EXPLORING THE BIOMEDICAL REVOLUTION

Published by the Howard Hughes Medical Institute
Distributed by The Johns Hopkins University Press, Baltimore and London

This book is dedicated to Purnell W. Choppin, M.D., president of the Howard Hughes Medical Institute from 1987 through 1999, on the occasion of his retirement. Not only did he lead the Institute to maturity as one of the world's premier biomedical research organizations, but his personal concern for furthering the public's understanding of the biomedical revolution made it possible to publish the material in these pages. His profound knowledge of science, his high standards, and his keen editorial eye added immeasurably to the quality of the Institute series upon which this book is based.

FOREWORD

· ·

By Robert A. Potter
Director of Communications
Howard Hughes Medical Institute

*E*xploring the Biomedical Revolution takes the reader on an exciting excursion into the world of biomedical research, where scientists are making startling new discoveries that have far-reaching effects on the lives of people throughout the world. The pace of this progress is breathtaking—hardly a day goes by without a new discovery in biology or medicine being reported on television and in the newspapers.

The chapters in this book are based upon a series of publications prepared for the general public and published by the Howard Hughes Medical Institute.* Although the publications were not originally intended for the classroom, the colorful writing and dramatic illustrations that are their hallmark have made them favorites of many thousands of teachers and students worldwide. Well over a million copies of the reports have been printed, and more than 7,500 classroom sets have been distributed in response to requests from high school and college teachers in the United States, Japan, India, Pakistan, Scotland, Australia, and other countries. In addition, the individual publications have received many awards for writing and design.

The editors would like to express particular gratitude to the patients who have allowed us to tell their stories. It is through their difficult, often heartbreaking experiences that we are able to illuminate the human dimensions of the problems being addressed by biomedical scientists, directly or indirectly, in their everyday work. Some of the patients have become special friends, and one, Jeff Pinard, has graciously responded to repeated inquiries about his well-being from many teachers and their students during the past few years.

In his introduction, Philip Leder, a distinguished geneticist, describes how the book fits into the context of contemporary biomedical research. In the process, he also captures the spirit of adventure that characterizes the scientists and their work. Each chapter has a new introduction and a brief report on recent developments. Material that was clearly outdated in the original publications has been removed, an index has been added, and appendixes are included to indicate where each of the scientists mentioned in the text is now working and what each of the contributing writers is now doing.

Maya Pines wrote many of the articles upon which this book is based, and she assigned and edited all of them. The series and the book itself were designed by Rodney C. Williams. These two creative individuals are primarily responsible for the remarkable quality of the series and of this book, although they had the help of many talented professionals along the way.

Finally, I would like to express my appreciation to Hanna H. Gray, chairman of the Institute's Trustees, for her support of this project, and to Stephen M. Cohen, vice president and chief financial officer, and Joan S. Leonard, vice president and general counsel, for their enthusiasm and encouragement.

Finding the Critical Shapes (1990); *Blazing a Genetic Trail* (1991); *From Egg to Adult* (1992); *Blood: Bearer of Life and Death* (1993); *Seeing, Hearing and Smelling the World* (1995); *The Race Against Lethal Microbes* (1996); *Arousing the Fury of the Immune System* (1998).

The Howard Hughes Medical Institute was founded in 1953 by the aviator-industrialist Howard R. Hughes.
Its charter reads, in part:
"The primary purpose and objective of the Howard Hughes Medical Institute shall be the promotion of human knowledge within the field of the basic sciences (principally the field of medical research and medical education) and the effective application thereof for the benefit of mankind."

For further information, please contact the Howard Hughes Medical Institute, 4000 Jones Bridge Road, Chevy Chase, Maryland 20815-6789.
www.hhmi.org

ISBN 0-8018-6398-8

EXPLORING THE BIOMEDICAL REVOLUTION

A look at the work of front-line scientists and how they are changing medicine.

INTRODUCTION

· ·

Philip Leder, M.D.
Senior Investigator, Howard Hughes Medical Institute
Professor and Chairman, Department of Genetics, Harvard Medical School

It is only modest overstatement to say we all need to know something of modern biology. For millions today who struggle simply to survive, it is true, mere knowledge can seem an unimaginable luxury. But for many other millions who hear DNA, bone marrow transplants, cloning, and AIDS cited on the evening news; who make personal health care decisions rooted in the work of yesterday's researchers; and who may face troubling new choices arising from the work of tomorrow's, biological literacy cannot be dismissed as an intellectual frill.

This book throws an arm around readers and gently nudges them toward that distant, desired goal. Along the way, it offers a revealing look back at what future generations will view, quite rightly, as the opening shots of a biomedical revolution.

Once, long ago, the science of genetics, my own field, amounted to little more than noticing that maple trees begat maple trees, not dogwoods, that children are something like their parents, and that sometimes disease runs in families. Then, a century and a half ago in a monastery garden in Brno, in today's Czech Republic, Gregor Mendel found strikingly predictable patterns of inherited traits among the peas he so lovingly cultivated; from that came the idea of recessive and dominant characteristics familiar to generations of high school biology students.

It was not until 1953, however, that James Watson and Francis Crick discovered how the very shape and constitution of a particular molecule, DNA, gave precise physical form to the then still vague notion of a gene. Today, we understand that the genetic make-up of virtually every living thing lies encoded in its DNA. One gene—that is, one sequence of DNA base pairs—codes for the collagen in your skin, another for a fat-metabolizing enzyme, and so on; in all, fifty to a hundred thousand genes in every human cell. And each, today, is on the threshold of being deciphered through the Human Genome Project.

Once, the deluge of data spewed out by this massive international program would have seemed almost useless, by reason of its quantity alone. But today, computers make sense of it, rapidly analyze and compare it with, say, the genomes of other organisms. Today, we snip apart DNA molecules at particular spots. We track down the DNA sequences responsible for human diseases. We use mice with known flaws in particular genes to study cancer.

If this is not a revolution, what is? And today it rumbles through every area of biomedicine—immunology, brain chemistry, parasitology, developmental biology—granting an intimacy with nature's workings, at a deep, molecular level, once unimaginable. Technologies like x-ray crystallography, mass spectrometry, and the polymerase chain reaction have proven not simply useful tools, adjuncts to the work of science, but powerful drivers of it. The pace of discovery has accelerated, and across an ever-widening front.

To a scientist, perched precariously on the divide between ignorance and comprehension, to at last understand some small piece of nature's exquisite machinery grants enormous pleasure. One of my own most deeply satisfying moments as a scientist came while working with Marshall Nirenberg at the National Institutes of Health during the early 1960s. Our work helped show that the genetic code was expressed in "triplets"—three nucleotide bases for each amino acid in a protein, and not, for example, two; that one particular base sequence, G-U-U, codes for the amino acid valine. ...Our contribution to the solution of this important problem afforded us great satisfaction.

Much of the research recounted in this book gives similar intellectual pleasure. But as I've grown older and have seen insights gleaned at the lab bench become medical advances that prolong human life and relieve human misery, I've come to derive equal satisfaction from this other side of the research enterprise, too. The revolution of molecular medicine is not just an intellectual one, but reaches out to the clinics and examining rooms of the nation and the world.

Today we know that many diseases are caused by particular genetic anomalies we can identify at a molecular level. Sickle cell anemia, for example, is caused by a single DNA "mistake." One wrong nucleotide base leaves one amino acid where another should be, alters the shape of the resulting protein, and thus disrupts its normal functioning. Today we realistically contemplate remedying this and other diseases through a kind of genetic engineering that—far from the scare scenarios the word conjures up to some—promises relief for those today consigned to pain, misery, and early death.

Over the years, most biomedical research has been carried out on lower organisms like bacteria and yeast; or in worms; or in mice (in which, for example, models of sickle cell anemia have been developed). But today, humans themselves are becoming the biological paradigm. This, inevitably, raises ethical issues. Medical records could conceivably include an individual's entire hereditary make-up—her genome, base pair by base pair, genetic defects and all; but the release of such molecularly intimate information exacts a price in privacy. We may one day be able to genetically change behavior; that could be good, or very bad indeed, depending on the behavior, and who decides it needs changing. We may learn to extend human life; but sustaining a much older population incurs problems, too. Even metaphysical issues intrude: Why are we here? How did we get here? A more penetrating grasp of our biological identity will color our sense of ourselves as human beings.

But while we dare not overlook such issues, we ought not belabor them, either. Molecular medicine is, after all, medicine. And it's to serve medicine—to treat and cure disease, to promote health, to do human good—that we turn to the new biotechnology. Research on the molecular front lines promises enormous, tangible human benefit.

Robert Koch's 1882 discovery of the mycobacterium responsible for tuberculosis helped found the science of microbiology. Antibiotics helped control this dread wasting disease for a time, but drug-resistant strains of the mycobacterium have caused a resurgence. Today's scientific counterattack relies on new molecular tools. One researcher we meet in this book used gene-splicing technology to develop a more rapid screening test for drug-resistant TB strains. The spliced gene is one for luciferase, the protein that makes fireflies glow; drug-resistant samples literally light up. Another researcher found that TB victims in one city had not all come down sick from long-dormant infections contracted years before, as was long assumed; rather, many had acquired new infections, the disease erupting in flurries and clusters around the city. To reach that key conclusion meant recording the molecular fingerprints of almost every TB patient in town.

Today, half of children born deaf or with significant hearing loss suffer from genetic disorders. In one Bedouin tribe in northern Israel, for example, a quarter of its children are born deaf. Which proteins, misshapen and enfeebled by mutation, account for this tribal tragedy? Or, for that matter, those responsible for more than thirty other distinct forms of hereditary hearing loss? Such questions could scarcely have been asked a few years back. New treatments will emerge when scientists find answers.

Common to the research assault against these and other scourges is science's determination to penetrate the realm of individual molecules—molecules that mark off one form of TB from another; that disrupt normal hearing; that promote aggravated immune responses; that interfere with normal embryonic development. All, today, are becoming better understood. And with understanding will come more realistic hope for treatment and cure.

The growing inventory of powerful molecular tools will confer new blessings we cannot today imagine. Human beings are better off thanks to the revolution in biomedicine of the past fifty years, and will be better off yet fifty years from now; of this I have no doubt.

This book first appeared in slightly different form as a series of seven "reports" published by the Howard Hughes Medical Institute over the past decade. But the word "report," I must say, fails to do them justice; they are not the withered, flat, recital of fact the word often implies. The individual chapters of this curious mixed breed of a book have been written by science writers—professional writers adept at taming the sometimes formidable complexities of science and revealing the passion scientists themselves feel for their research.

In the hands—or rather through the words—of these gifted writers, we see the haunting mystery of smell evoked by Marcel Proust's *Remembrance of Things Past*. We see the organization of the brain's billions of nerve cells likened to the vast network of streets and highways that link the nation's homes. We see the immune system's B-cells as microscopic drug factories, its T-cells as impresarios of the immune response, its neutrophils as sandbag-fillers at a cresting river.

Well, T-cells are *not* impresarios. And yet, images and metaphors like these suggest the complex, subtle things they really are. Lingering in the mind, literary devices like these stir interest and promote understanding—which may be why so many high school and college teachers embraced the original reports on which this book is based. *Exploring the Biomedical Revolution* is in no ordinary sense a textbook; it is not intended explicitly to *teach*. And yet, through its pages, a great deal of modern molecular biology can be *learned*.

One scientist cited here likens mammalian sexual development (where the gonads can become either male or female sexual organs) to a play in

which, with all characters in place and all parts learned, the audience decides how the play turns out; thus he represents one particular and fascinating slice of nature as science today understands it. But how was that understanding wrested from the shadows of ignorance? The authors of this book are not content to describe what is today known of embryology, or immunology, or genetics. They tell also *how scientists came to know what they know*—how their sometimes groping, inchoate explorations of nature yielded to understanding.

Inevitably, then, this book tells stories—beguiling human stories of disappointment and hope, frustration and triumph, in which scientists themselves are the main characters. ...Late one evening, Indiana University postdoctoral fellow Matthew Scott compares the sequences of two seemingly unrelated fruit fly genes and, to his astonishment, finds sixty-amino-acid segments common to them both. This segment proves identical to what other researchers had termed the "homeobox" (for how it seemed set apart from neighboring DNA sequences, as if confined to its own box) which, in species after species, plays a crucial role in development.

Research teams from Yale and Harvard, each at first dismissive of the other, collaborate on a vaccine against Lyme disease. When an experiment with infected deer ticks seems to go wrong, one of the Harvard men wonders how to break the news to his Yale colleague. And yet the failure proves no failure at all. On reviewing the data, they realize their vaccine works not as they'd assumed it did but by a mechanism until then entirely unknown.

By citing these stories, I do not wish to suggest, pejoratively, that *Exploring the Biological Revolution* ranks as "mere journalism." This book has not been written by scientists. But scientists have closely scrutinized its contents, as has a team of experienced editors at Hughes. Yes, the picture of the natural world these stories impart is less rigorous and nuanced than might be found in, say,

Science or *Nature*; "popularized" treatments like these do, inevitably, lose something. But they *gain* something, too. They gain sensuous detail. They gain human context. They gain feeling.

Good science writing plainly demands lucidity and accuracy. But by my reading, the contributors to this book hold out for a higher standard. Science writing, their work attests, must be infused with feeling, too—with the thrill of discovery, with reverence for nature's beauty and complexity, with hope for the human good that unraveling its secrets confers.

Exploring the Biomedical Revolution can be seen as an experiment in biological literacy. But biological literacy is no longer, if it ever was, one of words and numbers only; today, more than ever, pictures count. This book, replete with photos, illustrations, dramatic fold-outs, and posters (not to mention the 3-D glasses tucked into a pocket between pages 410 and 411), has a powerful visual presence; and from a biologist's point of view today, nothing could be more appropriate, for in so doing it marches in tandem with modern biomedicine. Hitched to the computer and other powerful means of visualization, research itself has become ever more oriented to the eye.

When scientists finally understood how T-cell receptors could simultaneously recognize antigen and MHC molecules, it was a single x-ray crystallographic image that blew away the last lingering uncertainty. "I think the first picture, the original Bjorkman and Wiley 1987 paper, just opened up everyone's eyes to what was going on," an immunologist recalls. He refers to an image, of a salmon pink peptide wrapped within the folds and grooves of a phosphorescent blue MHC molecule, that in this volume occupies most of a page. "It explained visually, I think, something that we'd all been having a very hard time grasping."

Cameras slipped into blood vessels photograph the yellow plaque that obstructs blood flow, cap-ture capillaries so fine that red blood cells line up in single file to pass through them. A scanning electron microscope image and a simple schematic sketch together show how hair cells in the ear transmit sound vibrations. PET scans come close to revealing the act of thought itself, as different parts of the brain light up depending, for example, on whether one reads words or hears them.

One chapter, "Finding the Critical Shapes," is given over almost entirely to the tools of x-ray crystallography and computer graphics. A famous photo from the 1950s shows Watson and Crick (whose work was aided by early x-ray crystallographic images) gazing up at the mechanical model they'd used to help unravel the structure of DNA. Their model, these pages remind us, "covered a tabletop and was held together with wire and clamps; today's structures can be viewed on the screen of a laptop, rotated with a knob, and colored with every hue of the rainbow." It took Max Perutz twenty-two years to establish the structure of the hemoglobin molecule. Today, new protein structures enter the Protein Data Bank at Brookhaven National Laboratory almost monthly.

It's not abstract understanding alone that the images in this book serve; many of them, like the book's words, carry emotional and aesthetic weight as well. In one two-page spread, a backlit sea of human red blood cells seems to convey in its gleaming redness all the sheer vaulting power of life. What do we "learn" from it? Perhaps nothing that could answer a multiple choice question on a biology quiz. But this intimation of nature's splendor counts for something, too.

Exploring the Biomedical Revolution celebrates the new molecular medicine that has changed our lives. It doesn't "celebrate" through empty fanfare, though, but by peering deep into its subject matter, life itself.

BLAZING GENETIC TRA

BLAZING

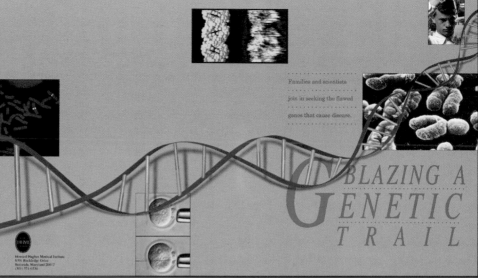

Families and scientists
join in seeking the flawed
genes that cause disease.

BLAZING A
GENETIC
TRAIL

Howard Hughes Medical Institute
6701 Rockledge Drive
Bethesda, Maryland 20817
(301) 571-0530

A GENETIC TRAIL

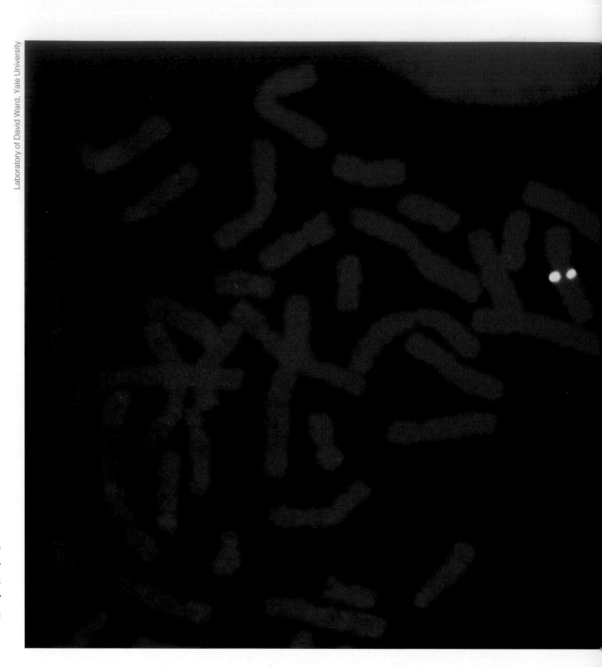

Fluorescent yellow dots show the location of a human muscle gene on a pair of chromosomes in a human cell.

BLAZING A *G*ENETIC TRAIL

Families and scientists join in seeking the flawed genes that cause disease.

INTRODUCTION

. . . .

"There's lots of zip in DNA-based biology today," declares James D. Watson, who, with Francis Crick, won a Nobel Prize in 1953 for discovering the structure of DNA. The speed of progress in understanding human genes since that time has been dizzying. Yet it is bound to accelerate even more in the new millennium, when genome projects reveal the sequence of all the genes that specify a human being.

The six feet of colorless, astonishingly thin DNA filaments that lie coiled inside each of the cells in our bodies contain all the genetic instructions necessary for our growth, development, and health. And now, for the first time, scientists are learning what these instructions spell out.

As researchers become fluent in DNA's language, however, they keep finding new copying errors in the precious texts. Every time a human cell divides, six billion base pairs of DNA must be reproduced in the correct sequence. Considering that this happens several billion times a day in an adult, our DNA is remarkably free of errors. But mistakes do occur. In some cases we are born with DNA typos that we inherited from our parents. Other defects arise anew inside our cells. Whatever their source, genetic flaws are common causes of pain, illness, and death.

This report focuses on the struggle to *conquer* inherited diseases. It sketches out a trail that, if completed, will lead to means of prevention or cure. To progress along this trail, scientists must answer three basic questions about each inherited disease. These questions and their answers—most of which depend on the use of recombinant DNA technology—are illustrated in a foldout guide that can also serve as a wall chart.

Easiest to analyze are disorders caused by errors in a single gene, such as cystic fibrosis or Duchenne muscular dystrophy. As shown in the story of Jeff Pinard, a gifted young man who was seeking his own mutation, scientists have now answered most of the early questions about cystic fibrosis: What faulty gene causes the disease? What kind of protein does the normal gene instruct the cell to make—in what quantities, and in what specific places? Does the flawed gene produce too little protein, the wrong kind of protein, or no protein at all? And, finally, what treatment will counteract this error? Scientists are still grappling with the problem of treatment, but they have considerable hope of success.

Although single-gene disorders are relatively rare, their clarity allows researchers to uncover mechanisms that may apply to other, more widespread diseases. For instance, as a result of identifying the gene defect that causes Duchenne muscular dystrophy, a muscle-wasting disease, scientists discovered a previously unknown protein that plays an important role in all muscle function. This gave them a richer view of how muscle cells work, enabled them to diagnose other muscle disorders with greater precision, and suggested new approaches to treatment.

Finding the causes of such complex ailments as heart disease, breast cancer, colon cancer, diabetes, arthritis, or schizophrenia presents much tougher challenges. Yet researchers have already identified a number of DNA errors that predispose people to developing these conditions, as well as some genetic variations that protect against them. Now scientists are zeroing in on the genetic pathways through which the flawed genes interact with other genes, or with the environment, to produce disease. Each pathway offers several possible targets for drugs, so this work raises the hope of developing new, individualized treatments for many widespread ills.

Thanks to the growth of powerful databases, researchers who identify a new gene can usually find a match for it right on their own computers by searching through a wealth of DNA sequences or proteins from animal models, as well as from humans. Additional databases are being set up to help identify the genetic variations that underlie disease and to speed the development of genetic tests and treatments. Every day, it seems, more pieces of the genetic puzzle fall into place.

Maya Pines, *Editor*

Under ultraviolet light in a cramped darkroom, Jeff Pinard wears a mask to protect his eyes as he examines how far DNA fragments (marked with a fluorescent dye) have moved through a gel. The fragments' size is revealed by how fast they move in an electric field.

A
GIFTED YOUNG
PATIENT SEEKS
HIS OWN GENETIC
FLAW

by Maya Pines

In the summer of 1992, 20-year-old Jeff Pinard set out to find the flaw in his genes that causes him to have cystic fibrosis.

He already knew quite a lot about genetic diseases, especially his own. Cystic fibrosis (CF) is a fatal disorder that clogs the lungs and other organs with a viscous, sticky mucus that interferes with breathing and digestion. It is the most common lethal inherited disease among white children and young adults, attacking about 30,000 Americans. Until recently, most patients died before reaching the age of 30.

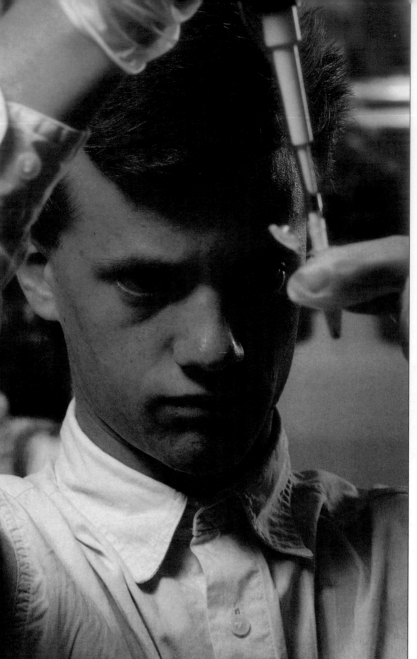

Pinard adds fragments of a CF patient's DNA to a dye that will help track these fragments as they move through a porous gel.

But Pinard, a microbiology major at the University of Michigan, was full of hope and could hardly contain his excitement at the thought of working with top scientists at the cutting edge of research on CF.

A year earlier—in one of the triumphs of molecular genetics—a research team headed by Francis Collins of the HHMI unit at the University of Michigan and Lap-Chee Tsui and John Riordan of Toronto's Hospital for Sick Children had discovered the errant gene that is responsible for CF. The researchers also identified the specific mutation—a missing snippet of deoxyribonucleic acid (DNA), the genetic material packed in the chromosomes within the body's cells—involved in most cases of CF.

As many as 1 in 25 Americans of north-ern European descent—some 10 million people—carry a gene with a CF-causing defect. Because CF is a recessive disorder, people who inherit a defective gene from only one parent remain healthy. Those unlucky enough to inherit a defective gene from both parents develop the disease (see p. 22). But in some patients the specific defects that cause CF differ from the one originally found by the research team. This was the case with Jeff Pinard.

When Pinard was a child, his parents could not understand what ailed him— why he had two serious bouts of pneumonia before he was six years old and why he was in and out of hospitals all the time. No one else in their family had had such problems. They were especially worried about their son's inability to gain weight.

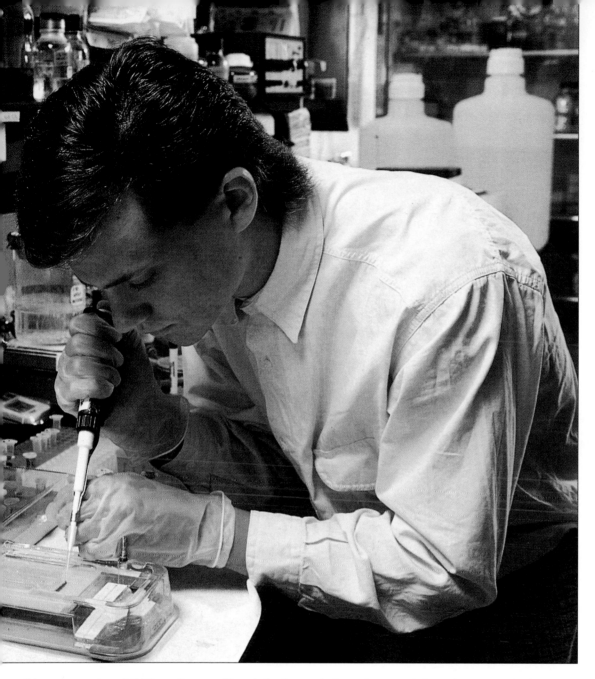

He carefully inserts DNA mixed with dye into a well at the starting end of the gel. A different well is used for each patient.

"I was emaciated," Pinard says. "I weighed only 30 pounds at age six."

They questioned their family doctor, who blamed Pinard's mother for "not feeding me properly," Pinard recalls. "It wasn't until my little sister nearly died of dehydration at one and a half that a different doctor, in Grand Rapids, diagnosed her—she has cystic fibrosis, too—and asked about others in the family." A test that measures salt content in sweat established that Jeff Pinard also had CF.

Thunderstruck—and without any information about what the disease meant—his parents went to the library to look it up. They found just a little paragraph in an encyclopedia that said, 'children with cystic fibrosis die by the age of six,'" Pinard says. "It was devastating for them. But in spite

of what they read, I got better and I lived."

In fact, Pinard is so full of life it is difficult to remember that he constantly faces the threat of a fatal flare-up of his disease. He shares an apartment near campus with another student. He often plays racquetball with his friends. He rides a bicycle, plays the trumpet. He has been singing in choirs since he was in sixth grade, and recently he toured the East Coast for two weeks with the University of Michigan Men's Glee Club, which gave him a chance to see the ocean for the first time.

Most of his waking hours, though, are focused on genetic research. As far back as he can remember, he has always been interested in science, he says. He turned to genetics in high school, where he wrote a 40-page report on the subject and became increasingly engrossed in it. When he entered the University of Michigan as a freshman and heard about the CF research going on there, he asked if he might do volunteer work in the lab. Soon he proved so useful that the Cystic Fibrosis Foundation gave him a small grant to work in the lab part time during his sophomore year. And last summer he received a larger grant so that he could work full time.

Pinard's presence "has had a fairly profound effect on my lab," said Francis Collins. "He's all enthusiasm, good humor, smarts. Many people in the lab had no direct contact with anyone who had the disease before. Now it's not so easy for them to see their work as just an intellectual exercise. They're much more driven by the urgency of it."

Every time Pinard coughs—often in mid-sentence—he brings a sobering reminder of the fragility of his life and that of other CF patients. At mealtimes, he must take five pills of pancreatic enzymes so he can digest his food. The pills cost $20 a day. Pinard remembers that "I had terrible pains and stomach troubles when I was little—really bad—until they found out I had CF and gave me these pills. Now I take them every single time I eat"—even when he eats just a single M & M.

He suffers from severe bacterial infections in his lungs several times a year. These are treated with antibiotics, but Pinard hates to take the drugs because

then he can't stay out in the sun (they cause photosensitivity) and there is a chance that the antibiotics may affect his liver. While 95 percent of CF patients die from lung complications and heart failure—"the heart works so hard and has so little air coming in," Pinard explains—the other 5 percent die of liver damage. "So it's kind of a tightrope," he says. He adds, "My medications run about $10,000 a year when I'm healthy."

To keep healthy, he must be pounded on the chest for an hour every day to loosen the mucus in his lungs. "There are eleven positions for the pounding," he says, "including the lower rib cage and the back. Since I moved here, I've been doing it myself, but it's not nearly so effective as having someone else do it. My arms get extremely tired, and I can't do my own back." He points out that most other patients with CF have to have their chests pounded two or three times a day. "And in bad cases, they need it four times a day, a full hour each time, just to get rid of that awful stuff—that mucus that lodges there, like a thick web," he says. "You've got to jog it loose."

"I've had many friends die of the disease," he adds quietly. "I used to go to a CF camp in Battle Creek, and I always felt guilty because I was healthy while others had it really bad."

Pinard attributes his relatively good health to the unusual mutation in one of his CF genes. His DNA was among the first to be tested last year, after the team isolated the gene. The researchers extracted his DNA from the nuclei of white cells in his blood and rapidly found the known mutation—called Delta-F508—on one of his CF genes. But his other CF gene contained a yet-unknown, probably milder, defect.

"My little sister seems to have exactly the same type of disease as I do," says

- - - - - - - - - - - - - - - - - - - -

Postdoctoral fellow Theresa Strong gives Pinard some pointers as they look for new mutations in the CF gene. They are comparing the sequence of bases in patients' DNA with the sequence of the normal gene.

Pinard. "It'll be very interesting for me to see how my mutation correlates with the severity of the disease—I can't wait to find out. One thing I hope is that the mutation we have will apply to a lot of other patients. That would cut down on the amount of tests that are needed to identify carriers. If there really are over 100 different CF mutations, it's going to be a bear! It'll be very expensive and a lot of work and harder to find treatments."

Last summer Pinard worked 40 to 50 hours a week in the lab, hunting for his own mutation and those of other atypical patients. He painstakingly analyzed hundreds of DNA fragments but could not locate his own flaw or any other unusual mutation. Then, just as he was examining some DNA that looked interesting, he fell sick and had to be hospitalized for ten days with a severe infection.

When he came out of the hospital, he learned that while he was away the lab had found two new mutations, from two different patients, in the very batch of DNA he had been working with. "It was so frustrating," he says. "I was happy for them, but wished I could have reaped the fruit of all that labor!" Meanwhile the mutation in his own gene remains a mystery. Pinard is still looking for it as he continues working in the lab on a part-time basis during the school year.

"It's really crucial to find the other mutations," says Collins. "Every new mutation tells us something more about the gene: which regions of it are essential for its function, specific details we need in order to develop effective treatments."

Some 85 groups of scientists in over 20 different countries are now collaborating in the search for new CF mutations through a consortium organized by Lap-Chee Tsui in Toronto. They have identified more than 75 new mutations, but most of these affect only a very small number of patients.

The scientists correspond by fax machine and share their findings months before their papers are published in journals. They understand the need for speed. As Collins puts it, "I fervently hope we'll find a cure in time to help someone like Jeff." ●

"His presence has had a profound effect on my lab. . ."

idwives used to lick the forehead of infants they delivered, believing that if the newborn's skin tasted abnormally salty it was a bad omen—the baby would soon grow sickly and probably die.

Their predictions were often accurate, for excessively salty sweat is a symptom of cystic fibrosis. Until recently, not much else was known about the deadly disease, even though doctors kept their patients alive longer by using antibiotics.

Since CF seemed to run in families, several teams of researchers tried for years to find the gene (or genes) responsible for it. But until 1978, the only way to identify which of the 50,000 to 100,000 human genes caused a disease was to find a physical clue: an abnormal chromosome that could be seen under the microscope, or a protein (the product of a gene) that was defective or missing in blood, urine, or body cells. Cystic fibrosis offered no such clues, and all attempts to find out which of the 23 pairs of chromosomes in human cells harbored the faulty gene hit a dead end.

In 1978, two scientists at the University of California, San Francisco—Y. W. Kan, who is now an HHMI investigator, and Andrée Dozy—discovered a harmless variation in DNA that was inherited together with sickle-cell disease, a blood disorder, within certain families. The variation could be used to detect the disease prenatally in such families. It also pointed the way to the guilty gene.

Since then, researchers have detected so many other variations in DNA that, in any family, a given gene defect is likely to be linked to one. This makes it possible to use what has come to be called "positional cloning." Instead of starting with a protein and then looking for the gene that produced it, scientists do the reverse. They find the general location of the gene for a specific trait by comparing the DNA of people who have the trait with that of close relatives who don't. Eventually, if the researchers examine the right DNA fragments, they zero in on the gene. Then they can deduce what protein the gene codes for and what flaw in the protein is responsible for the disease.

Using this approach, in 1982 Lap-Chee Tsui (pronounced "Choy") of Toronto's Hospital for Sick Children and his colleague Manuel Buchwald started looking for statistical links between CF and various genetic markers—recognizable variations in DNA that serve as molecular landmarks. The more consistently a marker is linked to a disease within a family, the more likely that the disease-causing gene is inherited together with the marker and, therefore, located close to it on the same chromosome.

Tsui contacted a large number of Canadian families whose children had CF and obtained samples of blood from more than 50 families in which at least two children had the fatal disease. He identified many previously uncharted variations in human DNA and looked for evidence that one of these was inherited more frequently by CF patients than by the general population. But for two years he did not find even a hint of linkage between any of these markers and his patients.

So when Tsui noticed a short report in an international journal from a group of researchers in Copenhagen who thought they saw a possible link between CF and one of their markers, he was very excited. The Danish group noted that they needed more families to confirm their link.

"I called them up at once and offered them my families," says Tsui. "By then they

"Walking" and "Jump

In Toronto, Lap-Chee Tsui analyzes a stream of faxed reports from CF researchers around the world and coordinates the search for new mutations.

ng" Toward the Gene

had already excluded that marker, but they had found another one. They applied our families to it and established a linkage."

Tsui's report aroused the interest of Collaborative Research, Inc., a biotechnology firm in Massachusetts. The company offered Tsui a deal: It would provide him with probes—labeled fragments of single-stranded DNA that search out and identify complementary DNA fragments—for some 200 additional markers that its researchers had developed, if Tsui would provide samples of DNA from the white blood cells of his CF families. Tsui accepted. Within three weeks, he found a definite link between one of the new markers and the patients' DNA.

Soon afterwards, Collaborative Research mapped this marker to chromosome 7. It was a key finding, for it meant that the CF gene, which was inherited along with the marker, was also located somewhere on that chromosome.

Meanwhile, two other research teams temporarily overtook Tsui. Raymond White, a pioneer in gene mapping who had been hunting for the CF gene at the University of Utah's HHMI unit, discovered a close link between CF and *met*, a cancer-causing gene that had been identified by scientists at the National Cancer Institute in Bethesda, Maryland. Robert Williamson of St. Mary's Hospital Medical School in London, one of the first researchers to seek the CF gene by studying affected families, discovered another close marker on chromosome 7 called J3.11. The findings of the three teams were published together in November 1985.

Up to that point, looking for the CF gene had been much like trying to find a particular house without any idea of its address, or even what continent it was on. Now researchers had a place to start. In fact, they could narrow their search to an area that was equivalent to a particular country, rather than a whole continent, since White's and Williamson's markers gave them a general location on chromosome 7: its long arm.

As a result, parents who already had a child with CF and were expecting

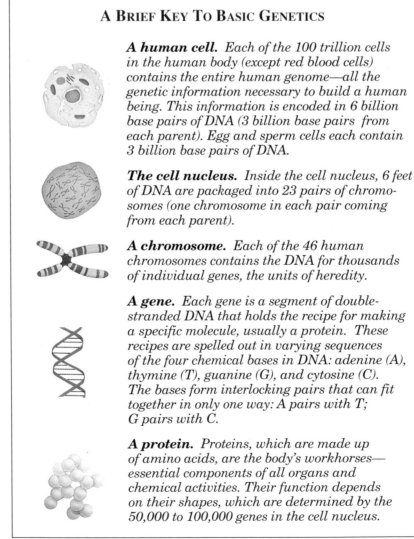

A BRIEF KEY TO BASIC GENETICS

*A **human cell**. Each of the 100 trillion cells in the human body (except red blood cells) contains the entire human genome—all the genetic information necessary to build a human being. This information is encoded in 6 billion base pairs of DNA (3 billion base pairs from each parent). Egg and sperm cells each contain 3 billion base pairs of DNA.*

*The **cell nucleus**. Inside the cell nucleus, 6 feet of DNA are packaged into 23 pairs of chromosomes (one chromosome in each pair coming from each parent).*

*A **chromosome**. Each of the 46 human chromosomes contains the DNA for thousands of individual genes, the units of heredity.*

*A **gene**. Each gene is a segment of double-stranded DNA that holds the recipe for making a specific molecule, usually a protein. These recipes are spelled out in varying sequences of the four chemical bases in DNA: adenine (A), thymine (T), guanine (G), and cytosine (C). The bases form interlocking pairs that can fit together in only one way: A pairs with T; G pairs with C.*

*A **protein**. Proteins, which are made up of amino acids, are the body's workhorses— essential components of all organs and chemical activities. Their function depends on their shapes, which are determined by the 50,000 to 100,000 genes in the cell nucleus.*

Francis Collins explains how a normal CF gene was pieced together in the lab out of smaller, overlapping fragments of DNA. Part of the gene's sequence is shown on the blackboard.

another child could be told with reasonable accuracy whether the fetus was destined to develop the disease. The verdict depended on whether the DNA extracted from fetal cells carried the same markers as the stricken child's DNA, showing that both parents had contributed a chromosome 7 harboring a CF mutation.

Finding the gene was still a long way off, however. The researchers did not even know what direction to take—was the gene to the left or right of the two markers, or in between? The only thing they could be sure of was that the three bits of DNA—the gene and the two markers—were quite close together on the chromosome. Therefore, determining the positions of the three pieces relative to one another, a statistically difficult task, would require DNA from an extremely large number of families in which some members had the disease while others did not.

Seven independent research teams had been racing to find the CF gene. Now they saw that none of them could do so large a study alone. At a meeting in Toronto, they decided to pool their families (211 in all), their probes, and their data.

"That was a really lucky break for CF research," says Tsui, "because in December 1986 Jean-Marc Lalouel (of the HHMI unit at Utah) and Ray White found that the two markers flanked the gene."

The search for the CF gene could then focus on a well-defined stretch of DNA— equivalent to an interstate highway, rather than an entire country. The scientists rejoiced. Nevertheless, the piece of DNA between the two markers was still very long—a whopping 1.6 million subunits, or

base pairs—long enough to hold perhaps 50 to 100 genes. It was far too great a distance for researchers to "walk" toward the gene.

For geneticists, walking along a chromosome means using partly overlapping DNA fragments to come closer and closer to the target gene, one step at a time, checking each fragment to see whether it is inherited together with the disease. But it is a slow and arduous task. The steps are very small and the path is strewn with roadblocks— repetitive sequences of DNA or other stretches that are difficult to cross. Tsui estimates that walking along 1.6 million bases would take about 18 years for the average lab.

They needed a shortcut. It was a time for imagination and ingenuity—as well as fierce competition. All thoughts of further collaboration vanished overnight as the seven research groups scrambled to be the first to find the CF gene.

At the beginning, Robert Williamson's team in London seemed far ahead of the pack. They had focused on a potentially interesting area 500,000 base pairs from the *met* oncogene, Williamson reported in April 1987, and found a "strong candidate" for the CF gene. "We really thought we had it," he said later. His report was so convincing that Raymond White and several other scientists dropped out of the race. But Williamson's group soon noticed some discrepancies. When they sequenced their candidate gene (that is, determined the sequence of bases in its DNA) and analyzed it to find out what kind of protein it coded for, the protein's profile looked wrong. The symptoms of CF suggested that the guilty protein would lie in or near cell membranes.

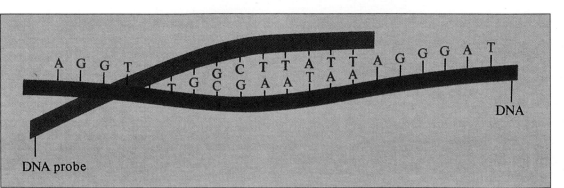

A probe—a strip of radioactively labeled DNA (black)—seeks out a complementary fragment of DNA and sticks to it.

Why So Many Errors In Our DNA?

As scientists learn to read the instructions in our genes, they are discovering that much of our DNA is riddled with errors.

Fortunately, most of these errors are harmless. Considering the difficulties involved—the 6 feet of DNA in a human cell consists of 6 billion subunits, or base pairs, coiled and tightly packed into 46 chromosomes, all of which must be duplicated every time a cell divides—our general state of health is something of a miracle.

We each inherit hundreds of genetic mutations from our parents, as they did from their forebears. In addition, billions of mutations occur in our DNA during our lifetime, either through mistakes in copying or through exposure to the environment. Bits of our DNA may be deleted, inserted, broken, or substituted. Most of these changes are harmless, usually because they affect parts of DNA that do not contain a gene's instructions. If they happen to affect genes that control cell growth, however, they may lead to cancer.

Problems arise only when an error in DNA alters a message that tells certain cells to manufacture a particular protein. Such messages are spelled out in varying sequences of the four chemical bases that make up DNA: adenine (A), thymine (T), guanine (G), and cytosine (C).

To stay alive and functioning, the human body requires a daily crop of billions of fresh protein molecules—about 50,000 different kinds of proteins that must be supplied in the right quantities, at the right times, and in the right places. We need hemoglobin to carry oxygen through the bloodstream, antibodies to fight foreign substances, hormones to deal with stress, neurotransmitters to evoke movements, emotions, and thought, and many other proteins to give structure to organs or speed up chemical reactions.

Our cells are kept extremely busy linking together amino acids—the building blocks of proteins—in the right order to produce these diverse proteins. The order is determined by the genes. According to the genetic code, which was deciphered in the 1960s, each triplet of bases in the genes' instructions either calls for a particular amino acid or gives a signal to start or stop making a protein.

An error in just one base can bring the wrong amino acid, altering the protein. And should one or two bases be missing, each succeeding triplet will be read in the wrong combination; such "reading-frame shifts" generally prevent cells from making the protein at all.

Actually the DNA's instructions are not transmitted directly; a copy made of ribonucleic acid (RNA) acts as an intermediary. The original DNA remains safely in the nucleus, somewhat like the printing block in a printing press, while the RNA copy is produced by transcribing just one strand of DNA, which carries the instructions for manufacturing protein.

Reading the DNA of humans and other mammals is complicated by the astonishing fact—discovered a little over a decade ago—that the genes' instructions are split into separate segments of DNA. These instructions must be spliced together before they can be carried out by a cell. Only about 5 percent of the DNA in mammalian genes actually contains the recipe for making a protein. The remaining 95 percent consists of intervening sequences, or "introns," whose function is unknown.

Splicing together the "exons"—the protein-coding sequences—is a very delicate, precise operation that involves snipping out the introns to end up with a much shorter strand of potent RNA. At the exon-intron boundaries are splicing signals, which researchers can now identify. Several genetic diseases have been traced to disrupted splicing.

Much of the recent progress in reading DNA has come from analyses of genetic errors.

James Wilson treats a CF patient's cells, shown growing in petri dishes, by inserting a virus that contains normal copies of the CF gene. The normal gene corrects the genetic defect within 24 hours.

Yet this protein did not resemble other membrane proteins. A few months after his announcement, Williamson regretfully admitted that he had made an error.

In Toronto, meanwhile, Tsui had continued his dogged hunt for the gene. His strategy was to bombard chromosome 7 with a large number of additional markers, which he created by cutting up and analyzing thousands of fragments of normal DNA from chromosome 7 "libraries." If he used enough markers, he reasoned, one of them was bound to be close to the CF gene. He referred to his method as "saturation mapping," or brute force.

When Williamson announced his candidate gene, "we didn't stop," Tsui says. "We figured that the basic defect in CF is really very subtle, since the lungs and pancreas of patients take years to deteriorate. It had to be a very small error—just one or a few base pairs. We thought it would be difficult for Williamson's group to find this error, and perhaps we would succeed."

Eventually he hit on a marker that was closer to the CF gene than any other except for Williamson's. But he still needed to get from this marker to the gene. Walking was still too slow. So he contacted Francis Collins of the University of Michigan, who had devised an imaginative technique for "jumping" along chromosomes.

Jumping is five to ten times faster than walking. It allows researchers to cover 100,000 to 200,000 bases of DNA at one time and simply leap over areas that might otherwise be difficult to cross. The technique involves snipping out a long segment

of the DNA under study—a segment that is labeled at one end—and letting it curl into a circle. This brings the labeled end next to a sequence of DNA that would otherwise be thousands of bases away. The circular segment of DNA is then opened up and used as a sort of bridge over very long stretches of DNA, while its far end is labeled for use as the starting point for another jump.

Joining forces, the researchers combined Collins' jumping technique with Tsui's marker. They walked, they jumped, and they walked some more. They knew they were making real progress. Nevertheless, "it took us one and a half years to find the gene," Tsui recalls, his eyes still shining as he relives the excitement of this period.

Each DNA fragment they used to walk or jump along the chromosome was examined by means of "zoo-blots," which compared the fragment with DNA sequences from animal species. One day they found a match with a sequence from a gene of chickens, mice, and cows. This implied that the fragment was important—only sequences that serve important functions in the body tend to be "conserved" in different species in the course of evolution. But they still had no evidence that it was the CF gene, and other conserved sequences had turned out to be dead ends.

Then another researcher at Toronto's Hospital for Sick Children, John Riordan, clinched it. Riordan, a biochemist, had been culturing and studying the membranes of sweat-gland cells—cells that are clearly involved in CF. Every time Tsui had a new segment of DNA to study, Riordan tested it to see if it corresponded to a genetic message in sweat-gland cells, and finally he found a match with a small part of this last conserved fragment. "That was the beginning of the gene," says Tsui happily. The researchers soon established that a gene containing this fragment was also expressed in other tissues that are specifically affected by CF. Furthermore, the gene looked as if it might code for a membrane protein.

"At that point," says Tsui, "his lab and ours began to sequence like crazy, looking for a difference between DNA from normal and CF cells."

In June 1989, as Collins and Tsui hovered over a fax machine in a freshman dorm room at Yale University, where they were attending a scientific conference, the reports from Tsui's lab brought the news they were waiting for: A small mutation in one particular DNA fragment had been sighted in 70 percent of the chromosomes from CF patients but was absent from normal chromosomes. Collins and Tsui realized they had found the CF gene. But they had to keep quiet about it because, as Collins recalls, "in the next room was one of our major competitors. The dorm walls were pretty thin, so we had to talk in whispers and not yell and scream."

Their reports were published in *Science* on September 8, 1989, to wide acclaim. "The one in 2,000 children born each year with a fatal defect now has a greater chance for a happy future," wrote Daniel E. Koshland, the journal's editor. "Until now, cystic fibrosis could not be studied in animals," he pointed out. The discovery of the gene would make this possible, "thus bringing the day of therapy and cure much closer."

Having the gene at hand would also make it possible to diagnose CF in the unborn, even in families that had no affected members. In addition, it would allow couples to be tested before they started having children, to see if they were carriers of CF (see p. 21).

As the researchers examined their new gene, they found it was made up of 27 segments of DNA that code for parts of a protein (known as "exons"), separated by a similar number of intervening sequences of DNA (known as "introns"). In the majority of patients, the error that caused CF was tiny: three of the gene's 250,000 base pairs were missing. This deletion led to the loss of just one amino acid out of the 1,480 in the protein for which the gene coded instructions. Yet this slight change was enough to radically disrupt the function of patients' lungs, sweat glands, and pancreas.

"Now that the gene has been identified, we have new possibilities for treatment," says Collins. He is encouraged by the fact

"I can't believe it worked!"

that the CF gene is not garbled and has only a small deletion. "The gene can still make a protein," he says. "The protein is there, but it doesn't function. Why?"

That is the question he and a dozen other researchers are now trying to answer. The new protein has been named CFTR, for CF transmembrane conductance regulator. It appears to work like a two-way pump, channeling vital compounds in and out of a cell.

When it functions normally, Collins explains, the protein somehow regulates the transport of chloride and sodium across cell membranes. But in CF this process fails, and the chloride channel stays closed. As a result, water is retained in the cell instead of going out into the airway; this leads to the buildup of thick, dehydrated mucus that characterizes the disease.

"Could we do something to tweak the mutant protein slightly, to make it work a little better?" Collins asks. "I think we'll figure out the basic biological defect in a year or two. I hope we'll find there's a drug that's already been used for something else that can stimulate the protein to function. That would be the nicest outcome. If not, we'll have to develop a new chemical treatment based on our growing understanding of the precise molecular defect."

Collins believes that eventually this will be feasible, and he has a suggestion about how it might be done: Aerosol sprays could be used to bring such treatments right to the patient's lungs, where they are most needed. "That's got to be an attractive delivery system," he says. "It could deliver either the normal CFTR protein or a drug that compensates for the faulty protein."

Looking even further ahead, Collins thinks aerosol sprays might deliver some form of gene therapy. The idea would be "to get the right gene into the right cell, and then get the cell to make the right amount of normal protein," he says.

In September 1990, two teams of researchers announced that they had actually corrected the CF defect—in a dish—by inserting normal genes into cells from CF patients. Michael Welsh, of the HHMI unit at the University of Iowa, and his colleagues "infected" cells taken from a patient's respiratory tract with a virus that had been engineered to act as a vehicle for the normal gene. The cells then began to manufacture the normal CFTR protein, and the chloride channel opened up. At the University of Michigan, HHMI investigator James Wilson and his associates corrected the same CF defect in cells grown from the pancreas of a CF patient. After they ferried the normal gene into these cells on a modified virus, the defective cells started to work normally.

Few scientists believed that success would come so quickly. When their experiment succeeded, Welsh and his co-workers were at first astonished. "There was a great deal of fun and excitement here," Welsh reported in a newspaper interview. "I remember one of my students running out of the lab room saying, 'I can't believe it worked!' "

If gene therapy is tried on CF patients, it may not rely on modified viruses to transport the normal gene because of concern that such viruses might eventually pose some danger. Instead, the normal gene might be packaged inside liposomes, small fat-coated bubbles that pass through cell membranes.

Gene therapy of this sort would raise no ethical issues, since it would not affect the genes of future generations. It would simply be a different kind of chemical treatment.

Many technical hurdles remain before scientists are able "to get the gene into the airways of a living, breathing person—and do it safely," Collins says. For instance, nobody knows whether the heavy mucus in the lungs of CF patients would get in the way of treatment. How many cells would need to be treated for the patients' airways to behave more normally? What is the precise function of the CFTR protein? How long would the inserted genes go on doing their job?

The answers to such questions may come when researchers succeed in developing an animal model of CF and then cure these animals of the disease.

WHO SHOULD BE TESTED?

Though ten million Americans carry an abnormal copy of the CF gene, hardly any of them know it or realize that they risk having children with the disease.

In 1989, right after the gene was isolated, the relatively small number of clinics that do genetic testing braced themselves for the expected onslaught of people demanding to be tested for CF. But testing was delayed because of the many mutations that can cause the disease.

Screening the general population with a test that identifies only 70 to 75 percent of carriers would be a mistake, a panel at the National Institutes of Health declared in March 1990. The panel pointed out that such screening would detect only about half the couples at risk of producing a child with the disease (couples in which both partners carry a recognizable CF mutation). Many people might be left feeling anxious or uncertain. If one member of a couple tested positive and the other negative, for example, the one who tested negative might actually have an as-yet-unidentified CF mutation. The panel called for pilot programs to determine how best to screen for CF, but recommended postponing more widespread programs until the test can detect at least 90 to 95 percent of carriers.

Since then, much progress has been made. It turns out that by taking into account ethnic origins and including the appropriate mutations in the test, one can greatly increase the test's ability to detect carriers—especially now that some new, moderately prevalent mutations have been uncovered.

"We're very close to being able to identify 85 percent of carriers," Arthur Beaudet, an HHMI investigator at the Baylor College of Medicine in Houston, Texas, said in October 1990. It is easy to test for several mutations simultaneously, he pointed out: "We can get four different mutations on a single test for as little effort as one." A test that identified 85 percent of carriers would detect 72 percent of the couples at risk, he said. Beaudet added that a reasonable goal for the next year or two would be to develop a CF test that can detect 95 percent of carriers, and emphasized that "population-based testing is an opportunity to prevent CF."

Some physicians are already offering a CF test to their private patients. Furthermore, several pilot projects for population screening have been started in various parts of the world. One of them, carried out by Robert Williamson's lab at St. Mary's Hospital in London, prides itself on its simplicity: Instead of requiring samples of blood, it uses a mouthwash.

"You take a small tube of saline solution, you wash your mouth out with it, and spit it back into a bottle. That's it!" explains Edward Mayall in the Williamson lab. "The great advantage is that you don't need a nurse to take a blood sample," says Eila Watson, the project's research coordinator. "So many people are terrified at the idea of giving a blood sample—they're much more likely to come if they don't have to give it. Plus the ease of handling. From the point of view of people in the lab, it's preferable not to handle blood if possible, so they don't have to worry about AIDS."

The mouthwash test costs less than $30, and the entire test can be done in one day. For children too young to do the mouthwash themselves, the lab uses a mouth scrape. In either case, only a few cells from the lining of the mouth are required. The cells are then heated, which makes them break down and release DNA from their nuclei. Extracting DNA from blood takes much longer.

"We've compared results from the blood test and the mouthwash," says Mayall, "and so far we have found absolutely no difference in the accuracy of the test. Even if the mouthwash sample is contaminated by bacteria or a bit of food, these are unlikely to have the same DNA

How Genetic Disorders Are Inherited

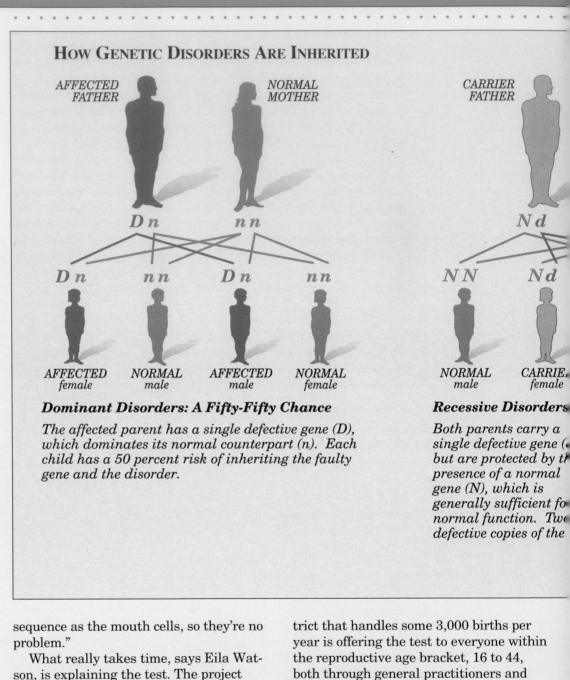

AFFECTED
FATHER

D n

NORMAL
MOTHER

n n

CARRIER
FATHER

N d

D n — AFFECTED *female*
n n — NORMAL *male*
D n — AFFECTED *male*
n n — NORMAL *female*

N N — NORMAL *male*
N d — CARRIE... *female*

Dominant Disorders: A Fifty-Fifty Chance

The affected parent has a single defective gene (D), which dominates its normal counterpart (n). Each child has a 50 percent risk of inheriting the faulty gene and the disorder.

Recessive Disorders

Both parents carry a single defective gene (... but are protected by t... presence of a normal gene (N), which is generally sufficient fo... normal function. Tw... defective copies of the...

sequence as the mouth cells, so they're no problem."

What really takes time, says Eila Watson, is explaining the test. The project provides genetic counseling in group sessions, eight persons at a time. Those who take the test pay nothing; the entire procedure is covered under the British national health service.

"We want to allow people to make informed decisions before conception," says Watson. This is why one health dis-

trict that handles some 3,000 births per year is offering the test to everyone within the reproductive age bracket, 16 to 44, both through general practitioners and through family-planning clinics. Whenever a carrier is identified, the test is offered to all members of the carrier's family.

If screening becomes practical on a wide scale, many questions will need to be answered. First, who should be screened? Should everyone be tested, including blacks (whose incidence of CF is only one-

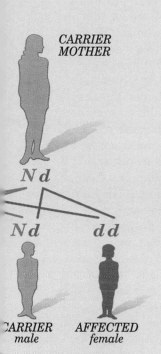

**CARRIER
MOTHER**

N d

N d   *d d*

CARRIER
male   AFFECTED
              female

ne Chance in Four

*ne are required to
roduce a disorder.
ach child has a 50
rcent chance of being
carrier like both
rents and a 25 per-
nt risk of inheriting
e disorder.*

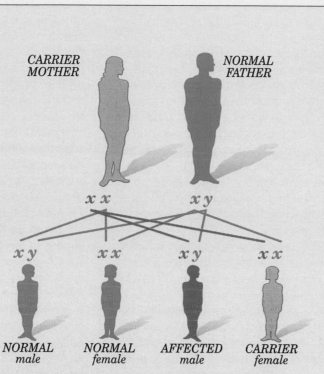

**CARRIER
MOTHER**   **NORMAL
                FATHER**

x x   *x y*

x y   *x x*   *x y*   *x x*

NORMAL   NORMAL   AFFECTED   CARRIER
male   female   male   female

X-Linked Disorders: Males Are at Risk

*One normal copy (blue x) of a gene on the X
chromosome is generally sufficient for normal
function. Women who have a defective gene (red x)
on one of their two X chromosomes are protected
by the normal copy of the same gene on the second
chromosome. But men lack this protection, since
they have one X and one Y chromosome. Each
male child of a mother who carries the defect has
a 50 percent risk of inheriting the faulty gene and
the disorder. Each female child has a 50 percent
chance of being a carrier like her mother.*

tenth that of whites) and Asians (who
almost never have the disease)? Or should
CF screening be limited to whites, as the
President's ethics commission recommended in 1983?

Another question is how to prevent
insurance companies from obtaining the
results of such tests and using them to
deny health-care coverage to people with
genetic defects. Discrimination of this sort
will grow unless new laws ensure that
genetic information is confidential,

according to Paul Billings, a medical ethi-
cist at Pacific Presbyterian Medical Center in San Francisco.

Finally, at what age should people be
tested?

Testing in the schools has some advantages—students are a captive audience
mostly in the pre-pregnancy stage—but
geneticists fear that confusion about the
test's results might lead to stigma. They
remember what happened in the early
1970s when the nation embarked on an

energetic program of screening for sickle-cell disease, which affects 1 in 400 American blacks.

At that time, many of the people who learned they were carriers of the sickle-cell defect mistakenly thought they had the disease (which requires inheriting two copies of the defective gene) and might die of it. Some were denied jobs or health insurance. And since no prenatal test was available then—it exists today—some carrier couples were told that the only way to prevent the disease was to avoid having children, advice that led to charges of racism.

When screening programs are not properly introduced or explained, they run the risk of being counterproductive. This danger is best illustrated by an often mentioned but perhaps apocryphal story about a Greek city's experience with screening for thalassemia, a severe blood disorder prevalent among Mediterranean peoples. Although people who carry just one copy of the thalassemia gene defect are perfectly healthy, parents are said to have looked askance at any prospective bride or groom who tested positive for it. Carriers thus became pariahs and ended up marrying each other—exactly the reverse of the desired outcome, since this greatly increased the chance that their children would inherit two copies of the defective gene and develop the disease.

By contrast, in Ferrara, Italy—a city that used to have about 30 new cases of thalassemia every year—screening was combined with extensive education about the program. This was so successful that hardly a child has been born with the disease in nearly a decade.

The most reasonable time to screen for CF might well be when a couple gets a marriage license, suggests Francis Collins. Testing at this stage would reduce the likelihood that the test results would influence a person's choice of marriage partner. But it would miss the large number of unmarried couples who have

children. On the other hand, it would be a mistake to wait until the first prenatal visit to the obstetrician because that would limit a couple's options. "Most couples would like to have this information prior to conception," Collins says.

There are, in fact, quite a number of options for people who learn that they carry a CF gene defect and want to make sure their children do not suffer from the disease. If they marry someone who is not a carrier, the problem does not arise. If both partners are carriers, they can ask for prenatal diagnosis and, if the fetus has CF (a one-in-four chance), they may choose to terminate the pregnancy. If they find abortion unacceptable and do not want to take the chance of having an affected fetus, they can opt for artificial insemination with sperm from someone who is not a carrier. They can adopt a baby. Or they can try *in vitro* fertilization. At the Illinois Masonic Hospital in Chicago, for instance, Charles Strom and Yury Verlinsky are developing a way to screen the eggs of women who are carriers, select only eggs that do not carry the CF mutation, and then use those eggs for *in vitro* fertilization with the husband's sperm; once fertilized, the eggs are reimplanted in the woman's body.

Everyone agrees that screening for genetic diseases should be voluntary and confidential and that counselors should not impose their own values. "If a patient who is struggling with a terrible decision asks me what I would do, I don't answer," says Francis Collins. "I must not answer. It is a sacred and unbreakable rule."

What counselors can do is explain the options. It is hard to predict how long CF children who are conceived today will live with good medical care, or how they will respond to new treatments. However, "there is still no cure for CF at this stage," Lap-Chee Tsui points out. "So if it is possible to avoid the disease, one should be given the chance."

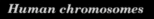
Human chromosomes

HOW TO CONQUER A GENETIC DISEASE

Nearly 4,000 genetic diseases afflict human beings. Given enough time and effort, scientists can learn to prevent or treat a great many of them. This requires answering three questions—major landmarks on a trail of genetic discoveries:

◆ Which altered gene causes the disease?

◆ What protein does this gene normally produce?

◆ Can the altered protein or gene be fixed or replaced?

Two different strategies may be used. Researchers may find the altered protein first (if it can be detected chemically in tissues that are affected by the disease) and then locate the gene that codes for it. When this is impossible, they use positional cloning: They find the gene first (by zeroing in on the DNA inherited with the disease, or by locating a similar gene in a mouse) and then identify the protein the gene makes.

The trail shown here illustrates positional cloning. This strategy recently led to spectacular progress toward diagnosing or treating cystic fibrosis (CF), Duchenne muscular dystrophy, neurofibromatosis, and other inherited disorders.

A BRIEF KEY TO BASIC GENETICS

A human cell. Each of the 100 trillion cells in the human body (except red blood cells) contains the entire human genome—all the genetic information necessary to build a human being. This information is encoded in 6 billion base pairs of DNA (3 billion base pairs from each parent). Egg and sperm cells each contain 3 billion base pairs of DNA.

The cell nucleus. Inside the cell nucleus, 6 feet of DNA are packaged into 23 pairs of chromosomes (one chromosome in each pair coming from each parent).

A chromosome. Each of the 46 human chromosomes contains the DNA for thousands of individual genes, the units of heredity.

A gene. Each gene is a segment of double-stranded DNA that holds the recipe for making a specific molecule, usually a protein. These recipes are spelled out in varying sequences of the four chemical bases in DNA: adenine (A), thymine (T), guanine (G), and cytosine (C). The bases form interlocking pairs that can fit together in only one way: A pairs with T; G pairs with C.

A protein. Proteins, which are made up of amino acids, are the body's workhorses—essential components of all organs and chemical activities. Their function depends on their shapes, which are determined by the 50,000 to 100,000 genes in the cell nucleus.

WHICH GENE IS AT FAULT?

1 *A child (blue box) develops a currently incurable genetic disease. Scientists who wish to find a specific treatment or a means of preventing the disease in other children must trace it to its cause: an altered gene.*

2 *Various clues, such as a visibly missing piece of a chromosome, may reveal the gene's rough location on a chromosome. When there are no such clues, researchers look for "markers" of the disease by comparing the DNA of the affected child to that of parents, relatives, and persons in other families. When placed on a "genetic map," these markers reveal which chromosome carries the altered gene. As more markers are added to the map, the location is narrowed to the space between two known markers.*

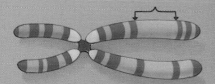

4 *"Walking" or "jumping" toward the gene, scientists create a physical map, or a chain of overlapping segments of DNA in space between the flanking markers. One of these segments must contain the altered gene. In the future, when researchers have covered all the chromosomes with overlapping fragments of DNA, any fragment they want will be available from a computer database.*

3 RESULT: Scientists may be able to diagnose the disease prenatally by following the inheritance of markers in an affected family. They may also recognize healthy carriers of the altered gene (light blue boxes). The family shown here has a recessive disease that develops only when a child inherits the altered gene from both parents.

5 *Zeroing in on the altered gene, scientists analyze each segment: Is it different from normal DNA? Finally they find the guilty gene and determine the error in its sequence of bases. The most common error in the CF gene is a deletion of three DNA bases (top right) out of a total 250,000.*

A mutant CF gene sequence:

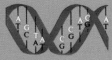

ATCATTGGTGTT

A normal CF gene sequence:

ATCATCTTTGGTGTT

(Some letters are hidden behind the DNA twists.)

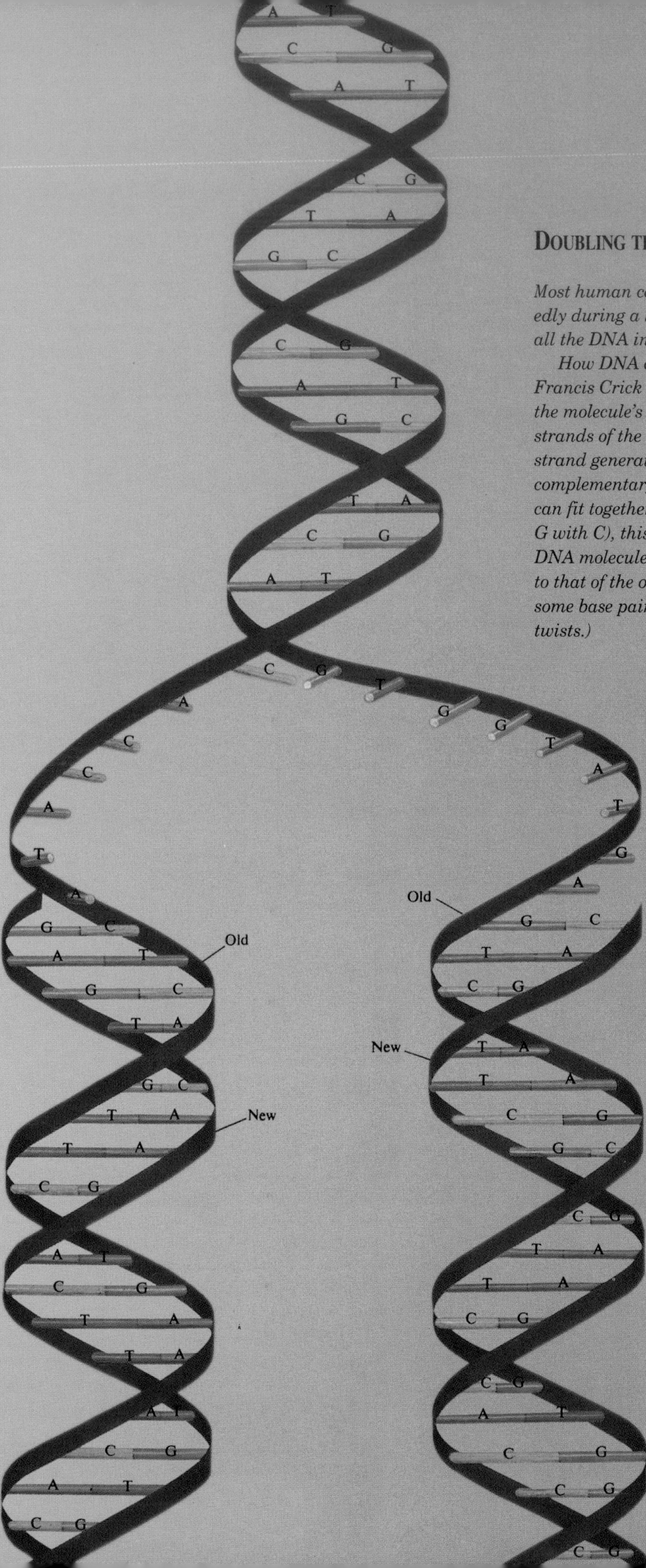

DOUBLING THE DOUBLE HELIX

Most human cells divide and multiply repeatedly during a lifetime. Before each cell division, all the DNA in the nucleus must be duplicated.

How DNA copies itself was revealed when Francis Crick and James Watson deciphered the molecule's structure in 1953: The two strands of the double helix unwind, and each strand generates a mirror image made up of complementary bases. Since DNA base pairs can fit together in only one way (A with T, and G with C), this process results in two daughter DNA molecules whose sequences are identical to that of the original DNA. (In this illustration, some base pairs are hidden behind the DNA twists.)

THE OTHER KINDS OF GENETIC DISORDERS: ABNORMAL CHROMOSOMES, ABNORMAL CELLS

Some genetic accidents go far beyond inherited errors in a single gene. Large chunks of a chromosome sometimes break off or attach themselves to the wrong place during the formation of egg and sperm cells. An entire chromosome may fail to separate properly, and an egg or sperm may end up with two chromosomes of the same type instead of one. The most frequent cause of Down's syndrome is an extra chromosome 21— three instead of the normal two per cell—as detected in the ominous picture to the right.

A totally different kind of genetic disorder— neither inherited nor inborn—results from accidents in a cell. For example, cancer is a genetic disease that can start in any somatic (body) cell, such as a breast, lung, or blood cell. If radiation, chemicals, or some other agents produce a mutation in the DNA of a somatic cell, the mutation will be transmitted to daughter cells as the cell divides. This may lead to cancer if there are other contributing factors.

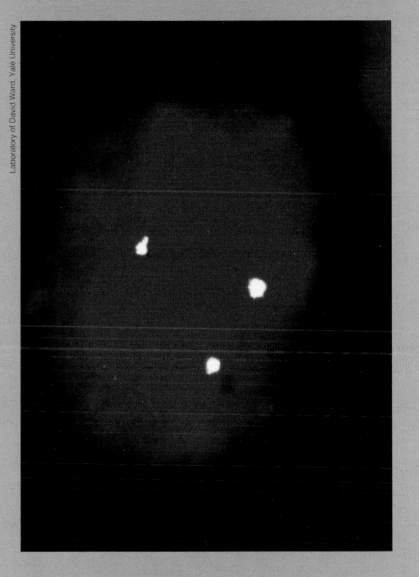

Laboratory of David Ward, Yale University

WHAT PROTEIN DOES IT MAKE?

7 *Knowing the gene's sequence, scientists use the genetic code to determine which amino acids make up the protein. Then they study the protein and find out what it is supposed to do. The tens of thousands of proteins in the body have different shapes and do different jobs, depending on instructions encoded in the genes.*

The elongated shape of collagen protein, shown in this molecular model, allows collagen to provide structural support to cells and organs.

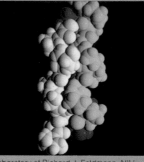

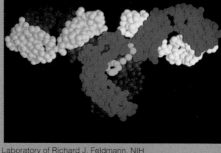

Laboratory of Richard J. Feldmann, NIH · Laboratory of Richard J. Feldmann, NIH

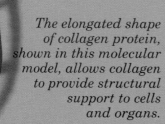

Laboratory of Harel Weinstein, Mt. Sinai Medical Center

Hemoglobin is the protein in red blood cells that delivers oxygen to all parts of the body via the bloodstream.

6 RESULT: Scientists can test for the disease directly in patients (blue boxes) and prenatally. They can identify healthy carriers of the altered gene (light blue boxes) in the general population— not just among members of an affected family. They can study the disease process in cultured cells and in animals, with a view to developing new treatments.

CAN THE PROTEIN OR GENE BE REPLACED?

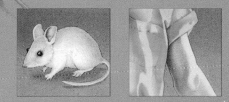

10 To make up for the genetic error, scientists may try to replace a missing or ineffective protein with a drug or with the normal protein. Such experiments are usually carried out first in cultured cells in the laboratory, then in animals, and finally in humans.

Antibodies, such as this human immunoglobulin, recognize and attack viruses, bacteria, and other foreign substances.

11 Another option is gene therapy. Some scientists "infect" cells with a virus into which they have inserted normal genes. Others use non-viral methods or even inject DNA directly into cells. Experiments that work in cultured cells are tried in animals and then in humans. For example, a patient's bone-marrow cells may be removed, treated with normal genes, and returned to the patient.

12 RESULT: Treatments are being developed for some genetic diseases. People will always carry genetic alterations, but in the future, prevention and treatment will vastly reduce suffering from genetic diseases.

8 Does the sick child's altered gene produce too little protein, a flawed protein, or no protein at all? Scientists need to understand just how the protein change causes the disease.

9 RESULT: As the mechanism of the disease becomes clear, scientists can devise new approaches to treatment involving either the protein or the gene. Understanding a relatively rare inherited disorder may also bring important insights into more common and complex diseases.

Illustrations: Wood Ronsaville Harlin, Inc.

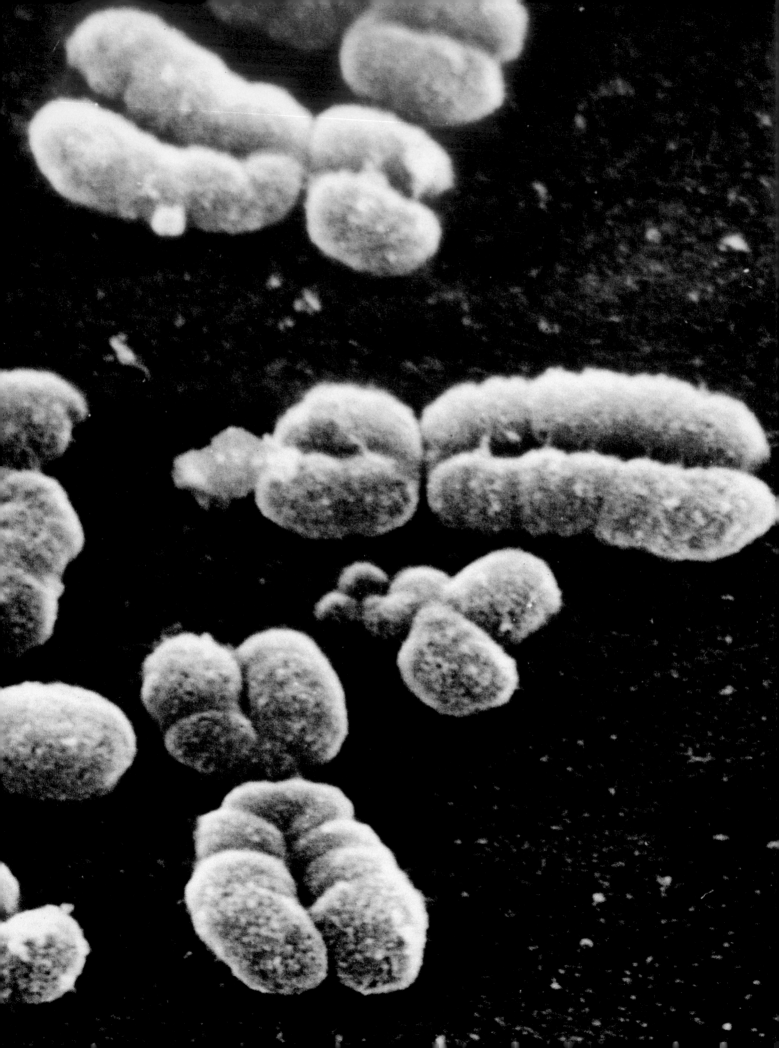

·IN·
SEARCH OF
L·A·R·G·E
FAMILIES

Like Southern novelists, geneticists love to chart the destinies of large, interrelated families with inherited disorders. But geneticists don't have the luxury of literary invention. They must find these kindreds in the flesh—and they will go to great lengths to do so.

Close-knit, isolated communities are as valuable to medical geneticists as inbred strains of plants are to botanists, points out Victor McKusick, who started the department of medical genetics at The Johns Hopkins University School of Medicine. Since these family groups—in obedience to religious or political beliefs, or because of geographic obstacles— have not mingled with other populations for generations, the genes of a small colony of founders are often passed on intact, and a variety of traits can be followed through parents, children, cousins, and aunts.

The lure of such families has brought geneticists to unlikely settings. It took Nancy Wexler from the National Institutes of Health in Bethesda, Maryland, to a remote shore of Lake Maracaibo in Venezuela, and it led Janice Egeland from Yale University to the sequestered Amish community in Lancaster County, Pennsylvania. However, any family can be valuable to geneticists if it is large and complete enough. In search of truly large families, Raymond White moved from the University of Massachusetts in Worcester to Salt Lake City, Utah, where families with twelve children are not too rare.

Nancy Wexler's remarkable quest for the cause of Huntington's disease (HD) has become a legend in less than a decade. HD is a genetic time bomb. In 1872 a New York physician, George Huntington, first described the inexorable course of this disease, which usually starts in middle age and slowly destroys its victims both mentally and physically. He noted that when either parent had the disease, "one or more offspring almost invariably suffer

by Beverly Merz

from it, if they live to adult age." Later physicians realized that HD's inheritance followed the pattern of "dominant" traits (see p. 22) first described by the Austrian monk Gregor Mendel in 1866. This meant that any child of a patient stood a fifty-fifty chance of developing the disease.

Having watched the disease erode her mother's body, mind, and spirit over the course of a decade, Wexler joined her father Milton, a Los Angeles psychoanalyst, at the Hereditary Disease Foundation, an organization he had created to do research into the disorder. At one of the foundation's research seminars, Wexler was riveted by a film of a Latin American village in which dozens of people lurched about, seemingly afflicted with the very disorder that held her mother in its grip. When federal funding for HD research finally became available in 1978, Wexler knew where to head with the money—San Luis, the Venezuelan village on Lake Maracaibo in which the film was set.

The people in the village were unaccustomed to medical attention and wary of strangers. However, Wexler had a unique bond with them because she, too, was at risk for HD. She managed to charm, wheedle, and cajole them into donating blood samples from which DNA could be extracted, while at the same time sorting out family connections. She eventually compiled the entire pedigree—a family tree showing the presence or absence of a trait—of almost 10,000 people and collected 2,000 samples, which are immortalized in cell cultures.

Wexler collaborated with Harvard molecular biologist James Gusella, Indiana University's Michael Conneally, and the Massachusetts Institute of Technology's David Housman, who agreed to analyze the DNA samples she collected. They had probes to recognize markers on an assortment of chromosomes and hoped that one of these probes might detect a marker that traveled with the gene. Expecting to spend years in this trial-and-error enterprise, they were elated to hit the jackpot on the third try.

Their third probe bound to fragments of DNA from people with HD in a distinctive pattern that differed from that of their healthy relatives. However, the pattern of markers that signaled the disease varied from family to family, making it necessary in some cases to test a large number of relatives to establish which pattern correlated with the presence of HD.

While the researchers had not found the gene itself, they had made it possible to tell, in most cases, which members of an affected family had inherited the fateful DNA patterns and therefore were likely to develop HD in middle age. This prediction could be made years before the treacherous course of the disease became manifest. It allowed people to make informed decisions during their childbearing years. Another new option within these families was prenatal diagnosis. The discovery of the marker also paved the way for finding the gene itself in the future and eventually developing new treatments.

At the time Nancy Wexler first traversed the hemisphere in search of the HD gene, Janice Egeland, who is now at the University of Miami, already had two decades of experience as a gene hunter among the Old Order Amish in the Pennsylvania farmlands. Egeland had approached the Amish in 1959—with some apprehension. She knew that the strict Mennonite sect, which eschewed modern dress, electricity, and automobiles, did not eagerly embrace outsiders. Her quiet perseverance eventually broke down their resistance, and she was taken into their confidence.

Egeland found that the Amish were as meticulous in their family record-keeping as they were in adhering to their religious convictions. It took her nine years, but she traced every one of the 7,000 living members back to the original 30 pioneer couples who founded the community. This allowed her to see patterns of disease running through families.

With an eye on the records and an ear tuned to local gossip, she also became aware of striking behavioral deviations within the Amish community. Members of certain families were known to have aban-

They hit the jackpot on the third try.

In a house built
on stilts over a
Venezuelan lagoon,
researchers gather
pedigree information
from families with
many cases of HD.

doned their chores and spent days sitting
in a corner, staring at the floor, only to
rise a few weeks later and suddenly
become loud or boastful. They went on
sudden trips, ordered as much as a ton of
chocolates for their small grocery store,
dressed in a flashy way, or cut their hair
in defiance of the community's rules.
While these aberrations might have been

overlooked in a big city, they were as out
of place as neon lights in the staid Amish
setting, and Egeland came to recognize
them as hallmarks of manic-depressive
disorder.

By the late 1970s, Egeland had charted
the passage of the disorder through as
many as 6 generations of 32 families. Yet
she hadn't a clue as to where the gene that

governed these mood swings might reside. Then, in 1982, she decided to try positional cloning. She collected blood samples from 81 family members, including 19 who suffered from manic-depressive illness, and sent them to Yale and MIT for testing. Daniela Gerhard, then a postdoctoral fellow in David Housman's lab at MIT (she is now at Washington University in St. Louis), found a marker on chromosome 11 that was present in the DNA of all the family's manic-depressive members but absent in most of their normal relatives. The marker pointed to the guilty gene's location on the tip of that chromosome.

Other groups have failed to substantiate these findings, however. Some have linked manic-depressive disease in families to markers on the X chromosome, while others have implicated chromosome 5 or found no links at all. And in 1989, after a few more members of the Amish families had come down with symptoms of manic-depressive disease, Egeland's own research group published a "re-evaluation" that markedly reduced the likelihood of

their marker being linked to the disease.

These results illustrate the difficulty of tracking down the genes for complex diseases, says Raymond White. Complex diseases such as cancer, schizophrenia, and heart disease, as well as manic-depressive disorders, are probably produced by several genes acting together and by different combinations of errors in different families. However, environmental influences such as diet, smoking, chemicals, or viral infections also play a role in triggering these diseases.

In dealing with mental illness, the picture may be clouded even further by uncertainties in diagnosis and a variable age of onset. By contrast, heart disease, despite its complexities, at least provides some solid indicators that can be measured and is therefore much more amenable to study.

In 1985 Michael Brown and Joseph Goldstein of the University of Texas Health Science Center at Dallas won a Nobel Prize for their work on a genetic disorder that can lead to early death from heart attack: familial hypercholesterolemia, which involves extremely high levels of cholesterol in the blood. The two scientists found that, though the symptoms of this disease are fairly uniform, it is caused by a variety of mutations that alter different functions of the same gene in different families. Some defects prevent special receptors on the surface of cells from binding as they should to low-density lipoprotein (LDL), which carries cholesterol in the blood. Other defects allow binding but prevent the uptake of LDL into cells. Either way, the cells fail to clear cholesterol from the blood. As Goldstein explains, "The more families we looked at, the more defects we continued to find."

Because every family may have a different skeleton hidden in its genetic closet, the quest to locate more large families with documented histories is endless. One of the best places to find such families is Salt Lake City, headquarters of the Church of Jesus Christ of Latter-Day Saints, also known as the Mormon Church. The Mormon Church not only encourages its members to produce large families, but urges them to compile elaborate genealogies.

"Every family has a historian, and every historian can tell you what diseases the members have had," says Raymond White. However he has found that he doesn't need to spend time combing the family historians' records for interesting pedigrees. They come to him.

"From time to time a doctor will call us and say, 'I've been taking care of a family you might be interested in,' " he explains. One family referred to White had familial polyposis of the colon, a condition that predisposes people to colon cancer. This condition was also being studied by Mark Skolnick, a population geneticist who has worked with Mormon genealogical records at the University of Utah since 1974. The classic harbinger of malignancy in people with this trait is the appearance of more than a hundred polyps, usually before the age of thirty. To avert the development of colon cancer, surgeons remove the colon at the first sign of polyps, usually while the patients are still in their twenties.

Pathologists have found that the polyps contain a variety of cells in different stages of alteration, ranging from cells closely resembling normal colon cells to cells that are all but cancerous. Oncologists now believe that this orderly progression to malignancy is caused by an accumulation of genetic mutations in colon cells. One of the principal genes involved has been traced to chromosome 5.

The patients in this family seemed to fit the classic pattern. Over seven generations, 15 of the 115 members developed colon cancer and 12 died of the disease. Yet the disease didn't always follow its predictable course; some members of the family had only a moderate number of polyps and one member who had no polyps at all developed colon cancer and died.

When White and his colleagues Mark Leppert and Jean-Marc Lalouel analyzed DNA from 81 family members, they found that several who appeared healthy actually carried the marker that signaled the gene on chromosome 5. White says that if these people hadn't been members of the family he was studying, they would have escaped identification as carriers of the gene and as people at risk for colon cancer. Now they and their children can be examined periodically to determine whether they are developing polyps, and they may be able to avoid losing their colons if their polyps are removed early. "Without the statistical power of a large family, it is often difficult to determine whether polyps are due to an inherited gene defect or to a sporadic mutation," White says.

Finding the genes for specific diseases is all very well, but Raymond White's main goal is to develop a set of tools for all geneticists—a genetic linkage map. In a landmark paper in 1980, David Botstein of the Massachusetts Institute of Technology, White, Skolnick, and Ronald Davis of Stanford University had proposed making such a map with the aid of DNA variations. White came to Utah with the intent of blanketing the genome with markers that could serve as reference points, or landmarks, for gene mapping. To do so he collected DNA from 46 normal families of

◆

Every family

may have a

different

skeleton

hidden in its

closet.

◆

three generations—four grandparents, two parents, and at least six children. He then resorted to a shotgun approach, using every available probe to look for markers.

By 1991 he had 500 markers scattered throughout the genome—a nearly complete, continuously linked set of landmarks no more than 20 million base pairs apart and generally much closer. He has sent probes for these markers to geneticists engaged in tracking down genes for inherited diseases, as well as to researchers looking at the sporadic mutations that cause cancer. His markers were instrumental in guiding Francis Collins and Lap-Chee Tsui to the cystic fibrosis gene, in establishing the location of the gene for familial polyposis of the colon, and in directing his own team to the gene for neurofibromatosis (NF), a disfiguring disease that used to be called erroneously "elephant man's disease."

It turns out that the NF gene is one of a family of genes which, when normal, operate as anticancer genes. Proteins produced by these genes may work to block tumor formation in the brain and central nervous system, as well as in the lung, liver, colon, and breast. This discovery may open the door to entirely new drugs to block the development of cancer.

All too often geneticists remain frustrated in their search for the "right" families. Although several teams of researchers are tracking the gene responsible for Alzheimer's disease, for example, there are few large families in which the disease has been documented reliably over a period of time. As a result, there have been conflicting reports. For example, a team led by John Hardy of St. Mary's Hospital, London, announced it had discovered a mutation in a gene on chromosome 21 in members of two unrelated families who have Alzheimer's disease. This agreed with a Harvard team's finding of linkage to markers on chromosome 21, but disagreed with a Duke University finding of linkage to chromosome 19.

One way to speed up research, Nancy Wexler suggests, is for all families to follow the lead of the Amish and the Mor-

mons and chart their own pedigrees. "Virtually everyone knows what a family tree is and how to put one together. And people know what diseases run in their families—not necessarily rare diseases but heart disease, cancer, diabetes." She adds that people with unusual family trees—for example, several members with early heart attacks—as well as those with rare diseases should try to link up with researchers through such groups as the National Organization for Rare Disorders, especially if their families are large.

Owen Davison, a management consultant in Hershey, Pennsylvania, did just that after his son Richard, a young social worker, died unexpectedly during routine surgery for a broken ankle. Richard was diagnosed as having malignant hyperthermia, an allergic condition in which certain commonly used anesthetics and muscle relaxants trigger a lethal series of events, including muscle contractions, bleeding, an increase in body temperature to 108° F or more, and cardiac arrest. The syndrome affects at least 1 person in 40,000. When Davison learned that malignant hyperthermia was a dominant trait, he wrote letters to some 300 relatives, warning them that they might be susceptible to the disorder and suggesting they might want to have a muscle biopsy to find out. Several members of the family who followed his advice, including Davison's niece Suellen Gallamore, learned that they, too, might succumb if subjected to these anesthetics. They were advised to wear identification bracelets, notifying hospital personnel of their susceptibility, and to take special precautions if scheduled for surgery.

In 1980 Davison, Gallamore, and Henry Rosenberg, an anesthesiologist at Hahnemann University, founded the Malignant Hyperthermia Association of the United States. The group has attracted other families with the disorder, as well as anesthesiologists and physiologists who are studying the bizarre syndrome.

Like the Wexlers, the Davisons have demonstrated that finding the key to hereditary disease is very much a family affair. ●

◆

Finding the key to hereditary disease is a family affair.

◆

READING THE HUMAN BLUEPRINT

"All human disease is genetic in origin," Nobel laureate Paul Berg of Stanford University told a cancer symposium a few years ago.

Berg was exaggerating, but only slightly. It has become increasingly evident that virtually all human afflictions, from cancer to psychological disorders and susceptibility to infection, are rooted in our genes. "What we need to do now is find those genes," says James Watson, who shared a Nobel Prize for deciphering the structure of DNA and who helped launch the Human Genome Project.

Single-gene diseases such as CF are relatively easy targets. Disorders that seem to be caused by the interplay of several genes—hypertension, atherosclerosis, and most forms of cancer and mental illness—are much more difficult to track down. Having a map of the entire human genome will make it possible to identify every gene that contributes to them.

A gene map can also lead researchers to new frontiers in drug development. Once all the genes are identified and their bases are sequenced, it will be possible to produce virtually any human protein— valuable natural pharmaceuticals, such as tissue plasminogen activator, interferon, and erythropoietin—as well as new molecules designed specifically to block disease-producing proteins.

The Human Genome Project officially began in October 1990. But the map of the human genome has been in the making for a good part of the century. It started in 1911, when the gene responsible for red-green color blindness was assigned to the X chromosome, following the observation that this disorder was passed on to sons by mothers who saw colors normally (see p. 23). Some other disorders that affect only males were likewise mapped to the X chromosome on the theory that females, who have two X chromosomes, were protected from these disorders by a normal copy of the gene on their second X chromosome—unlike males, who have one X and one Y chromosome.

The other 22 pairs of chromosomes remained virtually uncharted until the late 1960s. Then biologists fused human and mouse cells to create uneasy hybrid cells that cast off human chromosomes until only one or a few remained. Any recognizable human proteins in these hybrid cells thus had to be produced by genes located on the remaining human chromosomes. This strategy allowed scientists to assign about 100 genes to specific chromosomes.

Map-making really took off in the early 1970s, when geneticists discovered characteristic light and dark stripes or bands across each chromosome after it was stained with a chemical. These bands, which fluoresced under ultraviolet light, provided the chromosomal equivalent of latitudes. They made it easier to identify individual human chromosomes in hybrid cells and served as rough landmarks on the chromosomes, leading to the assignment of some 1,000 genes to specific chromosomes.

Around the same time, recombinant DNA technology began to revolutionize biology by allowing researchers to snip out pieces of DNA and splice them into bacteria, where they could be grown, or cloned, in large quantities. This led to two new mapping strategies. In one, *in situ* hybridization, scientists stop the division of human cells in such a way that each chromosome is clearly visible under a light microscope. Then they use probes to find the location of any DNA fragment on these chromosomes. Originally these probes were radioactively labeled, but chemically-tagged probes that can be made to fluoresce have been found to yield far more accurate and rapid results (see p. 41).

The other strategy is to use DNA variations as markers on the human genome, as proposed by Botstein, White,

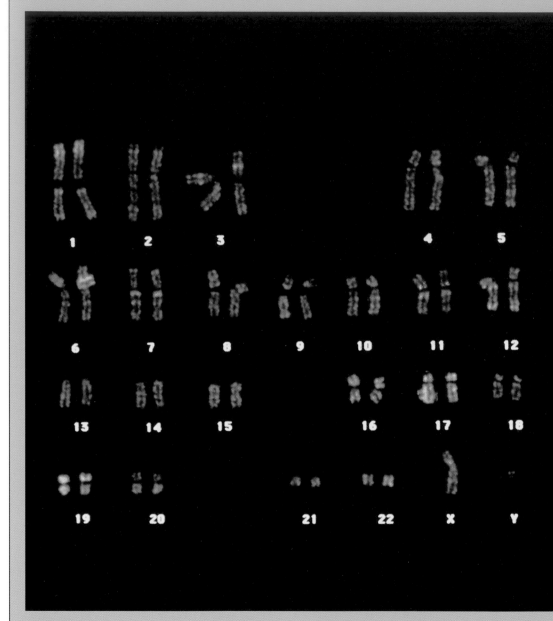

Skolnick, and Davis in 1980. This resulted in a flood of new markers and an explosion in the knowledge of genes' chromosomal whereabouts. The number of genes mapped tripled in 10 years. Gene mappers now update the map every day via electronic databases.

In the mid-1970s, when Frederick Sanger at Cambridge University and Walter Gilbert and Allan Maxam at Harvard University developed efficient new meth-

ods for determining the order of bases in a strand of DNA, scientists learned to sequence the genes they isolated. Automated sequencing followed in the 1980s. Now, once a new gene has been identified, it is immediately sequenced to understand the nature of the protein it codes for and to identify mutations that are related to disease.

Sequencing the entire genome, however, means sequencing at least 3 billion

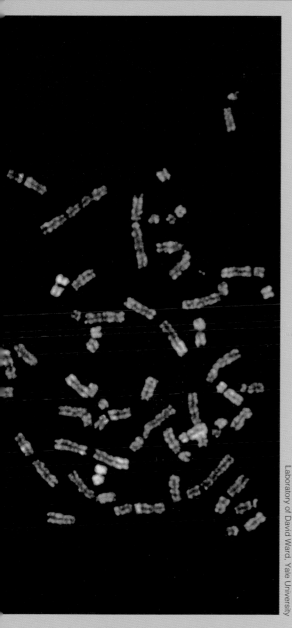

Laboratory of David Ward, Yale University

A previously uncharted fragment of human DNA is mapped to chromosome 11 with the aid of probes that glow in red. The 22 pairs of autosomal (non-sex) chromosomes in the genome, plus one X and one Y chromosome, are shown jumbled (right). After being identified by their banding patterns, they are arranged in order (left). Each chromosome appears double because it was caught in the process of division.

Scientists will be able to determine any gene's location relative to these markers.

In addition, a physical map will show actual distances along the chromosomes in terms of base pairs. The physical map probably will be constructed of long overlapping stretches of DNA cloned in yeast and known as yeast artificial chromosomes (YACs). Developed in 1987 by Maynard Olson, now an HHMI investigator at Washington University in St. Louis, YACs make it possible to clone and store much larger DNA segments than those cloned in bacteria. The technique has reduced the number of DNA pieces that need to be placed in the right order from about 100,000 to 10,000. Olson distributed his YAC library of the entire genome for the use of gene mappers.

Meanwhile, new strategies promise to speed up sequencing significantly. Automated sequencing and advanced computer software are already reducing sequencing time.

If a map of the genome and sets of overlapping clones had been available when researchers set out to find the cystic fibrosis gene, their task would have taken only a fraction of the time and cost, points out Thomas Caskey, of the HHMI unit at the Baylor College of Medicine. "The investigators could have simply reached into the freezer and pulled out two markers flanking the gene. The same would be true for many other diseases. And remember—once we make this map, we will never have to do it again."   —*B.M.*

"Once we make this map, we will never have to do it again."

base pairs of DNA—one chromosome of each type, or half the total number of chromosomes in a human cell.

Generally the most interesting or accessible genes have been located first, creating a disparity among chromosomal maps. The Human Genome Project is expected to produce a genetic linkage map in which the positions of genes for specific traits and diseases are superimposed on a grid of evenly spaced markers along the chromosomes.

OF MICE AND MEN

by Sandra Blakeslee

In the summer of 1980, Melvin Bosma, an immunologist at the Fox Chase Cancer Center in Philadelphia, began to examine four mice that were brought to his attention because of the peculiar results of their blood tests. The mice had none of the usual antibodies in their blood. In fact, they seemed to have no immune reactions at all.

The four mice turned out to be littermates, which suggested that their lack of immunity might be a genetic trait passed down from mother and father. Bosma soon realized that he and his coresearcher and wife, Gayle Bosma, had discovered an exceedingly rare and useful spontaneous mutation.

Named *scid* for *severe combined immunodeficiency*, the mutant mouse cannot make T cells or B cells, white blood cells

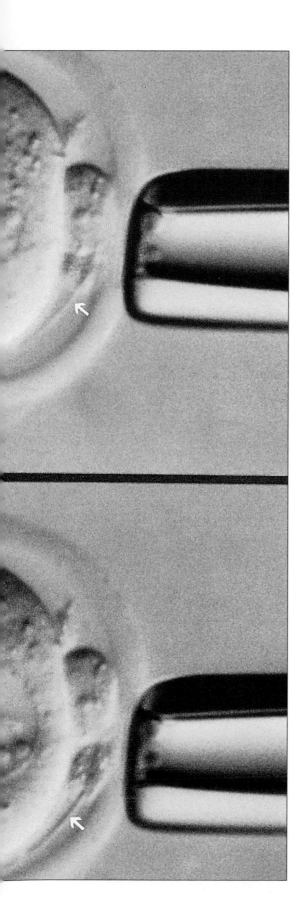

that fight off infections and foreign implanted tissues. It is one of the best animal models ever found for studying the basic biology of the immune system as well as several diseases.

The *scid* mouse "was a gift of nature," Melvin Bosma says. The Bosmas belong to a generation of modern scientists who use various strains of mice to study fundamental life processes.

Mice are small, handy, and remarkably fecund: three months after her own birth, a female mouse can produce a dozen new babies. Mice live only two to three years, allowing researchers to follow disease processes from beginning to end in a relatively short time. And their genes are remarkably similar to those of humans, despite an evolutionary distance of 75 million years between the two species.

As a result, researchers have come to realize that they can use mouse genes as a blueprint for finding and studying human genes, including disease genes. They can also use mouse models to test drugs, devise novel therapies, and study the physiology and biochemistry of genetic diseases in ways not possible in humans.

To create a mouse that models a human disorder, researchers inject a solution containing many copies of a defective gene into a fertilized mouse egg held by a large pipette (above). The male pronucleus, the genetic material that came from the sperm, swells up from the injection (below). The remnants of the sperm's tail can be seen (arrows) between the egg and the protective layer surrounding it.

They no longer have to wait for Mother Nature. . .

Most of the mutant mice strains under study arose spontaneously, and many of them are housed at the Jackson Laboratory in Bar Harbor, Maine, which maintains over 450 strains of mice with a wide variety of genetic afflictions and apt names.

The "stargazer" mutant, for example, has a neurological disorder that forces it to throw back its head and look at the sky. "Twitcher," "shiverer," and "quaking" mice have damaged nerve fibers and abnormal gaits. "Dwarf" and "little" are lacking growth hormones.

At least 20 new mutants are identified each year. Using various breeding tricks, the scientists are able to perpetuate the mutant strains in stable colonies.

Some of these mice have versions of common human diseases, such as diabetes, whose two forms—juvenile (Type I) and late-onset (Type II)—afflict millions of people worldwide. According to diabetes expert Edward Leiter of the Jackson Laboratory, these mice give researchers the opportunity to tease apart the genetic and environmental factors that underlie the disease. When fed high-fat diets, for example, some strains of mice become glucose-intolerant (as in Type-II diabetes), while others do not. By comparing the two strains of mice, scientists may be able to find the genes responsible for the different responses to diet.

The discovery of the *scid* mouse has allowed scientists to range farther afield and, for example, produce the first animal model of AIDS. It was generally believed that mice could not be infected with AIDS because many human viruses, such as HIV, which causes AIDS, attack only humans and higher primates. But in 1988, Michael McCune, who was treating AIDS patients at San Francisco General Hospital, had an inspiration: Since *scid* mice do not have a working immune system, they cannot reject human tissue, he reasoned; then why not implant human immune-system tissues into them, to see if that would allow the AIDS virus to infect the mice?

It was a wild idea, he admits. But to the utter astonishment of all his colleagues at Stanford University, where the experiment was carried out, it worked.

McCune now routinely infects his *scid*-human mice with the AIDS virus. The mice are treated with the drug AZT at different intervals after infection to test the advantages of early treatment. Hundreds of other antiviral compounds are also being screened, either alone or in combination, to see if they will stop the fatal march of AIDS.

In similar ways, a variety of fetal and adult human tissues can be implanted into *scid* mice to study cancer, brain disorders, and other conditions, as well as to develop gene therapy. McCune and other scientists have actually grown human lung, pancreas, intestinal, and brain tissues in these animals. The goal of research by Charles Baum of Systemix, a Palo Alto biotechnology firm, and HHMI investigator Irving Weissman of Stanford, is to isolate the human blood stem cell—a master cell that gives rise to all types of human blood and immune cells—which scientists will need in large quantities to make gene therapy become practical.

Researchers still lack animal models for most human genetic diseases. But this is changing rapidly as scientists learn they no longer have to wait for Mother Nature to make their mutants—they can create them to order.

By inserting foreign genes into animal embryos, they can produce "transgenic" animals, whose cells follow the instructions of the interloper genes as well as those of their ancestral genes. The result is an explosion of new information on how genes work in specific cells and how they go about promoting health and disease in both mice and humans.

In 1982, Richard Palmiter, an HHMI investigator at the University of Washington in Seattle, Ralph Brinster of the University of Pennsylvania, and their colleagues injected a modified rat growth-hormone gene into a fertilized mouse egg. The gene normally produces small quantities of growth hormone in the pituitary gland, but the researchers wanted to see what would happen if an animal had unusually high levels of the hormone.

Before injecting the gene into the mouse egg, therefore, they attached it to promoters, regions of DNA that control which tissue expresses a gene. The promoters would redirect the gene's expression to other cell types, such as liver cells, where the gene would be freed from its normal controls and would produce large quantities of hormone. Then they implanted the egg into a mouse foster mother. The mother gave birth to normal-sized pups that grew at an unusually rapid pace to become giant mice, nearly twice the size of their littermates. The picture of one of these super-mice was splashed across newspapers and magazines throughout the world.

This experiment paved the way for the first attempt to cure a genetic disorder—dwarfism—in transgenic animals by gene therapy. The mice "patients" were undersized because they lacked sufficient growth hormone. By inserting a modified growth-hormone gene into them, the researchers and Robert Hammer, who was then at the University of Pennsylvania (he is now a senior associate in the HHMI unit at the University of Texas), corrected the genetic defect. The correction was so good that the mice grew slightly larger than normal.

Since that seminal experiment, geneticists have been striving to find permanent cures for a variety of genetic diseases. But first they have to understand the basic biology of the diseases.

Five research groups—including the team of Palmiter, Brinster, Richard Behringer, then an HHMI associate in Brinster's lab, Tim Townes of the University of Alabama at Birmingham, and their associates—have succeeded in creating various mouse models of sickle-cell disease. A painful and often lethal genetic disease, sickle-cell disease affects about 2 million people around the world, mostly blacks. One in twelve American blacks carries the abnormal gene. More than 50,000 Americans suffer from the disease, which develops in children who inherit the gene from both parents.

Sickle-cell disease was the first genetic disease to be understood at the molecular level—it results from the mutation of a sin-

gle base in the DNA that codes for hemoglobin, the oxygen-carrying protein in red blood cells—yet there is still no effective treatment for it. Scientists have long been eager for an animal model on which to test new therapies.

To produce such a model, Richard Behringer injected a few hundred copies of the abnormal gene into fertilized mouse eggs. The pups born of these eggs were examined to see which, by chance, had incorporated the human gene into one or more chromosomes. Those with the gene were mated in the hope that the human gene had lodged in their egg or sperm and that the trait would be passed on to the next generation. By breeding these mice with a strain of mice that produced low amounts of mouse hemoglobin, the researchers have created mice with red blood cells that sickle.

However, these mice still have some of their normal mouse hemoglobin, Palmiter points out. Further experiments may yield mice with more abnormal human

A giant mouse (left) grown from an egg injected with rat growth-hormone genes weighs nearly twice as much as its normal sibling. This experiment was a major step toward creating animal models of human disease.

hemoglobin—mice that may be more like sickle-cell patients. Researchers could then test anti-sickling drugs in the mice and explore methods of gene therapy.

In other laboratories, cancers of the eye, breast, lymph tissues, pancreas, and other organs have been induced in mice by cancer genes and combinations of cancer genes. This work is fueling one of the most important revolutions in twentieth-century medicine—the ultimate understanding of cancer as a genetic disease.

It is now believed that cancer is caused by genetic mutations—most often, by a series of mutations, some of which may be inherited. Certain normal genes involved in cell growth, development, and differentiation can be converted into cancer-causing "oncogenes" by mutation. Other genes that normally prevent the uncontrolled growth of cells—"suppressor" genes—can also produce cancer if they are knocked out by genetic mutations.

Single mutations are generally not sufficient to cause cancer, but they produce changes that predispose cells to malignant growth. Additional mutations in other genes continue the cells' malignant transformation. Thus, cancer is a multi-step process involving the interaction between genes and their environment.

To test the hypothesis, researchers have put a variety of oncogenes into mice, using promoters to direct the genes to specific tissues. In this way, HHMI investigator Philip Leder and his co-workers at Harvard created and patented "onco-mice"—animals that reliably develop breast and lymph cancers. Onco-mice are being used worldwide to test drugs and therapies against those two forms of cancer, says Leder.

Aside from cancer, cardiovascular disease is the biggest killer in developed countries. Several researchers are using animal models in an effort to uncover the genes responsible for high levels of fats and cholesterol in the blood, to develop tests for identifying people with a genetic predisposition to heart disease, and to explore new therapies.

Michael Brown, Joseph Goldstein,

Robert Hammer, and their associates at the University of Texas recently created transgenic mice that never develop high levels of cholesterol in their blood, no matter what they eat, because they carry a human gene for receptor proteins that latch onto LDL, the carrier of "bad" cholesterol, and remove it from the bloodstream.

Most researchers make transgenic animals as Palmiter and Brinster do, by injecting foreign genes into fertilized eggs and then implanting the eggs in a surrogate mother. Others link the foreign genes to a virus and inject the viral combination into an embryo at a later stage of development; the virus then integrates itself into the animal's chromosomes, carrying the foreign genes along with it.

Either way, the researchers face a major problem: They cannot specify where in the animal's DNA the foreign gene will become integrated. If a gene is taken up at the wrong spot, it may disrupt a native gene or even cause a lethal mutation; if taken up in the right spot, it may cure a disease. Furthermore, one animal might integrate hundreds of copies of the gene, while another animal, under the same experimental conditions, might integrate but a single copy.

Another problem is that researchers can only add genes to the animal's own genome: They cannot get rid of an animal's unwanted or defective gene. The animal's own gene, equivalent to the inserted foreign gene, still functions in some form and could influence the experiment, as appears to have happened in mouse models of sickle-cell disease.

Recently a new technique has overcome these limitations and opened up a whole world of possibilities for making almost any wished-for animal model. It is called homologous recombination, or more loosely, gene targeting, and it is awesomely precise.

In homologous recombination, a desired gene finds an identical, or homologous, sequence of DNA in the animal's genome and swaps places with it. Several teams of researchers, including that of Oliver

Smithies, who is now at the University of North Carolina, have devised ingenious ways to achieve this precise exchange.

In 1987, Mario Capecchi and Kirk Thomas of the HHMI unit at the University of Utah showed that such strategies could be particularly effective when used with mouse embryonic stem (ES) cells. These cells, which were first isolated from very early mouse embryos by Martin Evans at Cambridge University and Gail Martin at the University of California, San Francisco, in the early 1980s, are unspecialized precursors of other cells. Each one is capable of giving rise to an entire animal.

When Capecchi introduced new genes into ES cells, most of the genes integrated into the chromosomes randomly. But in 1 out of 100 or 1,000 times, depending on the gene, the new gene found its exact counterpart in the mouse genome and integrated itself there, Capecchi says. It was as if the foreign gene had cruised up and down the mouse chromosomes and found its home address.

The challenge, then, was to find those cells in which the genes had landed on their home targets. If they could be identified, a powerful new genetic tool would be available.

Capecchi and his associates have pioneered a double-barreled system for selecting these rare cells. They start out by adding two other genes to the gene that is seeking its homologous site in an ES cell.

One of the added genes (the herpes virus thymidine kinase, or *tk*, gene) makes a cell vulnerable to the antibiotic gancyclovir. It is placed just outside the homologous sequence, so when the inserted DNA hits its precise target the *tk* gene is discarded because its sequence does not match up. The cells in which homologous recombination has occurred, therefore, remain invulnerable to gancyclovir, but those in which the DNA integrated randomly still have the *tk* gene and are killed by the antibiotic.

The second added gene, which confers resistance to the antibiotic neomycin, is placed right in the middle of the homologous sequence. When the cells are bathed with neomycin, those that failed to integrate the homologous gene are weeded out.

The cells that survive both treatments have the homologous gene right on target. These cells can then be used to recreate a mouse. Whole generations of mice can be raised from these cells, permanently carrying the inserted gene in exactly the right spot.

Homologous recombination allows scientists to carry out a new type of research with transgenic mice: knockout experiments. By introducing defective mouse genes into ES cells, researchers knock out the native gene, and the resulting mouse expresses the defective gene.

The technique should prove important for studying human diseases, such as cystic fibrosis, which until now have had no animal models. But there's a rub, says Capecchi. To do knockout experiments, researchers must already have cloned the human disease gene and its mouse homologue. So far, only a few hundred such genes have been deciphered.

Nevertheless, scientists are reveling in their new freedom to manipulate genomes in many ways.

"With gene targeting, you can knock out a gene, replace it, or change it more subtly," says Capecchi. "For example, you can put a gene under the control of a switch so that if you inject a certain drug, the gene is turned on—or turned off." Almost any disease process can now be studied in animals in this way.

Mice remain the favorite subjects for such studies—not only because they are convenient to work with, but because recent advances in mapping both the mouse and human genomes have made comparisons between the two especially fruitful. Scientists who work with transgenic mice confess astonishment and glee at the rapid progress in the field. The ability to add or replace mouse genes at will is leading to a medical revolution, they say, and new experiments are limited only by the human imagination. ●

Gene
targeting
is awesomely
precise.

THE FUTURE OF GEN

by
*Harold M.
Schmeck, Jr.*

The eminent British molecular biologist Sydney Brenner once got a hearty laugh from his audience by describing how some future graduate student will define a mouse—"ATC,GCC,AAG,GGT,GTA,ATA…"

But every year the idea of defining an organism by the sequence of its DNA bases seems a little less farfetched.

Victor McKusick, of The Johns Hopkins University School of Medicine, notes that scientists' growing ability to read and write in the language of the genes has already explained some of the once-mysterious basic concepts of genetics.

The difference between dominant and recessive traits as causes of genetic disease used to be just an abstraction based on a great deal of observation. If a genetic defect expressed itself only in patients who inherited the trait from both parents, it was called recessive; both copies of the gene coding for the trait were presumably defective, resulting in disease. If the trait was dominant, on the other hand, it meant that one defective copy of the gene was sufficient to spell disaster. But why should some disorders require two mistakes, while others resulted from only one? Molecular biology has given a concrete and remarkably simple explanation.

"It now appears that these two categories (recessive and dominant) correspond pretty closely to the two fundamental categories of proteins: enzymatic and structural," McKusick said in a review of genetics research. Recessive disorders tend to result from failures in genes that code for enzymes, the biological catalysts that do much of the body's chemical work. A person who has inherited the defective gene from only one parent often goes disease-free because the normal gene inherited from the other parent produces enough of the enzyme to serve the body's needs. The disorder appears only when the person inherits the same defect from both parents and therefore lacks any working copy of the normal gene.

If the genetic defect affects structural proteins, however—for example, collagen, a key component of connective tissues and bones—only one copy of the faulty gene is usually enough to cause disease. It is easy to see why. A four-engine airplane can still fly even if one of its engines fails, as long as the other engines provide enough power, but a single faulty strut that makes a wing fall off will cause the plane to crash.

The reason some genetic disorders are relatively common while most are extremely rare has also proved to be almost ridiculously obvious. The bigger the gene, the greater the chance that something will go wrong with part of it. In many cases, it seems as simple as that.

ETIC RESEARCH

Sometimes rather subtle differences in the defects of a single gene can make a profound difference in a patient's fate, as Louis Kunkel of the HHMI unit at Harvard University learned after he and his team discovered the gene for Duchenne muscular dystrophy (DMD) in 1986. Major flaws in that huge gene result in the presently incurable DMD, a muscle-wasting disease that leaves young boys wheelchair-bound by age 12 and may kill them by age 20 because the muscles that control breathing fail. By contrast, lesser defects in that same gene produce a much more benign disease, Becker's muscular dystrophy.

A year after discovering this gene, the team identified the protein it codes for—a previously unknown protein, now named dystrophin, which occurs in muscles in such small amounts that it would never have been found by ordinary means. Dystrophin plays a key role in muscle cells and may be involved in many other muscle diseases. Researchers are now analyzing how dystrophin functions, what other proteins it interacts with, and whether it might be replaced to interrupt the course of disease.

Experts see many more insights such as these in the future, as research in molecular genetics opens some of the "black boxes" of biology.

"I think we are going to have an explosion of understanding," says David Valle, of the HHMI unit at The Johns Hopkins University. For example, the causes of mental disorders certainly include environmental factors, but biological psychiatrists believe the genes are whispering an important message, if only it can be heard.

Genetic research will illuminate many disorders of single organs such as the eye, teeth, skin, and cochlea (the hearing apparatus of the ear), Valle believes. The deafness of about two-thirds of patients with serious hearing problems has a genetic basis, he says. Molecular biologists can find genes that are expressed only in the cochlea and therefore are probably important in hearing. Once such genes have been identified, several strategies exist for determining their functions and suggesting treatments.

Valle's current research focuses on a rare genetic disorder of the eye, gyrate atrophy, which leads to blindness through degeneration of the retina. The basic fault is an enzyme defect that causes an abnormal buildup of the amino acid ornithine. Surprisingly, some 35 different mutations in a single gene are able to produce the disease. The excess ornithine is found almost everywhere in the body—blood, urine, tears, spinal fluid—but the serious ill effects are limited almost entirely to the retina. As yet, nobody knows why.

Understanding the genetic cause of the

disease has led to a medical treatment that seems effective: severely restricting the patient's diet to bring the ornithine levels down to nearly normal. Scientists have compared the effects of this treatment on children in whom it was started early and on siblings who did not receive it until an older age. The studies confirm that the dietary restriction minimizes damage to the retina, Valle reports. But the diet is only a stopgap solution. Geneticists are searching for more effective remedies, including possible treatment for the gene defect itself.

"One of the really exciting things about modern molecular genetics is that we now have opportunities to make animal models of these diseases and to study what happens at the tissue level in a direct way," Valle says.

Philip Sharp, director of the Center for Cancer Research at MIT, divides the benefits of genetics research into two categories: those that generate knowledge and those that generate treatment. He sees animal models as extremely important to both. Deliberately produced genetic diseases in animals will have pathologies like those of human diseases. "We will learn how to recognize them, treat them, and analyze them in animals," he says. "That is going to be the forefront of biomedical science, in one area of it at least."

In addition, many aspects of human development will be clarified by work with mice, flies, and worms, he says. Scientists

Louis Kunkel and his team found the genetic flaws that cause two forms of muscular dystrophy.

have discovered that genes which are developmentally active in both *Drosophila*, the fruit fly, and *C. elegans*, the nematode worm, have direct counterparts in mammals, although the functions of these genes in humans are not yet entirely clear.

When genes of species that separated from each other many millions of years ago show so much similarity, there is every reason to believe they are related. Many molecular biologists have noticed that nature is quite frugal in preserving devices that have proved biologically effective. As an example, Valle points out that the human enzyme ornithine delta aminotransferase, which is defective in gyrate atrophy, is 54 percent identical to the comparable enzyme that functions in yeast.

"I think one of the real themes of biology is that Mother Nature uses things over and over again once she figures out how to solve a problem," Valle says.

This concept offers scientists a great opportunity, says Eric Lander of the Whitehead Institute. He thinks there is hope of compiling, eventually, a complete thesaurus of protein parts that function in Earth's myriad species. "That would be spectacular," Lander says. "If we had the thesaurus of all the moving parts, then we would understand life in a remarkable way."

Gene mapping and cloning are key to the assembly of the thesaurus, and progress in these areas is clearly accelerating. However, most of the 50,000 to 100,000 human genes remain totally unknown, and there is still a long way to go.

To date, most of the progress in understanding the genetics of human disease has involved relatively rare conditions, such as cystic fibrosis or Duchenne muscular dystrophy, that are caused by errors in single genes. But science is also stalking the genes that contribute to heart disease, cancer, diabetes, and mental illness—the big killers and cripplers of mankind.

It may soon be possible to tell some people that they have certain genetic predispositions to a specific major illness and suggest that they tailor their lifestyles accordingly. Similarly, the use of drugs to treat some of the important diseases could be tailored to the genetically varied needs of patients, with benefits for them and for the health care system in general—"Different strokes for different strokes," as one scientist put it.

On the other hand, some scientists fear that people might be stigmatized or become uninsurable because of genetic traits, such as carrier states, that don't in themselves have any appreciable effect on health.

Genetic research is advancing steadily, often rapidly, on many fronts. It has long been known that some disorders affect males, others affect females primarily, while still others may appear in either sex. But a few years ago researchers discovered that, even in some of the latter disorders, the gravity and sometimes even the nature of disease may depend on which parent provided the faulty gene. This phenomenon is called imprinting. Although it has been detected only in rare human conditions, imprinting is a subject of intense study as researchers look for other examples.

Other scientists have forsaken the genes that reside in cell nuclei and are finding new clues to disease in the genes of what are probably our oldest and most entrenched "parasites"—the mitochondria, tiny, energy-generating organs inside every cell. Mitochondria are thought to be the descendants of ancient bacteria that

"Different strokes for different strokes"

not only found a home in animal cells, but also adapted so thoroughly that they became indispensable functional parts of those cells.

We inherit mitochondria only from our mothers; sperm leave their mitochondria behind when they enter the egg. Flaws in mitochondrial genes have been found to lead to certain types of blindness and epilepsy and may also contribute to some degenerative disorders, such as dementia, that are associated with aging.

"Mitochondrial DNA gives us a whole new way to think about genetic transmission of diseases," said Douglas Wallace of Emory University, a specialist in those vital intracellular power stations.

In even more fundamental ways, discoveries in genetics have led to novel strategies for treating disease. DNA is mainly the archive of genetic information. Its orders are translated into action by segments of ribonucleic acid (RNA), which serve as the working blueprints for all proteins. Today, chemists are beginning to create valuable new drugs by fabricating "anti-sense" segments of RNA, whose sequence is the exact opposite of an unwanted sequence, to combine with certain existing strands of RNA and thus block the action of specific genes.

The bottom line in any kind of biomedical research lies in the realm of treatment and prevention. The ultimate step in that direction is gene therapy—the deliberate transplantation of genes to treat or even prevent human disease. Many geneticists dismiss gene therapy as a distant prospect; others disagree.

Though gene therapy has not been too successful so far, it adds to the excitement that pervades genetics today. Studies of microbes, plants, animals, and many normal human beings are all contributing to the explosion of new knowledge. In recent years, molecular genetics has given important insights into the origin of life and its evolution, the emergence of humans, and our intimate relatedness to every other species on Earth. We can expect many more advances as geneticists continue to explore the wonder of life. •

Much of the excitement in genetics today comes from its lively offspring, "genomics." This newcomer specializes in large-scale analyses of all the genetic material in the genomes of organisms ranging from bacteria to mammals. Genomics is expected to provide the functional meaning of newly revealed DNA sequences: What do these genes really do? And that precise knowledge, in turn, heralds a revolution in the diagnosis, monitoring, and treatment of diseases.

Meanwhile, new DNA sequencing techniques using high-speed robots are flourishing. So are ingenious combinations of YACs (yeast artificial chromosomes), BACs (bacterial artificial chromosomes), PACS (fragments of DNA in a vector derived from a bacteriophage known as P1), and MACs (mammalian artificial chromosomes), which supply genetic fodder for the machines that do the sequencing.

A Gifted Young Patient Battles Cystic Fibrosis

Jeff Pinard, the young man with cystic fibrosis who was seeking his own genetic flaw as a college student, is now 29 and doing quite well. In May of 1998 he married Darcy Hartley, who is studying for her master's degree in elementary education. They live in a house of their own in Grand Rapids, Michigan. He does computer work for an electric utility, mostly from home. But he has had some very rough times.

Several bad bouts of illness landed him in the hospital for months and forced him not only to give up his research work, but to drop out of the University of Michigan. After going home to his parents in Grand Rapids, he was hospitalized again because of pancreatic problems that produced excruciating pain. These problems remain unsolved, despite a variety of treatments. However, Pinard feels generally better and, according to his mother, "he's had some periods of time when he has been without pain." How long? "Oh, a couple of weeks at a time ...."

Lap-Chee Tsui of the University of Toronto finally identified and sequenced Pinard's second, milder CF mutation, adding it to the list of more than 850 known mutations in the CF gene. But it is difficult to count CF mutations these days, Tsui says, because "many mutations are now found in atypical diseases, such as male infertility and pancreatitis."

A major study of pancreatitis led by Jonathan Cohn of Duke University Medical Center and published in the *New England Journal of Medicine* recently concluded that many adults who suffer from so-called "idiopathic" pancreatitis (pancreatitis of unknown cause) actually have cystic fibrosis. The authors add that these findings "will change how physicians treat patients with this condition."

A Natural Antibiotic

The major symptom of CF is lung infection. In 1996, Michael Welsh, an HHMI investigator who teaches medicine and physiology at the University of Iowa, discovered why these infections occur—and offered a new approach to treatment that is still being developed. Welsh had a longtime interest in epithelia, the sheets of cells that line the internal and external surfaces of the body, including those lining the airway. When the CFTR gene was identified, he was among the first to examine the role of the protein made by this gene.

"We found out that it's actually a chloride channel, through which salt moves across the membrane," Welsh says. "That was very satisfying, because then you could begin to tie together the physiology—defective epithelia—and the gene product, the chloride channel." Then his team made an intriguing discovery: Normal epithelial tissue can kill a large number of bacteria, while similar tissue from people with CF fails to do so, or even allows the bacteria to multiply.

Welsh guessed that the fluid covering the airway normally contains factors such as defensins, molecules that are part of our innate, nonspecific defense system. He wondered whether these were also present in people with CF. To his surprise, he found natural antibiotic substances both in healthy people and in those with CF. In CF patients, however, the substances' activity was greatly reduced by the abnormally high salt concentration resulting from the defective CFTR channel. When Welsh lowered the salt concentration, even the epithelia of CF patients became able to kill bacteria.

The team's conclusion: Drugs that reduce the salt concentration in airway fluid may help treat or prevent the sometimes fatal lung infections of CF patients. Other antibiotic drugs that resemble defensins may also be developed for this purpose.

As for his experiments with gene transfer, "they remain just that—experiments," says Welsh. "We can deliver the normal CFTR gene, but we cannot deliver it efficiently enough," he explains. "The problem is the delivery. We need to go back to the lab and try to make it work better."

Genetic Screening

"We are slowly moving closer and closer to implementation of genetic screening for CF," says Arthur Beaudet. New recommendations about such screening emerged from an NIH Consensus Development Conference held in the spring of 1997. The consensus panel declared that all expectant couples and those planning to have children should be offered the option to test for gene mutations that cause CF; so should couples in which one or both partners have a family history of the disease. The panel further recommended that insurance cover the cost of these tests.

Several labs around the country now provide such tests. "And we can get 50 different mutations on a single test for as little effort as one," Beaudet says. He adds that the tests have become so sensitive that "somewhat over 90 percent of CF carriers would be correctly identified." Therefore the tests would detect more than 81 percent of the couples at risk. Beaudet believes that newly married couples should be given a set of prepared mouth swabs to take home, so that they can test themselves at their leisure. Further tests would then be necessary only if both members of the pair are carriers of CF.

About 40 specialized centers worldwide offer *in vitro* fertilization to avoid genetic diseases. At the Illinois Masonic Hospital in Chicago, for example, Charles Strom and Yuri Verlinsky screen the eggs of mothers who are carriers of CF before fertilizing them with the husband's sperm. In this analysis they use only the eggs' polar bodies, which would be cast off anyway, Strom explains. If the test indicates the egg is free of the CF mutation, the doctors proceed with fertilization. "More than 16 healthy children have been born to CF carriers with this method," Strom reports. The method has now been extended to a variety of genetic diseases, including hemophilia, thalassemia, and sickle cell anemia.

New Findings About Brain Disorders

There was great rejoicing when Nancy Wexler's quest for the cause of Huntington's disease finally succeeded in 1993, after nearly eight years of effort described as "a nightmare of false leads, confounding data, and backbreaking work." The faulty gene was named *huntingtin*. The guilty mutation turned out to code for an extra-long, repeated stretch of glutamine, an amino acid in huntingtin, the protein made by this gene. But no one knew how the expanded glutamine repeats cause brain neurons to sicken and die.

Several other "triplet-repeat" diseases are known. They all attack some part of the nervous system, and all of them are still mysterious. Scientists have begun to search for clues to the function of huntingtin in the proteins that interact with it.

"The beauty of having the Huntington's disease gene in hand is that we are now able to place it in animals and learn its effects," says Wexler. In 1996, Gillian Bates and her team at Guy's Hospital, London, put fragments of the human HD gene into mice for the first time. The mice developed HD-like symptoms two months after birth and died soon afterward.

Fruit flies were also enlisted in the fight. In 1998, George Jackson of UCLA's Department of Neurology inserted fragments of the HD gene into the large nerve cells in the eye of a fruit fly. He found that, just as in human beings, the cells' fate depended on the number of glutamine repeats in the HD gene's DNA. The eyes of flies whose gene had only two repeats remained normal. Those with 75 repeats were normal for a month, but then began to degenerate slowly. When the flies had 120 repeats, their eyes suffered massive cell destruction.

Wexler is greatly encouraged by these findings. She points out that "the fly eye is a perfect laboratory to test the effects of drugs that will protect the eye and prevent degeneration." And because of many similarities between HD and other neurodegenerative conditions, including Alzheimer's, and Parkinson's diseases, scientists hope that the findings from one of these areas will advance research in the others.

The Viking Genes

Other gene defects uncovered in recent years include mutations predisposing people to such widespread ailments as breast cancer, familial polyposis of the colon, Alzheimer's, and Parkinson's.

Many family groups have helped in these searches. Scientists now look forward to working with the biggest genetic trove of all—the Viking gene pool, which can be found in very pure form among the 170,000 people of Iceland. Some of these families can be traced back for hundreds of years, and their records will soon be available to researchers. As DNA-based biology expands, so does the need for large groups of people with detailed and accurate family trees.

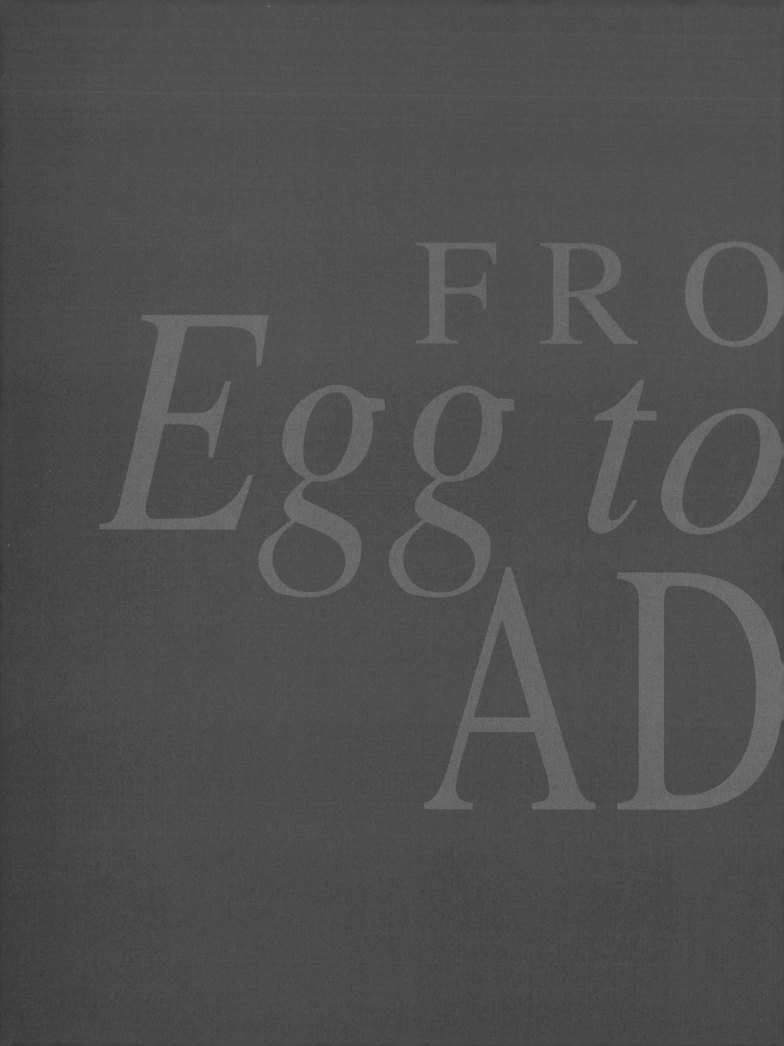

FROM Egg TO ADULT

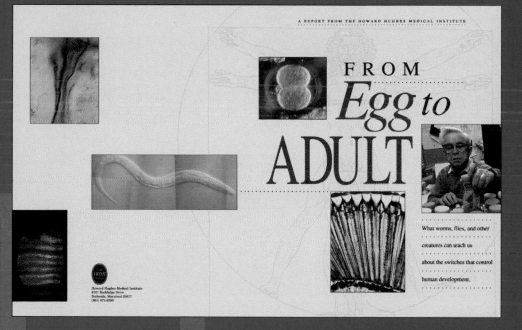

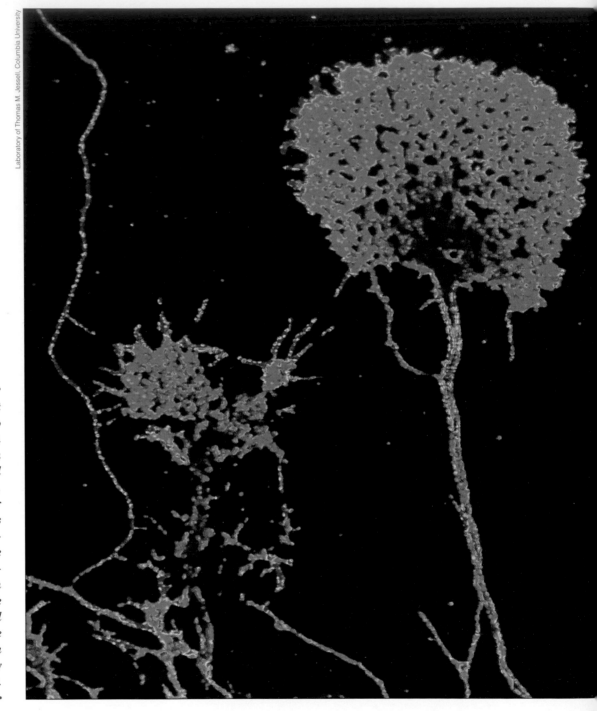

Axons, the fibers that connect nerve cells to nerve cells, can reach a length of nearly a meter in humans and 10 meters in whales. How they find their way to their targets during early development is under intense study. In this photograph, an axon from a rat embryo nerve cell shoots out toward a target cell. At the tip of the long fiber, a growth cone (bushy mass) leads the way.

FROM *E*GG TO *A*DULT

What worms, flies, and other tiny creatures can teach us about the switches that control human development

INTRODUCTION

Whe take it for granted that most babies are born with eyes, toes, and brains in the right places, and that some are boys while others are girls. But what actually controls these critical developments?

The answers are now pouring in from studies of fruit flies, microscopic roundworms, and other tiny creatures that reproduce rapidly. Scientists started observing and testing these small animals long ago, realizing that they could never examine human development in the same way. When recombinant DNA technology made it possible to isolate genes that become active in these animals' eggs or embryos at particular times, researchers suddenly gained the power to pinpoint the origins of developmental changes.

Would a discovery in one organism extend to others? Could a finding made in fruit flies apply to humans in anything but the most general sense? At first, the answer seemed to be "no." Today, to the astonishment of many researchers, studies are showing that different species, including humans, share the same genetic mechanisms.

Major chemical pathways determine the development of all species. In each pathway, genes produce proteins that switch on other genes, which then produce other proteins, and so on, with each step further defining the destiny of individual cells. Different cells respond to such switches in different ways. Some migrate. Others specialize—becoming, for instance, liver or muscle cells. Many of the cells divide. And some cells die.

Sometimes, cells have the potential to act in either of two ways and must "select" one of them. Researchers describe these cells as "making a choice" or "deciding" their fate. In fact, the cell's "decision" is driven either by its lineage (which cell it comes from) or its environment (what signals it receives from neighboring cells), according to Robert Horvitz, an HHMI investigator at the Massachusetts Institute of Technology.

"We have to trace some of these decisions back further and further, to the single-celled egg," Horvitz points out. "And ultimately, that takes you back to the mother." For instance, in the fruit fly, *Drosophila melanogaster*, the mother provides the future embryo with spatial information (such as which way is up) long before the egg is fertilized, storing this vital information in the egg in the form of molecules specified by her own genes.

Some of the most important mechanisms of early development have been discovered in the fruit fly. Three scientists who used the fruit fly to make fundamental contributions shared the Nobel Prize in 1995. They were Edward Lewis, who showed that certain genes direct individual body segments to become specialized organs; and the team of Christiane Nüsslein-Volhard and Eric Wieschaus, who identified the genes that determine the body plan and the formation of body segments. Since then, scientists have found similar genes in humans. This strongly suggests that the basic genetic mechanisms controlling early development have been conserved through millions of years of evolution.

Most of the genetic pathways involved are used not only in embryonic development, but also at various times of life. When they fail, the results can be catastrophic, ranging from birth defects to cancer or the degenerative diseases of old age. Researchers hope that a more complete understanding of these pathways will make it possible to prevent many such errors.

That prospect hinges on scientists' growing ability to apply to humans what they learn in insects and worms. For instance, while the development of a fly's eye might seem unrelated to human cancer, scientists who study the former have been amazed to find that they are also illuminating the latter, since the genetic pathways are the same. An equally surprising connection with cancer has turned up in investigations of sex-organ development in *Caenorhabditis elegans*—a roundworm so small that 10,000 of them can fit into a single laboratory dish.

These worms are treasured by developmental geneticists, who know them thoroughly, down to the fate of every single cell (see p. 88). Since December 1998, when an international team announced that it had deciphered the DNA sequence of every gene in *C. elegans*, the worms have become even more valuable: They are the first multicellular creatures to have their genetic blueprint completely spelled out. Their DNA encodes a total of 19,000 genes—roughly one-fifth the estimated number of genes in a human being.

"Before that, we knew what genes are required to make an individual cell, like a yeast cell," says Cornelia Bargmann, an HHMI investigator at the University of California, San Francisco. "Now we'll be able to find out what it takes for a group of cells to get together and communicate with each other."

Despite all the recent progress, developmental biology is still in its own embryonic stage. This is particularly clear when scientists wrestle with the nearly insurmountable problem of how the brain, the world's most complex system, develops and grows. Yet even here, some exciting discoveries have been made, and new genetic techniques are laying the groundwork for deeper understanding.

Maya Pines, *Editor*

How does a fertilized egg turn into a fly, a chicken, you or me?

Allan Spradling takes a vial full of tiny swarming flies, stills them with a squirt of anesthetic, and shakes them onto a glass plate. As seen under a microscope in his Baltimore laboratory, the flies, a quarter inch from wingtip to wingtip, are all monstrous bristles, segmented abdomens, and big red eyes made up of elements aligned as neatly as transistors on a computer chip. This is *Drosophila melanogaster,* the common fruit fly, a mainstay of biological research for most of a century.

In fruit flies, as in humans, the intricate dance of growth and change that biologists call development usually goes just right. Most fruit flies look like, and behave like, fruit flies. Most humans have two arms, two legs, and a heart with four chambers. The fertilized egg successfully develops into fly, chicken, or human; then the young fly, chicken, or human develops further. In humans, massive hormonal changes signal the onset of puberty thirteen or so years after birth, the body quietly tracking the time. All in seeming obedience to some schedule, pattern, plan.

Sometimes, of course, things go wrong: A fruit fly's eyes are white, not red. Its antennae project from the wrong point along its body. It grows an extra set of legs. Biologists have catalogued some 3,000 such genetic mutations in *Drosophila*. Comparable mistakes sometimes arise in humans, causing great suffering. A baby is born with the two large arteries of its heart reversed (as happens in about 25 of every 100,000 live births) or with attenuated limbs, like those babies whose mothers took thalidomide, a sedative prescribed against morning sickness, 30 years ago. About one in twenty infants has a birth defect, often a serious one, while half of all pregnancies never reach term because of fatal errors in

By Robert Kanigel

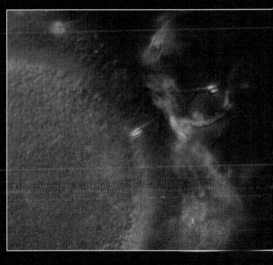

A human sperm breaks through an egg's inner membrane during studies of fertilization. Before reaching this membrane, the sperm must penetrate the egg's outer coat (insert).

development. Even years after birth, missteps in cell and tissue development can occur: lung, or breast, or liver tissue can go on a rampage of unrestricted growth, forming cancers.

When things go right, how is it that some cells grow up to become skin, others brain? Do organisms start out preformed, or do they take their form from internal and external cues? How do we wind up with five toes, not six or four? Why is the thumb where it is, the pinkie where it is?

Fruit flies such as this one (magnified 11 times) are attracted to fermenting fruit and feed on the yeast.

Laboratory of Edward B. Lewis, California Institute of Technology

How does cardiac tissue form the intricate pattern of ducts, chambers, and valves that make up the heart? What produces the leopard's spots, the filigreed intricacy of the body's arteries and veins, the porcelain skin of Meryl Streep?

"Developmental biology is not so much a discipline as a set of questions," says Donald Brown, director of the Carnegie Institution of Washington's Department of Embryology in Baltimore, Maryland, where Spradling, an HHMI investigator, studies the egg's role in fly development. These questions have tantalized thinkers—they have tantalized everyone—for centuries.

To be sure, an answer of sorts has been around for years. It was *Drosophila* pioneer Thomas Hunt Morgan who established, in the 1930s, what has become the fundamental principle of developmental biology: Growth and differentiation result from genes being turned on and off.

Morgan advanced this idea long before anyone actually knew what genes were made of. He died a decade before Watson and Crick unraveled the structure of DNA, before it became known that certain sequences of subunits, or bases, in the DNA double helix contain the code for making proteins, which are the body's workhorses. Such sequences determine a protein's biological role, whether as collagen in our skin, hemoglobin in our blood, or enzymes that control digestion.

A gene is the DNA corresponding to any one protein; some 50,000 to 100,000 different genes constitute the blueprint for a human being. But while each cell, even in the earliest embryo, carries all that DNA, each cell "expresses" only some of its genes—that is, produces only some proteins—at any given time; the rest lie silent. And which genes are on or off, in which cells, at what time, dictates the development of the egg into what it is to become.

These facts represent an understanding of sorts, but it is an incomplete one. As Gerald Edelman, a professor of developmental biology at the Rockefeller University, New York, has written in his book *Topobiology,* "It is very difficult to account for the forms, patterns, or shapes of complex animals simply by extrapolating from the rules governing the shapes of proteins" at the molecular level. None of these rules comes close to explaining the miracle of a human infant or, for that matter, a fruit fly. As Sydney Brenner of Britain's Medical Research Council has observed, the notion that all development is "simply a matter of turning on the right genes in the right places at the right times" is "while absolutely true...also absolutely vacuous. The paradigm does not tell us how to make a mouse but only how to make a switch. The real answers must surely be in the detail."

And it is just this sure and satisfying detail, the product of work with flies, mice, worms, and other organisms, that has begun to emerge from laboratories around the world.

An Assembly Line for Making Eggs

Allan Spradling can practically see *Drosophila* genes turn on and off. In color.

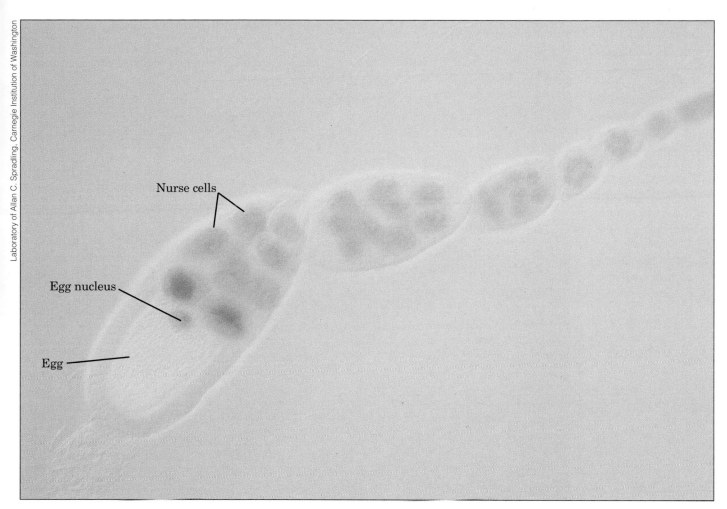

Nurse cells

Egg nucleus

Egg

He has looseleaf binders stuffed with photographs of fruit fly eggs in various stages of development, both before and after fertilization, in which cell nuclei are tinted a bright blue. The blue means a particular gene is being expressed in that cell at that time, which means the gene is making a protein the cell needs.

The female fruit fly, laden with eggs that account for more than half her weight, runs what amounts to assembly lines in which eggs develop into mature form. These little factories consist of seven work stations—teardrop-shaped egg chambers, strung together like beads on a necklace, in which the developing egg gradually changes form, clouding over into milky white along the way. The seventh chamber, at the far end of the assembly line, periodically ejects an egg ripe for fertilization and then transformation into a larva, the next step on the way to adulthood.

All this is visible under a low-power microscope and has been known since the days of Morgan, who began his work with *Drosophila* in 1909. What's new today is that Spradling uses the powerful tools of molecular biology to pinpoint gene action in each egg chamber. In one photo, for example, only four cells, two at either end of an egg chamber, contain distinct blue dots. Nothing subtle here, no squinting to see it: one particular gene and no other, among *Drosophila's* five to ten thousand, is shown to be active in these cells. Because the egg chambers change continuously as they move down the line, a photo of the area as a whole reveals, through the blue patterning, seven distinct slices of developmental time.

Spradling's tools are "transposons," snippets of DNA that occupy more or less random spots in the fly's genetic material

Each of these seven egg chambers will eventually produce one Drosophila *egg (white area at lower left). The nucleus of the developing egg and the nuclei of supporting "nurse" cells are stained blue, revealing that a particular gene is active in those cells at that time.*

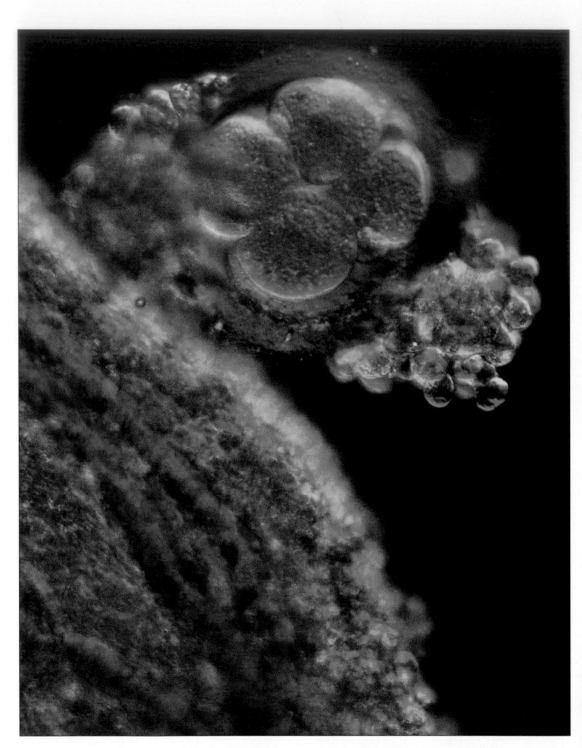

Three days after fertilization a human egg has divided into eight "totipotent" cells, four of which are visible here, surrounded by smaller cells that nourish them. Since totipotent cells are not yet committed to any particular developmental fate, none is essential to the embryo, and the clump occasionally breaks into two, producing identical twins.

and can move from one site to another. Transposons are found abundantly in nature, but those used in research have been elaborately doctored. The one Spradling uses now, genetically incorporated into the DNA of every cell of this line of fruit fly, has been furnished with a segment coding for an enzyme that makes a particular dye turn a telltale blue.

Wherever this transposon may land in the fly's DNA, some gene is likely to lie next to it. And when this gene's DNA is expressed, so is the transposon's. The cell floods with the enzyme, along with the product of the gene itself, and turns blue when the dye is added—thus revealing the gene's activity.

Getting the Basics Straight

The precision made possible by transposons and similar tools invites some scorn for the science of observation and description that preceded them. And yet, especially for embryology (as developmental biology was known until the 1950s, when the new, broader term caught on), that scorn is unwarranted. What we know today of the what-happens-when of embryos—never mind the why or how—reflects the prodigious work of centuries.

Early thinkers sometimes drew blanket assertions from thin air. The fifth century B.C. Greek poet and philosopher Empedocles, for example, asserted that infants are completely formed after only 50 days in the womb. An ancient Indian source claimed that the embryo is fashioned from semen and blood, with the firm parts of the body coming from the father and the soft ones from the mother. So simply getting the basics straight represented no small challenge to early embryologists.

Aristotle in the fourth century B.C. dissected animal embryos—and perhaps even an aborted human fetus. He was probably the first to broach one of the great questions of embryology: Does the organism simply unfold from preexisting structures ("preformation"), or does it develop afresh through interactions with its environment ("epigenesis")?

More than 2,000 years later, Karl Ernst von Baer, the first in a great line of 19th century German embryologists, found that the earlier in development he examined an embryo, the more it resembled those of other species. An early chick embryo looks like a mouse embryo or, for that matter, a human embryo (see p. 70).

The German embryologists following von Baer used frogs, salamanders, and chicks in their studies—mostly because they were near at hand. (In those days, the gentleman scientist normally paid for experimental animals out of his own pocket.) A little later, the American embryologists who took summer positions at the Marine Biological Laboratory (MBL) in Woods Hole, Massachusetts, began using eggs from fish, sea worms, snails, and other marine organisms. These eggs were typically small, but unlike those of frogs, for example, they had no heavy pigment, so scientists could see development proceed.

This was important because studies of cell lineage—the family tree of individual cells or regions of the embryo—were "the great overriding concern" around the turn of the century, says James Ebert, a senior developmental biologist at MBL. What was it possible for a tissue to become at a particular stage of development—its "potency"? What did it in fact become—its "fate"? In pursuing either question, continues Ebert, "you'd sit for hours and hours and watch things develop" under the microscope. Scientists would mark parts of the embryo with a dye and see where they wound up. Or, if the dye diffused too fast, they would dot the embryo with little specks of carbon and see where they went. The result was "fate maps" —elaborate, painstaking records of how embryos developed and grew.

In crude outline, animal development proceeds like this: Eggs and sperm form separately. Sperm fertilizes egg. The fertilized egg cleaves in two, initiating a series of rapid cell divisions. Early on, each cell has the ability to become any adult tissue—it is "totipotent." Cells in mouse embryos, for example, remain totipotent through the eight-cell stage. But soon these cells begin the march down the road toward what they will become; fate constricts them. First they form a hollow sphere—the blastula. Next an inner layer develops, the endoderm, and an outer layer, the ectoderm. Then, during a key process called "gastrulation," a middle layer, the mesoderm, begins to form, leading to furrows and folds, and thence the primordial organs. The ectoderm forms skin and nervous tissue, the endoderm the lining of the gut and the lungs, the mesoderm most everything else.

The descriptions, sketches, and drawings that recorded this knowledge were

"You'd sit for hours and hours and watch things develop."

The Dialogue Between Egg and Sperm

Eggs and sperm die within days if they do not meet. To improve their genes' chances of immortality, the two kinds of cells have worked out an intricate system of signals that seems to begin with a "come-hither" chemical from the egg.

David Garbers, of the HHMI unit at the University of Texas Southwestern Medical Center at Dallas, discovered such a chemical in fluid surrounding the eggs of sea urchins. When he placed a swarm of sea-urchin sperm in an experimental tank filled with ordinary sea water, they just floated about lazily. But as soon as he squirted a drop of the egg-related fluid into the center of their tank, the sperm became agitated. They darted forward, as in a race, and started swimming around in a frenzy.

The chemical proved extraordinarily potent. Garbers and a graduate student succeeded in identifying it and named it "speract" (for "sperm-activating"). His lab also identified the special receptors on the surface of sea-urchin sperm that recognize the egg's signals so the sperm can home in on them.

Human sperm appear to depend on similar signals. In cooperation with Michael Eisenbach of the Weizmann Institute in Israel, Garbers repeated his experiment with fluid taken from a human ovarian follicle, one of the cell clusters that surround developing eggs. The moment the human sperm had a whiff of this fluid, they sped up, giving "the first evidence that, in humans, the egg and sperm are really exchanging information even before they make con-tact," says Garbers.

An understanding of the dialogue between egg and sperm could lead to entirely new methods of treating infertility—as well as preventing conception. Fertilization is a relatively rare and complex event. A healthy man delivers about 280 million sperm to the woman with each ejaculation, but only a few hundred sperm manage to reach the part of the Fallopian tube that contains the egg. The sperm then respond to the egg only if it comes from a follicle with particular qualities.

In Eisenbach's lab, researchers examined how human sperm from the same donor reacted to fluids from 62 different human follicles. All the follicles had surrounded eggs from women who wanted in-vitro fertilization, a procedure in which sperm and egg are united in a laboratory dish and then injected into a woman's uterus; it succeeds only 20 percent of the time. In the experiment, the sperm were free to swim toward the follicular fluid through a porous membrane, but only half of the fluids attracted them.

"What was interesting was that there might be fluid from two or three different follicles taken from the same woman, and that one might attract but the other one might not," says Garbers. "We found an almost perfect correlation: the eggs from a particular follicle whose fluid attracted sperm were the eggs most likely to become fertilized in the test tube."

Now Garbers would like to identify the molecules that produce this attraction—the human equivalent of the sea urchins' speract—and also the receptors on the surface

---

stunning, fabulously detailed—a triumph of pure *seeing*. Historian of science Frederick Churchill gave this tradition its due when he wrote of "the towering edifice of classical descriptive embryology, solid in its discoveries, magnificent in its tracery and fine details, and as defiant of and removed from modern biology as a gothic cathedral is from today's secular world."

Then, too, as Carnegie's Donald Brown says, these early embryologists "phrased all the questions that people have asked for the next ninety years" as they groped for theories, explanations, and deeper

David Garbers injects a female sea urchin with a solution of potassium chloride to make it release its 100 million eggs for his experiments.

of human sperm that enable sperm to respond to these signals.

Other researchers are analyzing the next steps in the elaborate interaction between egg and sperm: the signals they exchange when they reach their rendezvous. These signals make it possible for sperm to pierce through the egg's thick outer coat, the zona pellucida. The first sperm to break through the zona reaches the egg's inner membrane, penetrates it, and propels its nucleus, containing its genes, into the egg. Immediately, thousands of little granules sent out by the egg's inner membrane change the structure of the zona pellucida, blocking it to entry by any other sperm.

The sperm's entry galvanizes the egg, which had been in a state of arrested development. The inner substance of the egg begins to move vigorously, forcing the sperm's nucleus towards the egg's nucleus until the two nuclei meet.

Twelve hours later, the male's 23 chromosomes mix with the female's 23 for the first time. The newly fertilized egg then duplicates itself and divides—a dramatic act that will be repeated billions of times in the cascade of cell divisions that marks the intricate, still perplexing journey from egg to adult.

—M.P.

It begins with a "come-hither" signal from the egg.

insight into what their pencils and pens recorded so faithfully. In the flush of excitement over Darwin's *Origin of Species* and theory of evolution, for example, came Ernst Haeckel's biogenetic "law." This stated that "ontogeny recapitulates phylogeny"—that the developing embryo goes through stages that mimic the evolutionary development of the species.

Haeckel had twisted von Baer's observation from earlier in the century to explain, well, *everything*. And he was a powerful propagandist for his views. As Jane Oppenheimer, a historian of embryology formerly at Bryn Mawr College, acidly portrays him, "he wrote lots of convincing popular books that were full of errors" and would brook no challenge to the overarching framework of his theory.

Even today Haeckel's "law" is occasionally passed off as biological truth. Yet "embryos do not pass through the adult stages of their ancestors; ontogeny does not recapitulate phylogeny," writes Lewis Wolpert, a British developmental biologist at University College, London. "Rather, ontogeny repeats some ontogeny—some embryonic features of ancestors are present in embryonic development. In any group of related animals, such as vertebrates, there is a stage in development which is common to all members of the group."

A student of Haeckel's, Wilhelm Roux, became the first real experimental embryologist. He and his followers didn't just look, speculate, or theorize—they *interfered* in development. They tried to see what happened if you cut off an embryo's tail; or shook an egg to pieces; or destroyed one cell of an embryo's two-cell stage with a hot needle. This tradition reached its height with Hans Spemann, who, with his young Ph.D. student Hilde Mangold, reported an experiment in 1924 that sent a shock wave through embryology.

Spemann and Mangold transplanted cells from a particular region, the dorsal lip, of an early salamander embryo just beginning gastrulation, to another embryo at the same developmental stage but from another species. Using species of different pigmentation so they could readily track the contributions of donor and host, they saw a new embryo—a second embryo—form at the transplant site. Plainly, something in the cells of the first embryo had triggered the new growth, organizing its development into a fully formed embryo. What was this something? What, as it came to be known, was Spemann's Organizer?

All of science set out in search of it, spawning what James Ebert calls "an embryological industry. People really did think it would be one substance, or one

kind of substance." Scientists ground up tissue, coagulated it, implanted it, froze it, thawed it, boiled it, passed it through superfine filters in order to establish its size. The literature devoted to finding the active ingredient mushroomed. "It was," as Ebert put it, "like the quest for the Holy Grail."

But in time, says Jane Oppenheimer, "people began to see that it was not just one thing that organized the embryo." Certain off-the-shelf chemicals, even a laboratory staple like methylene blue, could spawn some of the organizer effects. Prick the embryo just so and you could do it, too.

Spemann's Nobel Prize–winning work wasn't wrong; he had illuminated an important process, now called "induction," in which one group of cells signals an adjoining group and makes it switch to a different developmental pathway. A protein called activin, for example, has been shown to induce the growth of mesoderm, the middle of the three germ layers, in frog embryos.

Still, by today's lights, the great search was misguided. Scientists had sought *the* organizer, only to conclude that there was no such thing, only much more intricate and subtle mechanisms resistant to the methods of the day.

After World War II and into the 1970s, biochemists were using new chemical methods, says Brown, "while experimental embryology was going nowhere... People in the field were spinning their wheels on questions that were not ready to be asked." Not ready to be asked because the tools weren't there to get at the answers.

Then, in the early 1950s, came what James Ebert terms "that Watson and Crick thing"—the discovery of the structure of DNA and the opening salvos in the molecular biology revolution. Twenty years later came recombinant DNA technology, which allows scientists to study individual genes by snipping out pieces of DNA and splicing them into bacteria, where they can be grown, or cloned, in large quantities. "And that," says Brown, "is what really launched modern developmental biology."

"He wrote lots of convincing popular books that were full of errors."

Finding the Genes

The embryo that Ann Lawler has just removed from the uterus of a white mouse with two fine needles is a few hours too young for her purposes. She can tell because, peering through the microscope at it, she can see that its head still faces up, away from its chest. The swing down toward the chest—a developmental marker known as "the turning"—is still to come. This means that the RNA (ribonucleic acid, forming a single-stranded copy of DNA's instructions) she had planned to collect from the embryo would not have told her what she wants to know about how an invisible embryonic speck can become anything like the adult mouse gnawing on food pellets in the nearby cage.

This embryo is about 8½ days old. A female mouse in estrus placed in the same cage with a male will mate with it approximately halfway through the night. So if the mouse becomes pregnant, Lawler reckons the start of development at midnight and counts from there; without having someone stand watch, that's as accurate as she can get. But this time, it's not accurate enough. For while an adult mouse 8½ weeks old looks just like one of 7½ weeks, an embryo that is 8½ *days* old differs dramatically from one of 7½ days. Consider the somites, small blocks of cells on either side of the neural tube that runs along the back of the embryo. These will later give rise to connective tissue, bone, and muscle. At 8½ days, the somites number about a dozen; at 7½ days, zero. Things change fast in an embryo.

Lawler, a postdoctoral fellow in John Gearhart's laboratory at the Johns Hopkins School of Medicine in Baltimore, wants to know what genes turn on at each developmental stage. In the last few weeks she has collected about 300 embryos that are 8½ days old, as determined by somite count and the turning. From these she will extract messenger RNA.

This type of RNA is made only during the process through which a gene churns out protein. From the RNA she collects, Lawler will prepare a "library" of DNA—a sea of bacteria containing millions of cloned gene fragments which together represent all the genes that are active on day 8½.

How, she wonders, do the genes active on day 8½ differ from those active on another day, say 7½? Well, if you could take all the day 8½ genes and line them up beside matching day 7½ genes, the ones that found no matches would represent the difference. In principle, this is what Lawler does in the next step—called, fittingly enough, a "subtraction."

DNA is a double helix in which each strand is complementary to the other; if you know what is on one strand, you know what is at the corresponding position on the other. Indeed, a single-stranded version of DNA will, under the right conditions, naturally link up with its "missing" strand—if the missing strand is there to link up with. So in this next step, day 8½ messenger RNA (which is single-stranded) is allowed to find matching stretches on messenger RNA from day 7½. Any stretch that fails to find a match is thus unique to day 8½, and therefore includes genes that have switched on in the previous day.

It is this preoccupation with time that distinguishes developmental biology. The field asks not only what happened, but when? At what point does the mesoderm start to form? When does a particular gene turn on? How do genes active in the adult mouse heart differ from those active in the fetal mouse heart 10½ days or 12½ days after conception? Developmental biology, it has been observed, is "a science of becoming, not of being." Egg becomes adult only through a molecular dance far more intricate than any choreographed by Balanchine.

At Last, the Right Tools

But where, or what, is the choreographer?

Looming over developmental biology has always been that great, chiding question. As Allan Spradling puts it, "How do you get structure on a large scale, way larger than one individual cell? If individual cells knew only what they themselves were doing, they could never coordinate to make a large structure where every part is in the right place." What fashions the Big Picture?

Spradling turns to the metaphor of sig-

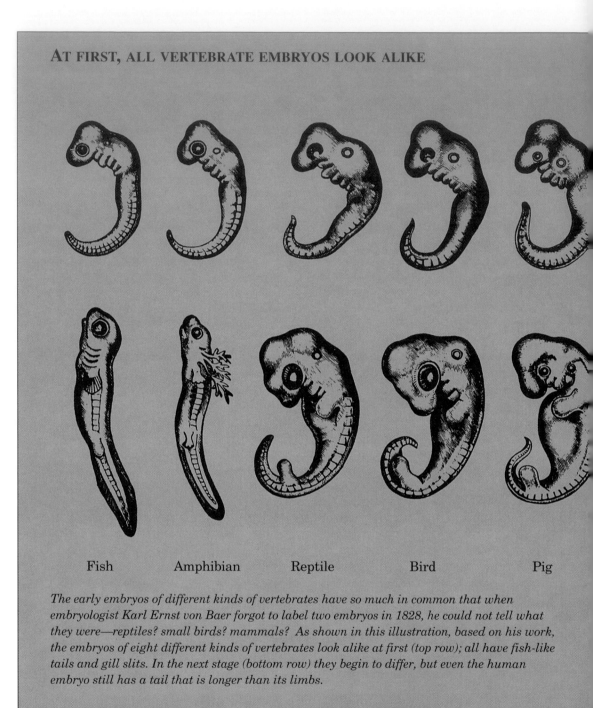

AT FIRST, ALL VERTEBRATE EMBRYOS LOOK ALIKE

| Fish | Amphibian | Reptile | Bird | Pig |

The early embryos of different kinds of vertebrates have so much in common that when embryologist Karl Ernst von Baer forgot to label two embryos in 1828, he could not tell what they were—reptiles? small birds? mammals? As shown in this illustration, based on his work, the embryos of eight different kinds of vertebrates look alike at first (top row); all have fish-like tails and gill slits. In the next stage (bottom row) they begin to differ, but even the human embryo still has a tail that is longer than its limbs.

naling systems—of one cell telling another, "Hey, you're not where you should be right now," of pattern emerging from loops of communication and control. Donald Brown speaks of signal cascades. A tantalizing development of recent years has been the discovery of genes that apparently play a crucial role in such cascades, serving as master switches, or gene controllers. The proteins these genes make don't become muscle or bone; rather, they attach themselves, or "bind" to the DNA of other genes, helping to express or silence these genes and thus directing the emergence of the body plan.

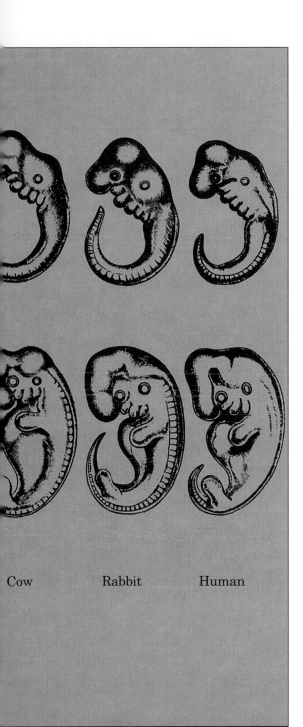

Cow   Rabbit   Human

Fruitful insight into the source of form and pattern has also come from the study of morphogens—biochemical substances presumed to act not locally, at a single point, but over a millimeter or two, and whose varying concentration influences development accordingly. If, for example, one applies retinoic acid, widely thought to qualify as a morphogen, to chick wing buds, the first wing digit forms at one concentration, the second digit at a concentration two-and-a-half times higher, and the third digit when the concentration reaches ten times higher than that. Activin, a substance thought to boast some of the properties attributed to Spemann's Organizer, has also been found to differ wildly in its effects at different concentrations.

With morphogens, it would seem, the dance of development proceeds without the choreographer. Imagine a weather map whose contour lines represent the concentration of a morphogen instead of temperature or barometer readings. A developing cell anywhere on the map that could "read" the strength of the surrounding morphogen would know better where it was and so, in principle, what it should become.

From this simple notion, complex patterns emerge quite naturally. Consider, says Donald Brown, morphogen A, densely concentrated at one point and simply trailing off to a low concentration as it disperses; no interesting pattern in that. But now add morphogens B, C, and D, radiating their effects from other origins, and genes responsive to their collective influence could presumably produce the peacock's feathers, the leopard's spots, or *Drosophila's* segmented abdomen.

None of these developmental models yet explains much in satisfying detail. "Patterning by positional information provides a relatively simple mechanism for making a wide variety of patterns," wrote Lewis Wolpert in a recent review. "Alas, that simplicity now seems more like simple-mindedness. Things seem, at this stage, much more complicated." "You've got to have tools commensurate with the complexity" of the organism and with the developmental process itself, points out Spradling.

Exactly. And now, says Brown, "we have all the tools"—at least for use in flies and other small organisms. As these tools are brought to bear on the great old questions, the answers grow surer and more detailed. Probing ever more precisely, scientists wrest common developmental mechanisms, themes, and patterns from their hiding places in nature's tangled variety.   ●

From this simple notion, complex patterns emerge quite naturally.

The Homeobox:
something very precious that we share with
flies

By Peter Radetsky

In late 1990 William McGinnis announced the results of an audacious experiment. He and his colleagues at Yale University in New Haven, Connecticut, had inserted a human gene into the embryo of a fruit fly, *Drosophila,* and discovered that the human gene fulfilled the same functions as the fly's own gene. Repeating the experiment with a comparable mouse gene, they obtained similar results.

"That was amazing," McGinnis says. "Although these human and mouse genes didn't act as powerfully as the fly genes, they still had the same effect on the organism." The effect was to control the development of the head of the fruit fly.

Actually, McGinnis's experiment mimicked a well-known spontaneous error in the fly's own gene—an error that made this gene turn on in the wrong parts of the embryo and produced visible defects in the fly's head. The human gene, which had been engineered to turn on inappropriately,

Edward Lewis, a pioneer of research on fruit-fly development, continues his work on the genes that control the formation of specific parts of the fly embryo. In his lectures he emphasizes that similar genes are active in the development of flies and humans—as shown in the cartoons of a human skeleton and a fly with matching colored dots for eight similar genes.

altered the fly's head in exactly the same way.

This provocative finding suggests that even though humans are separated from insects by over 600 million years of evolution, many of our genes function in virtually identical fashion and the differences between us and the lowly fruit fly are much smaller than we might expect.

It also illustrates the fact that nature has held on to certain genetic motifs that have proved to work, even in ostensibly unrelated animals.

The best evidence of nature's economy comes from the discovery of a powerful fragment of DNA called the "homeobox," which seems to play a key role in the development of nearly all species. This discovery grew out of the study of a class of genes that includes McGinnis's transplants: the "homeotic" genes.

Knocking Out Genes To Learn What They Do

One of the achievements of the early *Drosophila* pioneers was the observation that certain spontaneous mutations in the insects led to the bizarre appearance of perfectly formed body parts in decidedly inappropriate places. Some flies grew new sets of wings behind their normal pair, others developed legs in the normally legless abdominal section, and, most curious of all, still others sprouted legs from their heads instead of antennae.

In time, geneticists saw they could produce "homeotic" mutations of this sort (from the Greek word for "similar") by bombarding flies with x-rays, which broke apart their chromosomes. Perhaps one in a thou-

sand of the flies' abundant progeny, perhaps not even that many, would exhibit any particular mutation—it was up to the researchers to find it. The mutated flies could then be bred together to reproduce the anomaly. By discerning what went wrong, geneticists could deduce why things normally go right—and thus learn what normal homeotic genes do.

The gene that in its mutated form gave rise to the extra pair of wings was named *bithorax* because it seemed to produce a double thoracic segment. The one that caused legs to grow in the place of antennae was called *Antennapedia* (which means "antenna feet").

In the early 1950s, Edward Lewis of the California Institute of Technology discovered that *bithorax* and a number of other homeotic genes are linked tightly together in the third chromosome of *Drosophila*. He named the cluster the Bithorax complex. Later, he began to knock out the genes in this complex with x-rays. He wanted to find out precisely which body parts were affected by the genes' absence—which would show him where these genes normally functioned.

"That's what was really exciting, to take the genes away," says Lewis, whose work is

This fly grew an extra pair of wings because of three mutations in its bithorax *gene.*

It explained
how evolution
could have
created
different
animals by only
minor genetic
changes.

still a strong force in the study of developmental genetics. "If you don't take them away, it's hard to interpret what you have."

Though fruit flies look like mere specks to the naked eye, they are complicated creatures. Each part of the fly's body is composed of discrete segments—eight in the abdomen, three in the thorax, and at least three more in the head—almost as though a series of disks had been pressed together to make the whole insect. In one of Lewis's experiments, he recalls, "we made embryos in which each of the eight abdominal segments had turned into a thoracic segment." From such work Lewis concluded that the genes in the Bithorax complex control the rear half of the fly— half the thorax and the entire abdomen.

Then, to his amazement, he saw that these genes were lined up along the chromosome in exactly the same order as the parts of the fly's body that they controlled. Genes at the beginning of the cluster switched on development in the thorax; genes farther along the cluster controlled development in the upper abdomen; while genes at the end of the cluster controlled the lower end of the abdomen.

Lewis had long suspected that all these genes might have sprung from a single ancestral gene. Perhaps in some simple, unsegmented forerunner of the fruit fly, he suggested, an original homeotic gene had somehow become duplicated, each copy mutating over time to produce a divergent form in the organism. From such evolution came the segments of the fruit fly, each of which is different from the next, all of which are governed by homeotic genes.

It was an appealing theory because it explained how evolution could have created different animals by only minor genetic changes. And it seemed all the more appealing when in 1980 Thomas Kaufman of Indiana University (now an HHMI investigator) announced that he had tracked another homeotic cluster to the same arm of the same third chromosome. These genes were responsible for the fly's front half, from the middle of the thorax to its head. Here too, just as in the Bitho-

rax complex, the order of individual genes within the cluster corresponded to the order of the body segments that they controlled. Kaufman named the complex Antennapedia, after the gene that produced legs where antennae should be.

Although the genes in these two clusters made up less than 1 percent of the animal's estimated 15,000 genes, they specified much of its shape. And if Lewis's theory that such genes had evolved from an ancestral gene was correct, all the genes in these clusters should have some similar stretches of DNA.

The Homeobox

"I took Lewis's proposal very seriously," says William McGinnis. McGinnis later ran his own lab at Yale, but in the spring of 1983, he was a young postdoctoral fellow in the laboratory of Walter Gehring at the University of Basel, in Switzerland. Earlier that year, another Gehring postdoc, Richard Garber, had isolated a section of the *Antennapedia* gene. The Basel scientists put their heads together and decided to see if the fragment was similar to any other sections of homeotic genes in the fly. If so, it might be an indication that the genes came from the same ancestor.

They soon found that their probe—a labeled fragment of DNA that seeks out complementary fragments—sought out not merely one spot, but eight different sites on the fly's third chromosome. "Within a week or so we cloned all those genes," remembers McGinnis. "It was immediately apparent that they were very good candidates for other homeotic genes in the Antennapedia and Bithorax complexes." The researchers were even more intrigued to discover a short sequence of DNA that was virtually identical in all these genes. Any sequence that remains unchanged through millions of years of evolution is likely to perform some essential, irreplaceable function, so "we were tremendously excited," McGinnis says.

The sequence was only 180 base pairs in length (180 bases on two complementary strands of DNA), no more than a speck in the larger gene (*Antennapedia,*

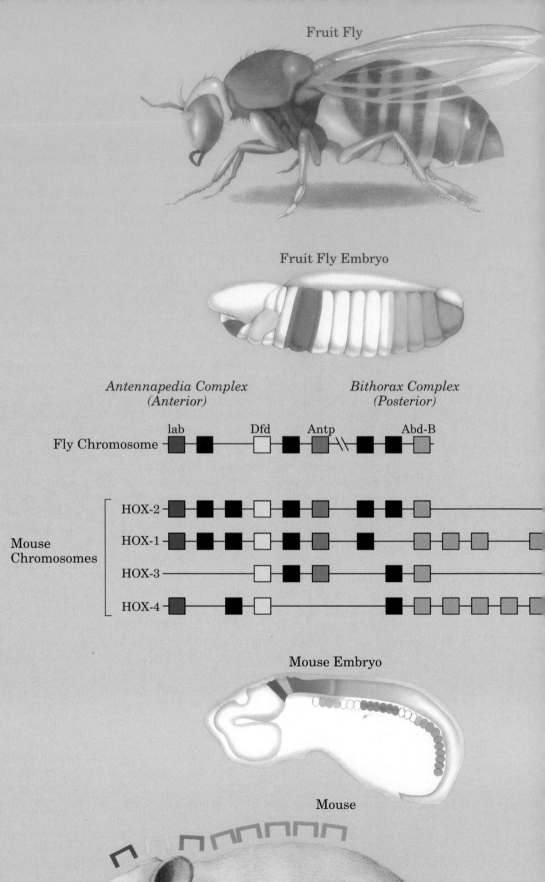

Fruit Fly

Fruit Fly Embryo

In insects and mammals, very similar genes—in the same order—control the development of anterior and posterior parts of the body. These homeobox-containing genes lie on a single chromosome in the fly (top row of colored squares) and on four separate chromosomes in mammals (lower rows of squares). In each case, the genes of the Antennapedia complex (left side of the chromosome) produce proteins that control the formation of the upper part of the body, while those in the Bithorax complex (right side of the chromosome) direct the formation of the lower part of the body.

The genes shown in color on the chromosomes of fly and mouse are especially similar to one another. In the Antennapedia complex, they are lab (labial, red), Dfd (Deformed, yellow), and Antp (Antennapedia, blue). In the Bithorax complex, they are Abd-B (Abdominal-B, green). The black boxes represent genes that are similar, but not quite as close, in fly and mouse. The genes are color-coded to match the parts of the body in which they are expressed in a 10-hour Drosophila embryo and a 12-day mouse embryo. Corresponding regions of the adult fruit fly and mouse are shown in the same colors.

Antennapedia Complex
(Anterior)

Bithorax Complex
(Posterior)

Fly Chromosome

lab   Dfd   Antp   Abd-B

Mouse Chromosomes

HOX-2
HOX-1
HOX-3
HOX-4

Mouse Embryo

Mouse

for example, stretches along for 100,000 base pairs). The tiny fragment was sharply, neatly defined, as though enclosed in a box. In an inspired moment, McGinnis and his colleagues called the sequence a "homeobox."

They were not, however, the only ones marveling at this small stretch of DNA. A quarter of the way around the world, in Thomas Kaufman's lab at Indiana University, another young postdoc named Matthew Scott had stumbled upon the same fragment. But he had come to it in quite a different manner.

Scott and his colleagues were trying to find out which parts of the *Antennapedia* gene are active in making protein. Sometimes as little as 10 percent of a gene consists of exons, the DNA segments that contain the recipe for making a protein. So they prepared a probe that would bind to the exons scattered among *Antennapedia's* 100,000 bases. But to their surprise, the probe also bound to something else: a stretch of a gene called *fushi tarazu,* or *ftz.*

Scott was puzzled. Although the two genes were neighbors in the Antennapedia complex, "it didn't make any sense," he says. "These are genes with utterly different functions. *Ftz* affects the *number* of segments that form in the early embryo; it's a segmentation gene. *Antennapedia* is a homeotic gene; it controls the *form* of a segment in the thorax of the adult fly. There was no reason why they should have been related. We figured the probe was contaminated with some *ftz* DNA or something."

So Scott went back to the drawing board, trying all the tests he could think of to flush out the contamination. He had no luck. The correspondence, or "homology," with *ftz* held up. He then decided to work out the sequence of DNA bases in the corresponding sections of the two genes so he could see how they compared. He was particularly interested in comparing the amino acids—the building blocks of protein—which these DNA sequences specified. It takes three letters, or bases of DNA (known by the letters A,C,T,G) to

code for a single amino acid. But some amino acids can be "spelled" in more than one way, by several different triplets.

"I had this wonderful moment," remembers Scott. "Late one night I was reading the sequences of *ftz* and *Antennapedia,* and suddenly I found a stretch of amino acids that was common to both of them. It was a stunning thing, because these genes are so different. The common stretch ran to 60 amino acids." The homologous gene fragments, therefore, were 180 base pairs long (since three base pairs of DNA code for one amino acid). Scott had found his homeobox—which proved identical to the one found by the Swiss group.

Meanwhile McGinnis uncovered hints that the homeobox was common to many other animals. The notion began to grow among certain developmental biologists that here was something very precious. "As developmental biologists, what we're looking for is not the bricks and mortar of life but the architect's plans," says Scott. "The homeobox appeared to be an important part of those plans." Many others, however, were unconvinced. If this homeobox were to be taken seriously, it would be necessary to *prove* its universality and show what role it played.

The first step was taken by Eddy De Robertis, now a professor of biological chemistry at UCLA, who was then running a lab in Basel next door to Walter Gehring. De Robertis specialized in working with the fat South African frog called *Xenopus laevis.* He was interested in finding a homeobox in the frog not only to investigate the importance of homeobox-containing genes, but to illuminate the genetic workings of vertebrates in general. For compared to all that was being learned about *Drosophila* development, vertebrates were relatively unknown territory. Using the fly homeobox as a probe, De Robertis and his colleagues set out to search for a similar homeobox in the chromosomes of the frog. "At the time, it was a completely unrealistic experiment," he says. "Frogs and flies are such unrelated species. Some of our students didn't even want to do it. But you have to have faith that you will find interesting things. It was a big conceptual leap."

It didn't take long to justify De Robertis's faith. Sure enough, the fly DNA probe bound to specific sites on the frog's chromosomes. "We cloned the frog gene," says De Robertis, "and then we tested its DNA for sequences that were complementary to three different homeobox-containing genes of the fly. It bound to all of them. This meant that it could only be the same thing as in the fly. We broke out the champagne at that."

The frog homeobox, then, was virtually identical to fly homeoboxes from *Antennapedia, ftz,* and another homeotic gene, *Ultrabithorax*. As Scott had done, De Robertis sequenced his homeobox and derived its amino acid structure. Here, too, the similarity was astounding. For example, the frog homeobox protein was identical to the protein produced by the *Antennapedia* homeobox in 55 of 60 amino acids.

When other researchers began to look for proof of the existence of homeoboxes in other animals, they found the same DNA sequences—in earthworms, beetles, mice, chickens, cows, and, yes, people. More recently, homeoboxes have been found in sea urchins, yeast, and plants, as well.

But What Do All Of These Homeoboxes Do?

A strong indication of the answer came from Scott's lab. Soon after finding his fruit-fly homeobox at Indiana University, Scott migrated to the University of Colorado in Boulder to head his own lab (he has since settled at Stanford University). One of his first postdocs was a molecular biologist named Allen Laughon (pronounced "Lawn"), who had been studying yeast at the University of Utah. When Laughon began looking closely at the sequence of amino acids in the homeobox, it reminded him of something familiar: certain proteins in yeast and bacteria that are known to control gene activity, the so-called DNA-binding proteins. By literally grabbing hold of the DNA double helix, these proteins are able to switch genes on and off.

This binding can take place in several different ways. Three-dimensional images produced by x-ray crystallography in the early 1980s revealed one intriguing design. A small section of the protein is twisted into the shape of a helix, with a sharp bend in the middle, like a hairpin. One side of this "helix-turn-helix" corkscrews itself into a groove in the DNA's double helix and, appropriately for a hairpin, the protein holds itself in place by pressing down on the other side. Thus anchored, it can go ahead and turn on the gene.

Laughon predicted that the amino acids specified by the homeobox would fold up into this hairpin structure. And since the homeobox appeared in the heart of key developmental genes, there could be no more logical role for it than to switch on other genes that take part in the developmental cascade. "So we stuck our necks out and said that this looks like a DNA-binding protein," says Scott.

Subsequent studies by Claude Desplan, who is now an HHMI investigator at the Rockefeller University, and Patrick O'Farrell, his colleague at the University of California, San Francisco, showed that the protein region, or "domain," made by the homeobox does indeed bind to DNA. (This region is now called the "homeodomain.") Then Kurt Wüthrich of the Swiss Federal Institute of Technology in Zurich proved that it has the helix-turn-helix structure. And in 1990, HHMI investigator Carl Pabo and his associates at the Johns Hopkins School of Medicine solved the 3-D structure of a fly gene's homeodomain while it was actually bound to a stretch of DNA (see p. 79).

"What's fascinating about structural studies of the homeodomain is that you can see this same regulatory system throughout evolution," says Pabo, who is now at the Massachusetts Institute of Technology. "It can be in a fly, in a mouse, or in a human being, and it's still the same fold of the protein, and the same helix still fits into the major groove of the DNA in an almost identical fashion. We've even seen it in yeast."

They found the same DNA sequences— in earthworms, beetles, mice, chickens, cows, and, yes, people.

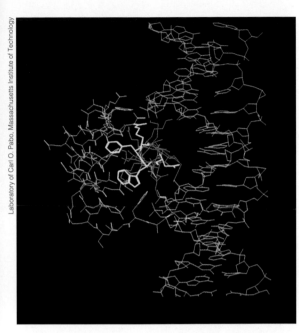

The homeodomain (orange) of a regulatory protein fits into the major groove of DNA's double helix (blue) to turn a gene on or off. The pattern of amino acids shown in yellow in this computer-generated image is found in all known homeodomains.

Over 100 homeobox–containing genes have been identified so far, in a wide variety of organisms. The precise roles of these genes and how they interact to produce normal development can best be studied in flies and worms. But some researchers have already begun to apply the findings from such work to vertebrates—particularly to mice, whose gene map is so similar to that of humans.

How Mammals Are Built

Interestingly, mice and humans have four strings of homeobox-containing genes, each on a separate chromosome. The strings appear to duplicate the fly's Bithorax and Antennapedia complexes (see illustration on p. 76). Not only do their genes match in number, but they also match in their order along the chromosomes. And—most impressive to scientists—all the genes in the strings obey the same rule that Lewis first noted in flies: The genes that are located near one end of the string are expressed in the upper half of the animal's body, near the head; those located near the other end of the string are expressed in the lower half of the body; and those in between follow the same pattern in an orderly array.

Researchers at the University of Utah have begun to reveal how these genes affect the development of mice. A few years ago Mario Capecchi and his associates in the HHMI unit there developed a revolutionary technique that allows them to knock out any known mouse gene and, if they wish, to replace it with a gene of their choice. Using this "gene-targeting" technique, they can breed mice that have a specific gene permanently knocked out. They can then study the results.

Capecchi has set himself the Herculean task of determining the function of every homeobox-containing gene in the mouse. He and his colleague Osamu Chisaka have concentrated on a homeotic gene called *Hox-1.5,* which resembles genes from the Antennapedia complex in flies. Knocking out the gene has striking results. Just as in *Drosophila,* the effects occur in the head and thorax of the mice. "Their musculature is affected," says Capecchi. "Fibers are disorganized and the wrong shape. The structure of cartilage is affected, and certain tissues are missing. The thymus, which is responsible for making antibodies, is totally missing, so the mice would be immune deficient. They die within a few hours of birth, probably of heart failure."

These defects, and others in the heart and face, are probably due to the fact that, because of lack of regulation from *Hox-1.5,* the section of the embryo that gives rise to these structures does not develop properly. The deficiencies in Capecchi's mice resemble those in a human genetic disorder called DiGeorge's syndrome. We humans have our own version of *Hox-1.5.* Might it be possible that the human syndrome is the direct or indirect result of a malfunctioning human homeotic gene?

"I'm really trying to answer two questions," says Capecchi. "First, what individual *Hox* genes are up to. And second, what the whole assemblage of *Hox* genes in these four strings is up to. That is much more difficult." Both questions are directly relevant to humans.

The genes in the four strings work together, "talking" to each other to produce a more complex creature than the fly, Capecchi says. He hopes to disentangle their functions and work out a blueprint of how mammals are built. But "there are 38 *Hox* genes involved in the mouse, and it takes two and a half years to make and analyze each mutation," he says with some awe. "It's a life's work." ●

DISCO

The Body Plan

Which end of an embryo will become its head? Which its tail? What determines this axis, and how are the body's basic segments formed?

A team headed by Christiane Nüsslein-Volhard, formerly at the European Molecular Biology Laboratory in Heidelberg, Germany, has provided some answers to these questions—at least for *Drosophila.* Starting in the late 1970s, the researchers undertook the mind-boggling task of mutating virtually every gene in the early embryo in order to observe its effects on the body plan, a mission they have continued to the present day (the team now pursues this research at the Max Planck Institute for Developmental Biology in Tübingen, where Nüsslein-Volhard is a director).

They made the mutations by feeding adult male flies sugar water that was laced with a DNA-altering chemical. When the flies were mated with selected females, their offspring exhibited a wide range of very specific defects that allowed the researchers to identify nearly all the genes and proteins that control the overall pattern of the fly's body.

This revealed that the embryo's body axis is actually determined by molecules slipped in by the mother fly. Long before fertilization, specialized nurse cells that help the egg to grow inside the mother move some of their own RNA into the end of the egg nearest them. This produces an initial asymmetry. Part of this maternal RNA (from a gene named *bicoid*) remains trapped near its point of entry, marking what will become the embryo's front end. After fertilization, the maternal RNA starts making *bicoid* protein, which diffuses in the now developing embryo and forms a gradient (see p. 82).

At the same time, from the other end of the embryo, protein from another maternal gene called *nanos* diffuses towards the center. (The names of these genes are inspired by their function—or lack of it, due to mutation. A mutant embryo without *nanos* genes lacks abdominal segments and so is small—the word "nanos" is derived from the Greek for "dwarf.") It is these opposing protein gradients that determine the fly's basic body plan.

Nüsslein-Volhard and her colleagues discovered the importance of *bicoid* when they saw that mother flies unable to make *bicoid* protein produced embryos lacking head and thorax. (They were, in effect, two-tailed embryos, or bi-caudal; thus the shortened name, *bicoid*.) Yet when *bicoid* RNA was injected into the front end of such mutated embryos, they developed normally. And when the scientists introduced the RNA into the opposite end of the embryo, a head and thorax developed at the point of injection. There was no doubt—*bicoid* is in charge of the anterior, or front, end of the embryo. Similarly, *nanos* was found to control its posterior end, its abdomen.

But just how do these protein gradients exert their influence? The mechanism is still being uncovered. It involves a series of genes that act in several overlapping zones within the embryo. For example, Nüsslein-Volhard's team in Germany and HHMI investigator Gary Struhl at Columbia University's College of Physicians and Surgeons in New York have determined how *bicoid* affects a lower-level gene called *hunchback*.

By Peter Radetsky

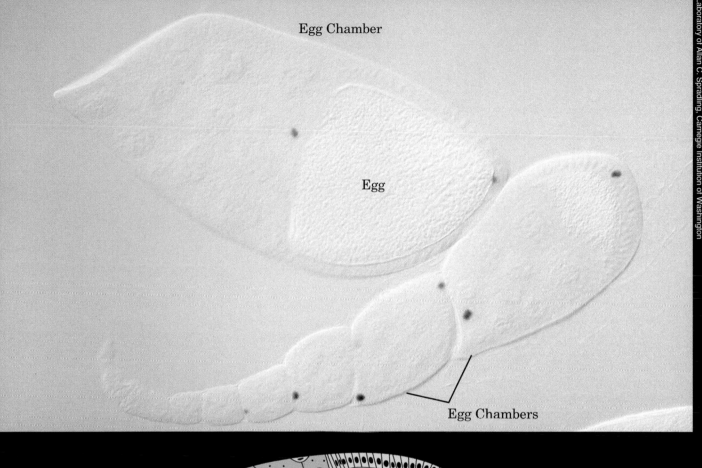

Egg Chamber

Egg

Egg Chambers

Nurse Cell

Nurse Cell
Nucleus

RNA →

Egg

Egg Nucleus

Follicle Cell

*An egg must know its front end from its back. An embryo must know its head from its tail. In the fly, the mother
supplies these compass points by means of molecules made in her egg chambers. Early in egg development,
for instance, a gene turns on in two ovarian follicle cells at the front of each egg chamber, as well as in
two follicle cells at the back of each chamber. This gene,* fasciclin III, *makes a protein that marks the two poles
of the egg. The cell nuclei in which the protein is present are stained blue in the picture above (two nuclei
look like one because of their small size).*

*Later on, the nurse cells (green) move some of their own RNA into the end of the egg nearest them, as
shown in the sketch. This determines what will be the embryo's front end. Only a few of the egg chamber's
15 nurse cells and approximately 1,000 follicle cells (blue) can be seen in this cross-section.*

(An embryo without *hunchback* genes lacks a thorax and so takes on a hunched shape.) Although *hunchback* RNA is distributed throughout the embryo, *nanos* prevents *hunchback* from making protein, as Gary Struhl and HHMI investigator Ruth Lehmann, of the Massachusetts Institute of Technology in Cambridge, Mass., have shown. The only place where *hunchback* is able to get away from the clutches of *nanos* and produce protein is in the anterior end of the embryo, the area controlled by *bicoid*.

Bicoid allows additional *hunchback* protein to be made at the front end of the embryo; the protein then diffuses into the rest of the embryo. As a result, "you wind up getting a very high level of *hunchback* protein at the anterior end and creating a second, different gradient that slopes off toward the posterior end," Struhl explains. These *hunchback* gradients are responsible for making the fly's thoracic and abdominal segments.

Bicoid also switches on other genes, which, in turn, activate still more genes, which, controlling ever smaller portions of the developing embryo, progressively section it off, much as a groundskeeper sections off a football field.

Approximately three hours after fertilization, the fly embryo's basic body plan is already drawn. *Bicoid* has successfully activated *hunchback* and other genes, which have turned on genes that define the embryo's segments, until finally the cascade crests with the switching on of the homeotic genes. These orchestrate the fate of the segments in the fly's more mature forms, creating wings, legs, and other structures. Altogether, at least 30 genes are involved in this development—and 14 more in another patterning process that takes place along the embryo's dorsal-ventral axis (backside to underside). This second axis provides a "latitude" to the anterior-posterior axis's "longitude."

What isn't known, however, is the extent to which this model applies to other animals. The early embryos of flies are very different from those of mammals: They are more nearly one cell with multiple nuclei suspended in cytoplasm, while mammalian embryos consist of multiple cells separated from one another by membranes. *Bicoid* and other morphogens—diffusible substances that affect body shape—can spread rapidly in the fluid mass of a fly embryo, but would have a tough time diffusing through the many cell membranes of an embryonic mouse or human. Though there may well be gradient genes involved in the development of mammals, they most likely work in a different manner—at least such is Gary Struhl's view. "The fact that there are cell boundaries in other organisms simply means that there must be mechanisms to get a signal to diffuse across many cells at once," he says.

So far, no such mechanism has been proved to be at work in other animals. However, Eddy De Robertis at UCLA suspects that he may have found a *bicoid* equivalent in his frogs. Named *goosecoid*, this gene may instigate the process that determines the anterior pattern of frog embryos. "Our hypothesis is that this gene is working through cell-to-cell signaling," says De Robertis. "It turns on a number of other genes, and in the end you get the ordered activation of head-forming genes. We think—we don't yet know—that this could be the vertebrate equivalent of the *bicoid* gradient."

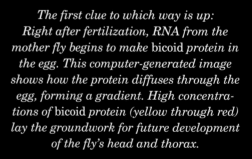

The first clue to which way is up: Right after fertilization, RNA from the mother fly begins to make bicoid *protein in the egg. This computer-generated image shows how the protein diffuses through the egg, forming a gradient. High concentrations of* bicoid *protein (yellow through red) lay the groundwork for future development of the fly's head and thorax.*

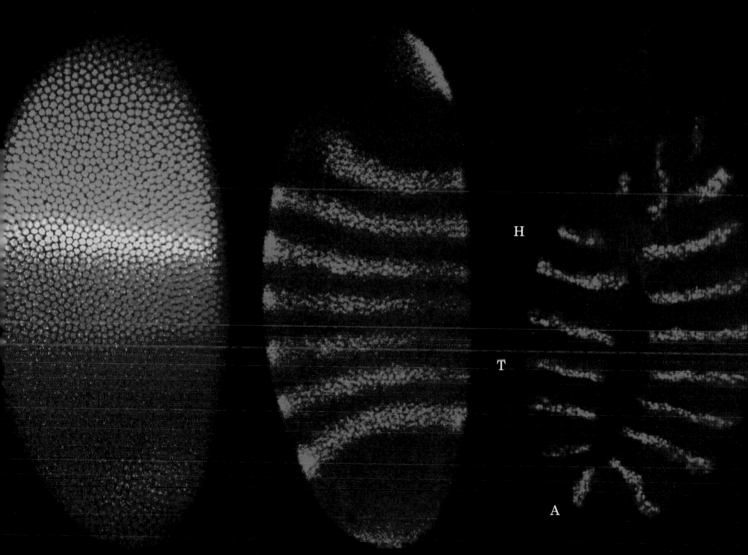

H

T

A

Brief signals from a cascade of genes then
split the fly embryo into ever smaller
and more specialized regions.
In this photograph the embryo is divided
into large blocks by proteins from
so-called gap genes—Krüppel (red) and
hunchback (green), which is turned on by
bicoid 2 $^1/_2$ hours after fertilization.
The region where the two proteins
overlap is yellow. The colors come from
fluorescent dyes in antibodies that
bind to these proteins.

About half an hour later, hairy, a
"pair-rule" gene that is regulated by the
gap genes, switches on and produces
seven transient stripes. As Francis Crick,
who shared a Nobel Prize for the
discovery of the double helix structure of
DNA, once remarked, "embryos are very
fond of stripes." These stripes act like
boundaries, dividing the embryo
into seven segments.

Finally the engrailed gene, a "segment-
polarity" gene, divides each of the
previous units into anterior and posterior
compartments. The fourteen narrow compartments shown here correspond to specific segments of the embryo. There are
three head segments (H, top left), three
thoracic segments (T, lower left),
and eight abdominal segments (A, from
bottom left to upper right).

(3) Laboratory of Sean B. Carroll, University of Wisconsin

DECISIONS, DECISIONS . . . WATCHING CELLS CHOOSE THEIR FATE

LEARNING FROM THE WORM

By Maya Pines

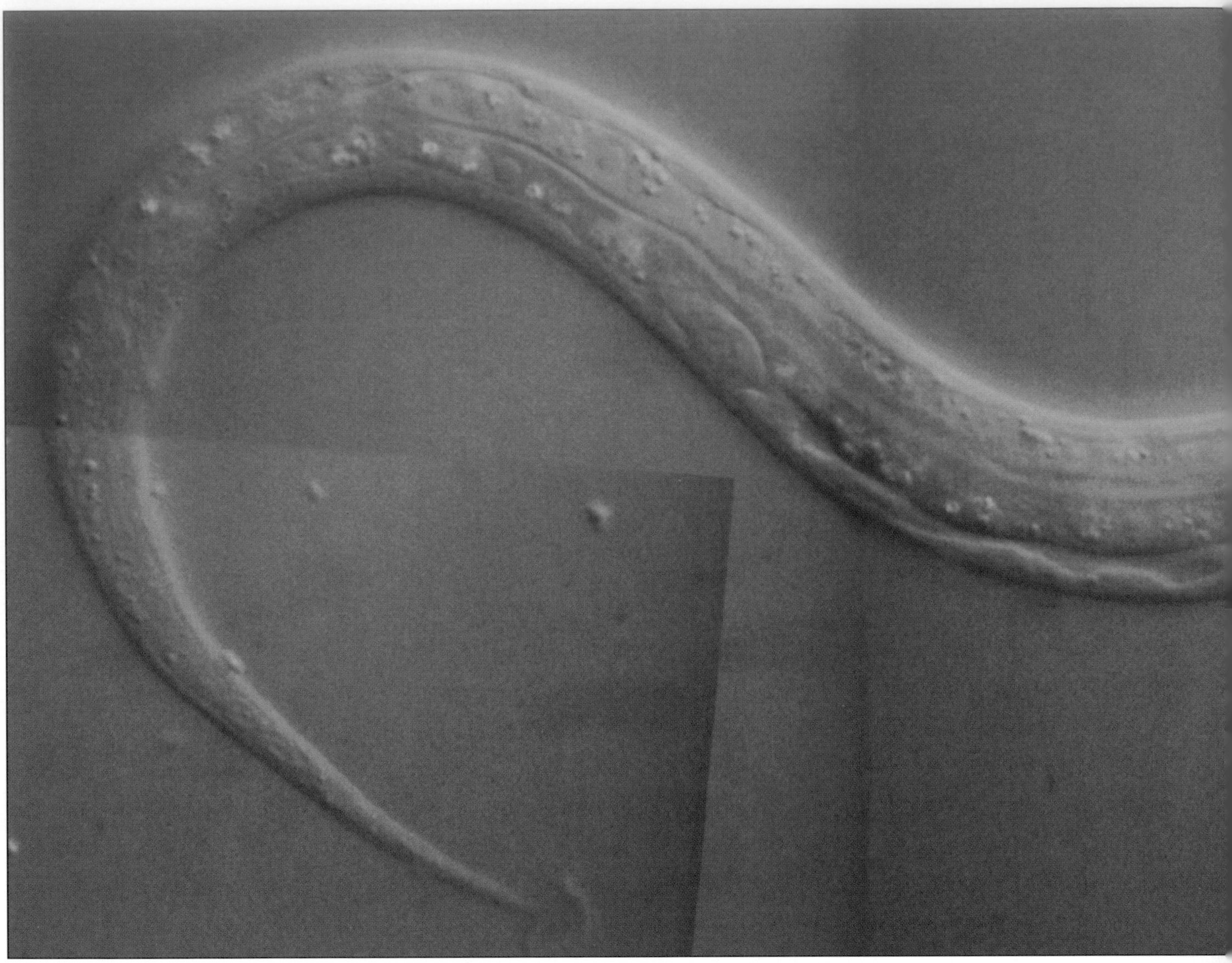

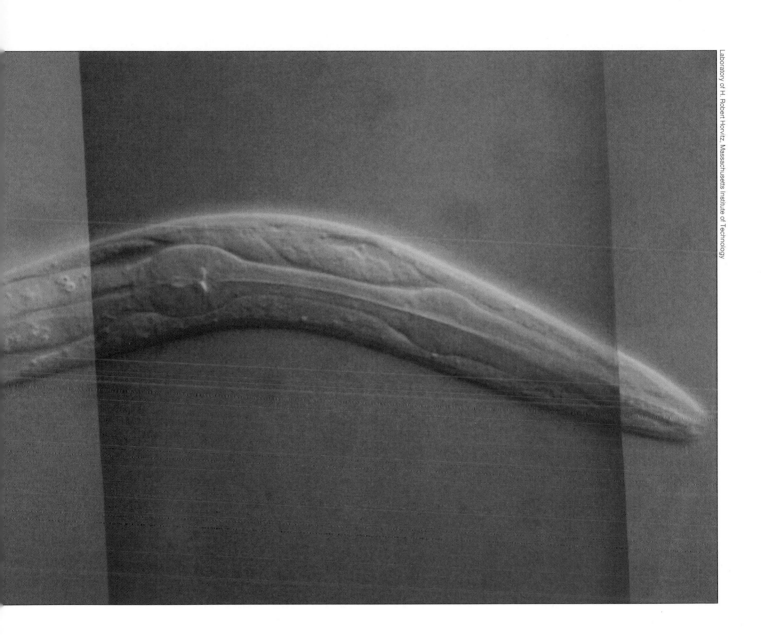

Midway through a film on the development of a small worm called *Caenorhabditis elegans,* people in the audience gasp. It happens when they realize that the embryo's cells—which were multiplying at a rapid pace—have stopped acting independently and that the whole embryo has begun to move sinuously, as one, inside the egg's transparent shell.

Even scientists who have witnessed this event many times before find it exciting. It is the first mark of unity in a multicellular organism. And it requires hun-

dreds of individual decisions, in the right order, by the cells that form the embryo. Every time a cell divides, each new cell must decide whether to be exactly like its "mother" cell or different, like its "sister," or unlike either of them. Each cell must decide whether to move to another part of the egg—and if so, where. Should it, too, divide? And what should it be when it grows up—a nerve cell, a sex cell, a skin cell, or a muscle cell?

How cells make these choices is the central mystery of development. All the cells in an embryo contain exactly the

Fresh out of its shell, this worm larva (magnified 2,200 times) is ready to start growing adult structures. Its tail is on the left, its head on the right.

same genes. What enables them to migrate and specialize in just the right way to create an integrated, working organism?

To find out, Sydney Brenner of the Medical Research Council in Cambridge, England, selected the little roundworm *C. elegans* as his experimental subject in 1965, even though, as he recalls, "people thought I was crazy." He wanted a really stripped-down animal—something as close as possible to the single-celled organisms on which scientists had done the experiments that produced the basic principles of molecular biology.

Little was known about *C. elegans* at the time, but in the intervening years the "worm project," as it is called, spawned several generations of researchers, and *C. elegans* may soon become the most completely understood multicellular creature.

To the naked eye, a petri dish filled with 10,000 of these worms looks almost empty, with just a faint shadow in its center. This shadow is a thin layer, or "lawn," of bacteria that serve as food for *C. elegans*. The

adult worms themselves are just 1 millimeter long. Clumped together, slithering along, eating, laying eggs, hatching, growing, or dying, they have few secrets from the scientists who observe their universe under a microscope.

The "worm people"—*C. elegans* researchers in more than 90 labs around the world—collaborate closely and share a real fondness for their tiny subject.

"Look at how many animals you can deal with," says Robert Horvitz, an HHMI investigator at MIT who is a "first-generation" worm person (he worked directly with Brenner in England). "You can put 10,000 worms on one petri plate. You can put 1,000 flies in one bottle. But you put only a few mice in one cage. So if you are looking for rare events, as we are—mutations or recombinations that occur only one in a million times—you are much better off with worms than with mice, both in terms of dollars and in terms of genetic power.

"Secondly, you want to do experiments very rapidly, with many generations. Look at

A burst of laser light (blue spot) neatly destroys the nucleus of a single cell whose outline is visible (arrows) near the larva's tail. Experiments of this sort reveal how the loss of a single cell affects the development of neighboring cells.

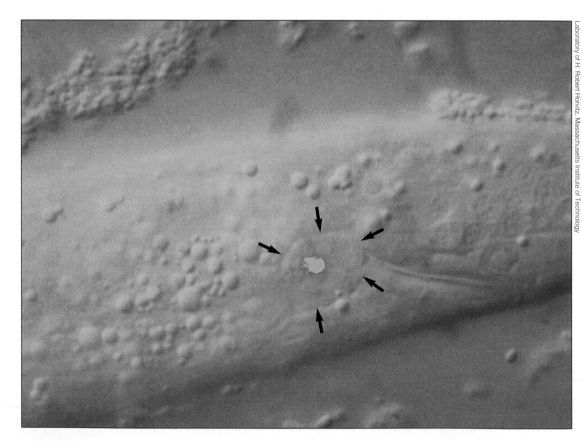

Laboratory of H. Robert Horvitz, Massachusetts Institute of Technology

the life cycle. In the worm, it's just three-and-a-half days from egg to sexual maturity.

"Thirdly, *C. elegans* is a self-fertilizing hermaphrodite; it makes both sperm and eggs, and it fertilizes internally. That turns out to have some striking technical advantages. For example, an animal doesn't have to move in order to reproduce, so mutants that would die out in a species that requires mating for propagation can have offspring anyway—they just lie there and make little ones. These are just some of the reasons why the genetics of *C. elegans* are very appealing."

Another trait that makes *C. elegans* "incredibly useful," according to Horvitz, is its transparent body, which allows scientists to see every one of the 959 cells in the adult worm. The researchers can observe food traveling down the worm's alimentary tract, eggs being made and fertilized, larvae hatching, and—most important—individual cells dividing or migrating.

That is what English scientist John Sulston did in Brenner's lab, using a light microscope equipped with so-called Nomarski contrast optics, which highlighted individual cell nuclei in the living embryo. After staring patiently at the worm's cells as the animal developed, he was able to trace the history of every cell.

Another technique was to "take the worm, slice it nose to tail like a sausage, and look at each slice under an electron microscope," says Horvitz. "In this way we can study every cell and find its connections." By now scientists know the anatomy of every neuron (nerve cell) in the worm and have deciphered all the connections in its nervous system. "It's the only animal for which the entire circuit of the nervous system is known," says Horvitz.

Having this kind of "unprecedented, incomparable information," as he puts it, allows scientists to do experiments analyzing the rules of development at the single-cell level. "We can knock out a single cell and ask, 'What does this cell normally do?'" he explains. "We use a laser beam. We focus it on a single cell in the animal while we view it with Nomarski optics, and we kill that cell. Then if another cell does

something different, we can say that the first cell must normally be affecting the second cell's development. So we can identify which cells are involved in induction—in signaling between tissues."

In humans, the fertilized egg from which we all began divides into two identical cells, then four, then eight, each one of which is totipotent, or capable of forming any part of an adult. But in the worm—a much more economical creature, with few duplicate cells and no time to waste—the very first division of the fertilized egg creates two different types of cells.

"How does that happen?" asks Horvitz. He points out that the mechanism for creating such differences must be either "intrinsic" (something that occurs because of what the cells are born with) or "cell-interactive" (dependent on signals from outside the cell). As some scientists say, the cells follow either the "European" or "American" plan of development. Under the European plan, the cells' fate is determined by their ancestry—by whether their parents are royalty or working-class. Under the American plan, it is determined by their current position and their neighbors.

To find out which mechanism prevailed during the worm's first division, the researchers removed each daughter cell in turn. As this did not alter the fate of its sister, the mechanism seemed to be intrinsic. They later learned that the difference between the cells was probably caused by an unequal distribution, within the egg, of specific molecules inherited from the mother worm.

Although some people jumped to the conclusion that the fate of all *C. elegans* cells is determined in this way, experiments have shown that, in many cases, the worm's cells acquire their identity through signals from other cells.

The destinies of two cells born in the worm's second round of cell division can be reversed simply by flipping their positions in the embryo, James Priess of the Fred Hutchinson Cancer Research Center in Seattle, Washington, discovered a few years ago. "They were born equivalent sister cells," Priess says. "They were identical.

Under the European plan, the cells' fate is determined by their ancestry...

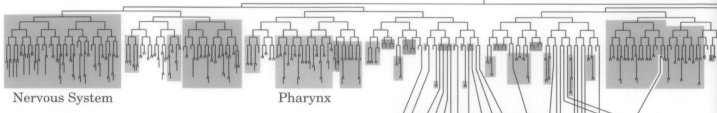

Nervous System   Pharynx

A Family Tree of Every Cell in the Worm

Scientists have learned where each of the 959 cells that make up an adult C. elegans *comes from, tracing it back to a single fertilized egg. As shown on this lineage map, the egg divides into two, then its daughter cells continue to divide. Each horizontal line represents one round of cell division. The length of each vertical line represents the time between cell divisions, and the end of each vertical line represents one fully differentiated cell.*

Some of these differentiated cells are "born" after only 8 rounds of cell division—for example, some of the cells that generate the cuticle, the animal's coat; other cuticle cells require as many as 14 rounds. The cells that make up the worm's pharynx, or feeding organ, are born after 9 to 11 rounds of division. Cells in the gonad require up to 17 divisions.

Exactly 302 nerve cells are destined for the worm's nervous system. Exactly 131 cells are programmed to die, mostly within minutes of their birth. The fate of each cell is the same in every C. elegans *nematode, except for the cells that will become egg and sperm. The major organs of the worm are color-coded to match the colors of the corresponding groups of cells on the lineage map.*

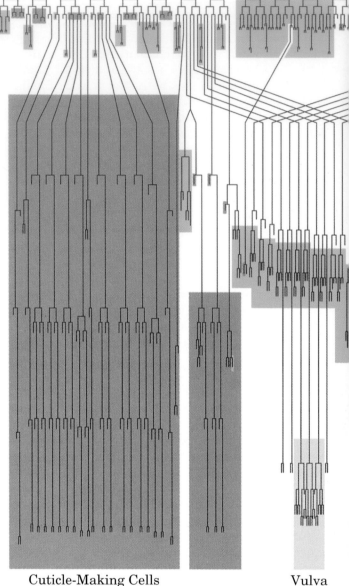

Cuticle-Making Cells   Vulva

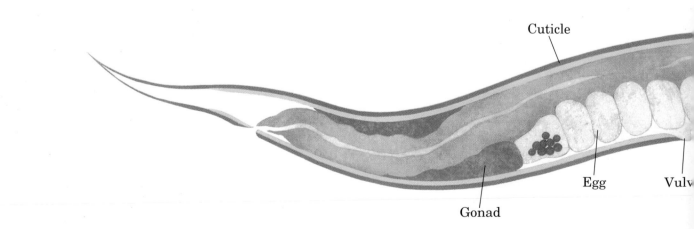

Cuticle

Gonad   Egg   Vulv

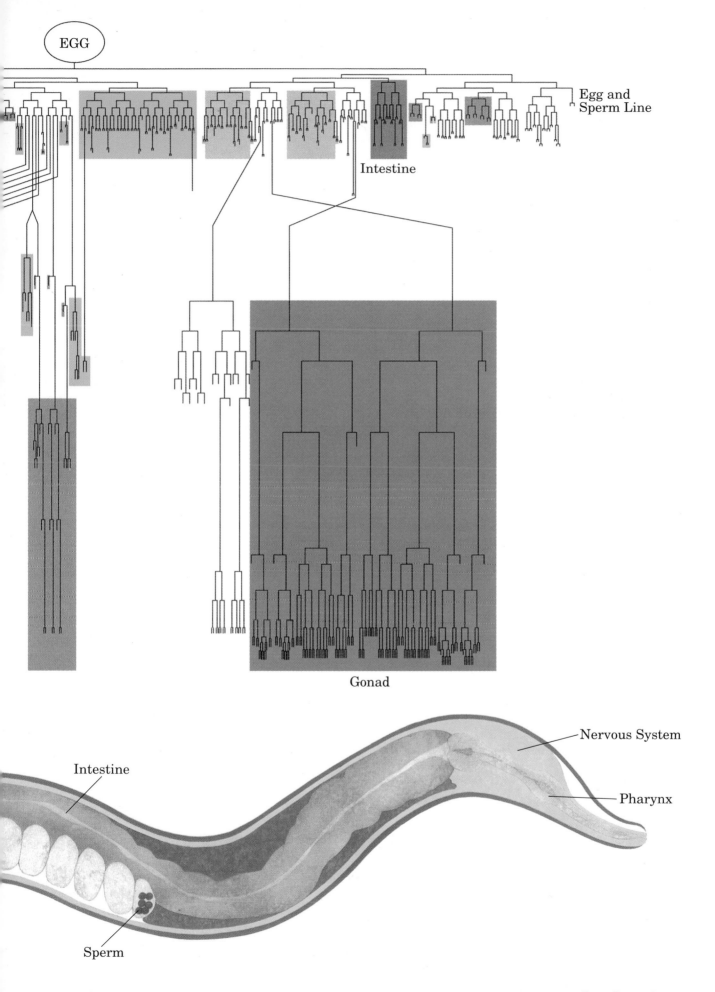

EGG

Egg and
Sperm Line

Intestine

Gonad

Nervous System

Pharynx

Intestine

Sperm

But only the one that interacts with a neighboring cell will, later on, form a particular type of muscle tissue. This is the earliest example of cell-cell interaction in the worm." Priess and other researchers also identified the protein that did the signaling. This protein—and the gene that codes for it, *glp-1*—turned out to be part of a family of molecules that are used over and over again at various stages of development in many species.

The Link with Cancer

Meanwhile Horvitz and his associates were examining how young worms develop the sexual organs they need for reproduction. This led to a discovery that is proving very valuable in medical research.

The scientists started out by looking for mutants that lacked the proper sexual organs. "It was very easy to recognize mutants that were defective in egg-laying," says Horvitz. Worms that cannot lay eggs because they have no vulva, or opening through which eggs are expelled, are still able to fertilize their own eggs internally. These eggs then develop and hatch within the worm's body. Using a microscope, researchers see an animal that is severely bloated with retained eggs or a "bag of worms"—a mass of larvae that rapidly consume their parent's body.

The researchers soon identified some of the signals that control the formation of the worm's vulva. The vulva develops rapidly in newly hatched larvae—within five hours—and the whole process can be observed directly in living animals, says Horvitz. Six cells have the potential to generate vulval tissue. They form an "equivalence group," and each cell can choose one of three fates, depending on the signals it receives.

A signal from a particular cell in the worm's gonad, the "anchor cell," is needed to start the process. "If the anchor cell is destroyed, all six cells in the equivalence group become skin, rather than vulva, and no vulva is formed," says Paul Sternberg, an HHMI investigator at the California Institute of Technology, who has been working on the vulva project ever since he

joined Horvitz's lab as a graduate student in 1979. "On the other hand, if the anchor cell is moved to a different position, it can induce other cells to become a vulva."

Upon receiving the anchor cell's signal, half the cells in the equivalence group become committed to making a vulva, while the other half go on to make skin. "Why? That's what we really wanted to understand," says Sternberg. "What type of molecules control this?"

Sternberg (a "second-generation" worm person) and Horvitz collaborate on this research by mail, sending worm-filled petri dishes across the continent in ordinary padded envelopes. They have focused on two kinds of mutant worms. "The simplest mutant produced no vulva at all; we called it 'vulvaless,'" says Sternberg. "Another mutant—'multi-vulva'—made one normal vulva and several additional vulvas—ugly-looking bumps. We have cloned several of the genes involved.

"So what do we find?" Sternberg asks with rising excitement, speaking volubly. "A series of three genes, some of which are related to human cancer!"

This finding was a total surprise, he says. One of the genes, *let-23,* codes for a receptor on cell membranes that binds to growth factors—chemicals that stimulate cells to proliferate. This receptor bears a strong resemblance to growth-factor receptors that have been studied in human cells and implicated in cancer. In the worm, this receptor appears to bind to the signal sent by the anchor cell. Then it interacts with another protein that is inside the cell, starting a cascade of interactions that eventually reaches the nucleus. The gene that codes for this other protein, *let-60,* belongs to a notorious family of proto-oncogenes (genes which, when mutated, can cause cancer)—the *ras* genes, which were originally found in human tumors. Mutant *ras* genes have been isolated in cancers of the colon, liver, lungs, and bladder, as well as in some leukemias.

Until *ras* was found in the worm, "nobody knew what ras proteins normally do in an organism," declares Horvitz. "And here was an answer. We could say, very

A Computer That Turns Genes On or Off

When cells must choose between two fates at key points in the development of an embryo, certain pivotal genes within the cells switch on and tell them what to do.

These genes, in turn, take their orders from intricate biochemical computers called transcription complexes, which scientists are now studying with great interest. Each transcription complex is formed by varying combinations of two elements: selected control sites on the DNA molecule (sequences of DNA, in the vicinity of a gene, that regulate the gene's activity) and proteins called transcription factors, which recognize these particular sites and bind to them.

"Transcription complexes are integrating systems," explains Eric Davidson, a professor of biology at the California Institute of Technology who has been using sea-urchin embryos to study the control of development.

"One transcription factor may tell a gene what kind of cell it's in. Another factor may tell the gene what time it is. A third may give information about signals received from other cells. But these factors will form a productive transcription complex only when all are in place. That's how different kinds of biological information get integrated."

Hundreds of transcription factors have been found in recent years—far more than biologists expected. For example, the homeobox proteins are transcription factors. So are some of the proteins produced by proto-oncogenes.

According to Davidson, half a dozen or more transcription factors must be present in exactly the right combination to switch a gene on—that is, to allow a particular stretch of DNA to be copied into RNA, the first step in making a specific protein. Depending on which DNA sequence a transcription factor binds to, it helps to promote—or repress—the transcription of a specific gene. The transcription complexes keep a running account of such inputs. In this way they decide which proteins should be made in a particular cell—and thus determine the cell's fate.

---

unambiguously, that ras proteins control the communication between cells that triggers specific changes in tissues during development. Not only that, but the very same mutation that is associated with the active form of the gene in human cancers is the mutation that we identified in worm development—same mutation, same amino acids. That says something pretty striking about conservation of function."

How Cells Tell Time

When *C. elegans* embryos grow normally, the lineage of every cell is predictable; cells from different embryos follow exactly the same pattern of division and migration, and every cell is always in the same place. Such "invariance" is not found in vertebrates, where the fate of individual cells is unpredictable after the first cell divisions. But in the worm, even those cells that depend on signals from outside the cell always have the same fate because they always receive the same signals, Horvitz points out. The worm has very few cells, he notes, and "if the interactions between cells are invariant, the fates will be invariant."

The timing of these signals is crucial, however. The cells must know what stage they are in so as to decide what signals to send out, or how to respond. The signals themselves may be just "stop" or "go." Everything depends on the timing.

Part of the time-giving mechanism "was discovered here, in the worm, back in 1984," says Horvitz. "We found some genes that tell cells to do things now or later." These genes could never have been recognized if the sci-

Sometimes
the genes
tell a cell
that it must
die soon
after birth.

entists had not known the complete pattern of normal cell divisions in the worm, he says, because "the way we found them was to look at mutations where the pattern was early or late. Some of the mutants were precocious, and others were retarded." The precocious ones made a particular adult structure too early, in the fourth larval stage instead of the adult. "Everything was out of sync," Horvitz says, "because the rest of the animal was developing normally, while these tissues just skipped one stage." As to the retarded mutants, "they never made the adult structure at all," he says. "They were stuck as larvae, repeating the immature pattern over and over again, never advancing. And what's interesting is that these can be opposite types of mutations in the same gene."

Horvitz made up a name for these mutations: "heterochronic," to parallel "homeotic," he says. "These genes do exactly what homeotic genes do—they specify cell fate." The homeotic genes tell cells where they belong spatially, in terms of their position in the embryo. The heterochronic genes tell cells where they belong in terms of developmental stage—in terms of time. Both types of genes make DNA-binding proteins that contribute to turning other genes on or off.

Sometimes the genes tell a cell that it must die soon after birth. Scientists call such events "programmed cell death," as opposed to death from accidental injury or disease. According to Horvitz, it is a normal and universal part of development.

"Cell death, on the face of it, is counterintuitive," Horvitz says. "Why bother to make these cells if you're just going to kill them off?" He points out that parts of the nervous system kill 85 percent of the neurons that are made. In the immune system, more than 90 percent of the cells die. "They are made and they die—some of them very early and some much later, depending on the region and the kind of cells," he says.

Although these deaths "seem like a waste," they are necessary because embryos start out with a vast developmental potential, he says. Some cells, for example, must die in order to eliminate sexual organs that would be inappropriate. Simi-

larly, "the way we make our hands is by sculpting," Horvitz says. Human fetuses have some webbing between their fingers that must be whittled away before birth.

There is nothing passive about programmed cell death, however. "It is an active process that requires the expression of specific genes," according to Horvitz. His lab has identified 11 of these genes. "Overall, I think it's fair to say we know more about naturally occurring cell death in *C. elegans* than in any other animal," he says.

Now the researchers are trying to understand "what mechanisms are involved, how cells decide which should live and which should die, and how the killing gets done," he says—not only in normal development, but also in disease. He emphasizes that "many human diseases involve cell death. For instance, Alzheimer's and other neurodegenerative diseases. Strokes, too. What happens with strokes, as well as some of these other diseases and also traumatic brain injury, as when someone falls off a motorcycle, is that many of the injured cells don't die right away. It's not an immediate 'squishing' caused by the fall, but rather a second-order effect that occurs some days later. Some physiological program is causing these cells to die. If we understand it, we may find ways of developing treatments."

In addition, Horvitz and other scientists are studying genes that appear to protect against cell death.

Watching the little worm's cells go about their business and choose their fate under the influence of these varied genes is "beautiful" as well as instructive, Horvitz finds. He realizes that "if someone had designed development in the most efficient way, this is probably not the way it would work." Scientists could not have predicted that the baroque patterns of cell divisions, migrations, and deaths they observed would produce a worm.

"But evolution is a tinkerer," he says. "It takes what already exists and adds to it, or shaves some of it away." Thus, human beings retain a large number of the mechanisms now being discovered so efficiently in *C. elegans*, and the worm can be used as a sort of microscope to search them out. ●

MAKING THE FLY'S
EYE

"People wonder, why study the fly's eye?" says Gerald Rubin, an HHMI investigator and professor of genetics at the University of California, Berkeley. "One answer is that it may be the shortest path to understanding basic mechanisms that relate to cancer in the human."

In fact, Rubin and his colleagues are in a race with the worm people to decipher the role of *ras* and other oncogenes. It is a friendly race. "We reinforce each other's work," he says. "If a mechanism were found only in flies—or only in worms—people could say, Who cares? But the argument is untenable if we find the same mechanism— even the same molecules— in both flies and nematodes, which are as far apart from each other as they are from humans."

Rubin picked the fly's eye to work on because "it's convenient," he says. "A fly

that becomes blind still lives; it can be bred and maintained in the lab." Furthermore, the fly's eye is made of 800 identical units, or ommatidia, each of which consists of 20 cells (8 photoreceptor neurons and 12 other cells) that are arranged in a precise, easily recognizable geometric pattern. These units form fairly late in development, at a time when the nearly adult fly has 3,000 cells "who know they're going to make up an eye, but haven't yet decided what part of the eye they'll be in," Rubin says.

How cells that were previously uncommitted decide to become eight specific photoreceptor neurons is now the focus of intense research.

The cells' decisions are always made in the same sequence, as was shown by Andrew Tomlinson and Donald Ready in 1986, when they were at Princeton University. First one cell is selected to be the linchpin—R8, the central cell of

Even a small segment of a fly's eye contains over 30 ommatidia, each with its own sensory bristle, as shown in this scanning electron micrograph.

Laboratory of Gerald M. Rubin, University of California, Berkeley

This cross-section of an adult fly's compound eye reveals different views of identical ommatidia. The solid lens at the top of each ommatidium is shown at the top of the picture. Pigment cells appear red. Seven of the eight photoreceptor cells (two are stacked on top of each other) can be seen at the bottom, in blue.

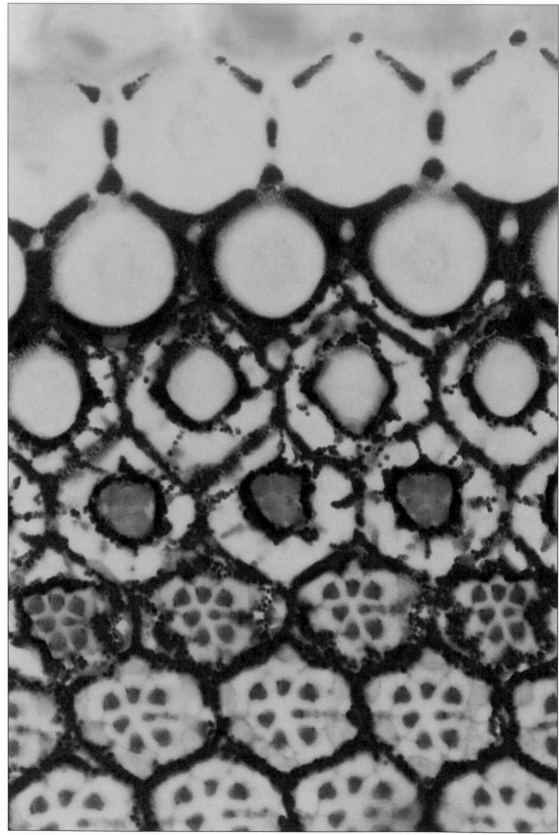

the wheel-like unit. At regular intervals, the unit then builds around the central cell in paired steps. To the immediate right and left of R8, two cells become committed to the unit. Three to four hours later, two additional cells join it, followed, after another three hours, by two more cells.

Finally, cell R7 completes the pattern. Meanwhile, other nearby epidermal cells are added on: the cone cells that form the lens, the pigment cells that insulate each unit from its neighbors, and the cells that will become eye bristles. According to Tomlinson, any remaining, unallocated cells then die, "to leave the whole structure as perfect and as precise as a crystal."

Some of the most interesting experiments on the fly's eye have centered on a mutant fly that never makes the final R7 cell at all. In 1987, Rubin and his associates cloned a gene, *sevenless,* which controls this cell's fate. Whatever cell finds itself in the pocket between R1 and R6 will become an R7 cell if its *sevenless* gene is normal and produces protein inside the cell. But if this gene is mutated and no *sevenless* protein is present, the cell will become a cone cell.

When the researchers analyzed the *sevenless* protein, they found, to their surprise, that it was a cell-surface receptor containing the enzyme tyrosine kinase—a type of receptor that biochemists knew well from previous studies of mouse and human cells. Receptors of this type bind with growth factors that play a role in the development of cancer. However, their role in normal development was obscure.

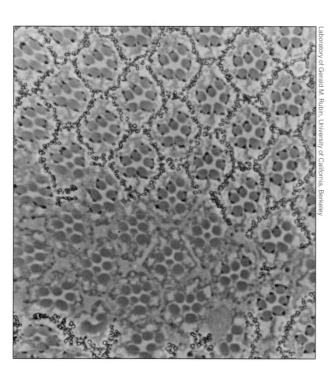

Laboratory of Gerald M. Rubin, University of California, Berkeley

The *sevenless* receptor sat on the surface of the uncommitted cell and touched the R8 cell. Shortly after this discovery, Larry Zipursky (who is now an HHMI investigator) and his associates at the University of California, Los Angeles, identified a signal sent out by the R8 cell that bound to the *sevenless* receptor; they called it *boss,* for "bride of *sevenless.*" Experiments in Rubin's lab have revealed the next step in the pathway: the *sevenless* receptor activates a *ras* gene—a proto-oncogene—inside the uncommitted cell.

"Many human cancers are caused by activated *ras* genes or tyrosine kinase genes," says Rubin. "*Ras* exists in two forms—active or inactive. Here we can start with an activated *ras* gene and try to fill in the steps. We can exploit this system to ask, What does *ras* do to affect the pathway from cell surface to nucleus?"

Rubin is struck by the similarity between the signaling mechanisms he found in the fly's eye and those that Sternberg and Horvitz found in the worm. "In one case, they were discovered in fly neurons; in the other case, in the worm's reproductive system," he says. "Yet the pathways are nearly identical and the *ras* sequences are nearly identical."

These mechanisms are very relevant to human cancer, Rubin declares. "If we want to devise drugs to interfere with some of the signals that cause cancer, we need to know what these signals are," he says. "And because of the kind of genetic studies we can do in flies and worms, we expect to find out here." ●

Evidence that the ras *gene is involved in transmitting signals from* sevenless: *In experiments where the signal from* sevenless *is weakened, ommatidia can make an R7 cell only when they have two copies of the* ras *gene (as in the central, grayish area). Those with a single copy of the* ras *gene lack the R7 cell (yellow area).*

BECOMING A MALE, BECOMING A FEMALE

Scientists are just beginning to uncover the dramatic events that make us male or female.

The fact that human beings come in two basic models is such a fundamental part of our lives that it comes as a shock to realize that embryos originally develop the characteristics of both sexes, becoming committed to one or the other only during the seventh week after conception.

The decision to become male or female —one of the most important decisions our cells ever make—is usually foreordained by the sex chromosomes we inherit from our par-

ents: generally, if we inherit one X and one Y chromosome, we will be male, and if we inherit two Xs we will be female. But nothing happens in the embryo to bring about this fate until certain cells make a gender choice during the critical seventh week.

Until then, embryos have the potential to go either way. In 1990, researchers in England and the United States discovered that everything hangs on the activity—or silence—of a gene on the Y chromosome. If this "maleness gene," *SRY* (Sex-determining Region of the Y), turns on in cells of the embryo's primitive gonads—the precursors of the ovaries or testes—the embryo

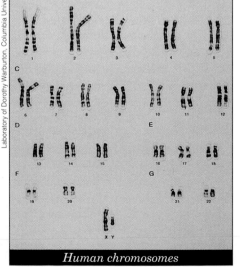

Laboratory of Dorothy Warburton, Columbia University

Human chromosomes

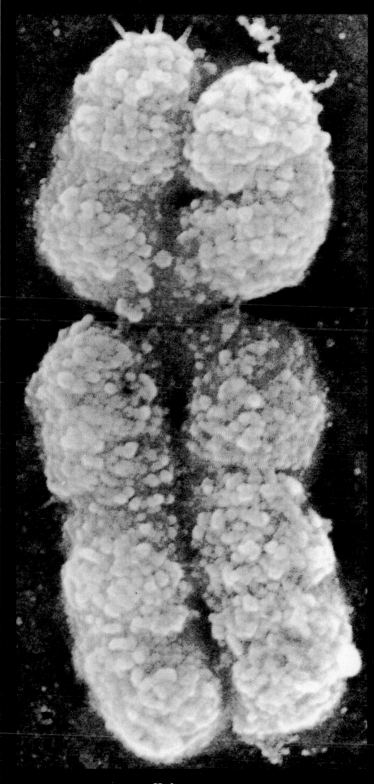

The differences between men and women can be traced to a tiny piece of DNA on the Y chromosome—just a fraction of the total Y chromosome carried by males. The Y itself appears rather stunted next to the much larger X chromosome. The critical piece of DNA is a recently discovered gene, SRY, that switches on the "maleness" program in a previously uncommitted embryo. Its location is shown with an arrow.

Except for the two sex chromosomes, X and Y, all human chromosomes (22 numbered pairs) are inherited equally from mother and father. Women carry two X chromosomes; men carry one X and one Y. Therefore, the mother can supply only an X, while the father can supply either an X or a Y. If an egg is fertilized by sperm containing the father's X chromosome, the child will be female. If the sperm contains the father's Y chromosome, the child will be genetically male. In order to develop as a male, however, the embryo must have functioning testes; these require the SRY gene.

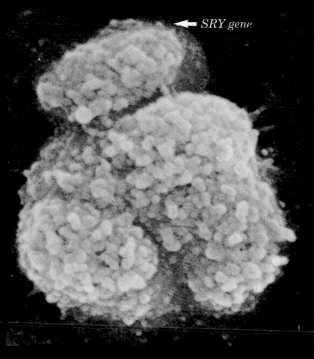

← SRY *gene*

X chromosome

Y chromosome

will develop into a male. If not, the embryo will develop into a female.

The primitive gonads in which this drama is played out are called "indifferent" because, at this early stage, they can become either testes or ovaries, says David Page, an HHMI investigator at the Whitehead Institute in Cambridge, Massachusetts, who has been tracing the development of gender differences. "The *SRY* gene turns on only in the indifferent gonads and—extrapolating from studies on mice—only in this particular week," he says. "In response to this switch, the testes begin to form. They churn out two hormones that control all sexual development in the male. One hormone, testosterone, stimulates the development of male organs. The other, Müllerian-inhibiting substance, destroys tissues that would otherwise become female organs."

If the *SRY* gene does not turn on during week 7, however, the indifferent gonads develop into ovaries. At the same time, the Müllerian ducts—precursors of the uterus, fallopian tubes, and part of the vagina—begin to grow, while tissues that would otherwise become male organs regress and disappear.

"This is the only situation in mammalian development where you have the option of what organs to make," Page points out. "Here you have two highly evolved, highly perfected pathways. Both are normal. And in any given embryo, a choice has to be made between the two.

"It's like those plays in which a story unfolds, all the characters are in place and all the roles have been learned, but the audience decides how the play should end," Page says. "Depending on its decision, some of the lines will remain unspoken on any given night."

Around week 8, the mammary cords and the nipples form, both in males and in females. "Why do men have nipples?" Page asks. He compares male nipples to "actors who know their role and are just waiting for a chance." Men have nipples but no functioning breasts, he notes. Around week 9, when breast tissue develops in females, the mammary cords die in males.

At this stage the genitals are still just a series of small swellings, containing features of both sexes. But around week 10 the sex organs begin to differentiate.

Meanwhile the primordial germ cells—precursors of egg and sperm—make a decision of their own. Having migrated from the yolk sac to the indifferent gonads around week 6, they choose their fate "according to what the indifferent gonad has become," Page explains. "They may say, 'I woke up this morning and found myself in an ovary, so I will become an egg.'" By week 16, five million immature ova (eggs) have been produced in the ovaries—the total for a lifetime—and immature sperm are in position in the seminal vesicles.

Becoming a female is not a "default" pathway, Page emphasizes. "The ovary is not the absence of a testis," he says. "There are two alternative paths, and following either of them is a very active process, requiring the coordinated activity of many genes and proteins."

Zeroing in on the Maleness Gene

Nearly everything that scientists have learned about these pathways has come from the study of mutations. In 1982 David Page became intrigued by some unusual cases: men who had the female genetic pattern (two X chromosomes) and women who had the male pattern (one X and one Y). "We had known since 1959 that maleness is determined by something on the Y chromosome," he says. "But these exceptions seemed to contradict the rule."

Page then started to test these people's genetic material with probes made of DNA from the human Y chromosome. "We found that most men with XX chromosomes actually carried a small fraction of the Y chromosome that was not detectable by older methods," he says. "The really exciting finding was that the bit of the Y chromosome these men carried was exactly the same piece that the XY females were missing." This meant that the presence or absence of a very tiny fraction of the Y was the key to sexual difference.

The Y chromosome is one of the smallest

human chromosomes. Page and his colleagues made a map of it on which they marked out the narrow band that contained the sex-determining gene—a segment consisting of 300,000 bases, or 300 kilobases (kb) of DNA, only half of 1 percent of the chromosome.

"We could say that the other 99.5 percent of the Y had nothing to do with sex determination," recalls Page. "That began the era of molecular research on this topic."

The race was on. Several labs, including Page's, set out to find the gene itself. "We scoured that 300 kb region of the Y," Page says, in an effort to pinpoint the smallest piece of DNA that XX men had in common. In 1987 Page thought he had it—a 140-kb section that seemed particularly important and contained a gene he called *ZFY*. "It looked like a very good candidate," he says. "But then Peter Goodfellow at the Imperial Cancer Research Fund in London found some XX males who didn't have this gene."

Eventually two British teams, headed by Goodfellow and by Robin Lovell-Badge of the Medical Research Council in London, jointly located the maleness gene within a 35-kb piece of DNA on the human Y chromosome. One year later, in 1991, they ran a striking experiment which showed that an even smaller piece of DNA was critical, at least in mice: Taking fertilized mouse eggs that would otherwise become females, they injected a snippet of DNA only 14 kb long into them—and thereby changed the embryos' destinies, turning them into males.

The protein made by the *SRY* gene appears to function as a transcription factor. "It probably regulates the activity of one or more target genes," Page says. "But what genes regulate *SRY* itself, to make it turn on at the proper time and place? And what are its target genes?" *SRY* is a switch, he points out. The subsequent steps in becoming a male depend on other genes, most of which are located on the autosomes—the chromosomes that both sexes inherit equally from both parents. "We can already show that the gene for Müllerian-inhibiting substance, which destroys cells that would become female organs, is located on chromosome 19 in

both males and females," he says. "Yet it is turned on only in males. It is regulated by the *SRY* protein, but we don't know if the interaction is direct or indirect. We don't know the nature of any of these additional genes."

Similarly, embryos whose *SRY* gene is switched on at the right time may still end up looking like females if they have a defect in a gene that codes for the testosterone receptor—a gene that lies on the X chromosome and is present in both men and women. No matter how much testosterone their testes pump out, they cannot develop male genitals or any other male sexual characteristics if their tissues do not respond to this hormone.

Many medical conditions, including infertility, arise from errors in the pathways that control gender differences, Page emphasizes, so it is important to discover how the pathways work.

Some of the best clues to the sex-development pathways may come from studies of flies, even though the mechanisms these insects use to choose their sex appear totally different from those of humans. All flies develop as males unless a special signal flips a genetic switch that starts a cascade of "female" developmental events. In 1974, Thomas Cline, who is now in the division of genetics at the University of California, Berkeley, identified this switch as the protein produced by a gene he named *Sex-lethal*. Since then Cline and others have been able to find some of the gene targets downstream of the sex-determining switch, putting them far ahead of researchers on human sex determination.

"Sex is a mechanism for mixing and reassorting genes," Cline says, "and sex determination is an example of a decision that every organism makes. We need to understand how the ratio of males to females is set. We also need to know how cells that become committed to a particular sex *remember* the choice they've made."

According to Cline, scientists are making rapid progress in understanding these mechanisms. "It's all coming together in flies now," he says, "and worms are close behind." —*M.P.*

Becoming a female is not a "default" pathway.

Building The World's Most Complex System: The Brain

By Larry Thompson

A couple of times a year, a boy will be born without a sense of smell. As an adult, he will be unable to father children. This combination of symptoms is so strange that for years no one could see how the two defects were related.

Now researchers have solved the mystery of Kallmann's syndrome, as the disorder is called. And in so doing, they have discovered a gene that is needed for one of the fundamental processes of brain development—the migration of nerve cells, or neurons, to their proper places in the growing brain.

It turns out that the cells that produce sex hormones originate in the same part of the embryonic brain as the olfactory cells, which relay information about smells to the brain. Rockefeller University scientists who examined the brains of mice learned that the two kinds of cells normally travel together along a common pathway during development. But in people with Kallman's

syndrome, the cells fail to migrate. Lacking olfactory tracts, these people cannot smell, and lacking sex hormones, they cannot develop normal genitals.

An international team of scientists zeroed in on the genetic error that is responsible for both defects: a missing gene on the X chromosome. The gene, which they called *KALIG-1,* normally produces a protein that seems to help nerve cells stick to each other and navigate. By studying this gene and its effects, researchers can now analyze how brain cells find their way to proper locations in the growing embryo.

This piece of detective work is just one example of recent attempts to understand how the most complex system in the world—the human brain—is put together from early after conception to birth, and how it develops thereafter.

It is a daunting task. The brain has so many neurons (over 100 billion), of so many different kinds (thousands—far more than any other organ), with such complicated

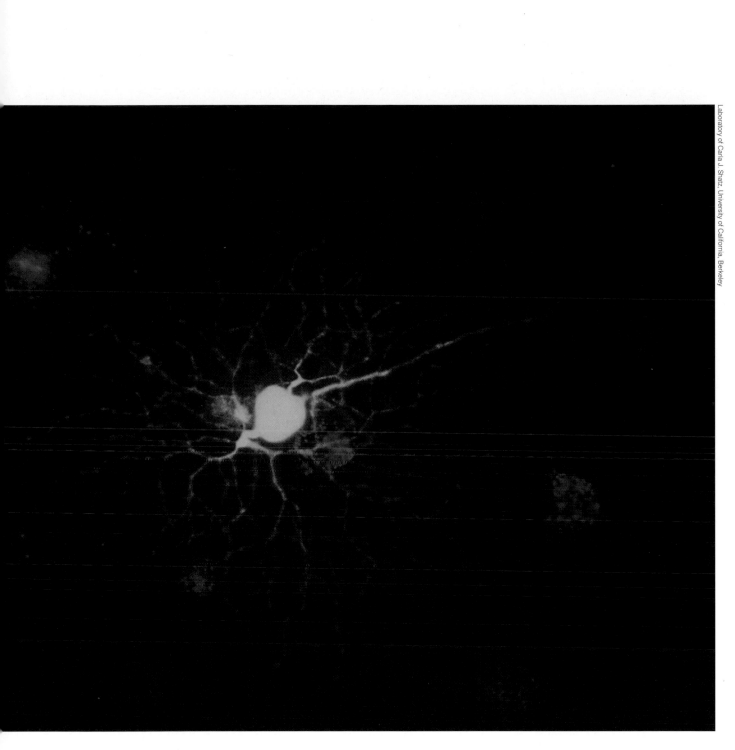

A developing nerve cell (yellow) from a cat's retina sends its axon to cells in the LGN, a visual relay station in the brain. The cell will survive only if it connects with its proper target there— as did the surrounding neurons (red), which gained their color from a red dye injected in the LGN.

interconnections (trillions), and so many support cells (such as the glial cells, which outnumber the neurons by 10 to 1) that it seems well beyond our ability to grasp. Somehow this mass of cells forms an exquisitely sensitive system that allows the individual to see, hear, smell, touch, feel, talk—and think. The brain also controls muscular contractions, and thereby all movements and activities. It releases hormones that regulate sexuality, stature, blood pressure, and even the immune system. It embodies our consciousness, our humanity itself.

While most major questions about the brain remain unanswered, scientists have begun to sketch some of the broad outlines of brain development that are established in the early embryo. They have also uncovered some of the mechanisms involved in more detailed aspects of brain development.

"The overall pattern is set up in the embryo in the space of less than a millimeter," says Thomas Jessell, an HHMI investigator at Columbia University. "Then the pattern is enlarged as the organism grows."

The process begins when a small group of cells in the early embryo decide to become brain cells, he explains. Shortly after conception, when the fertilized egg has become a ball of cells, gastrulation begins: a dimple forms and the ball starts to push into itself to form two layers. It is as if a finger poked into an inflated balloon until one half of the balloon pushed inside the other. The embryonic cells then rearrange themselves and decide whether they will become skin or bone, muscle or nerves. As German zoologist Hans Spemann showed in studies of frog embryos, some of the cells in the embryo "induce" the fate of other cells. The cells that are committed to forming the nervous system must then choose whether to become glial cells or immature neurons. And those that have chosen to be neurons must migrate to their final position in the brain.

Only then do they become full-fledged neurons. They grow short, branch-like extensions called dendrites, which act as antennae, receiving signals from other nerve cells. They also produce a single fiber

called an axon, which enables them to form connections with distant target cells and to deliver their own signals to these cells. Most dramatically, they lose their ability to divide and have daughter cells. The time of this irrevocable decision is called the neuron's "birthday." When all the neurons have reached this stage, no more are added to the brain. However, to ensure an appropriate number of neurons, the brain produces many more than will eventually survive. In some areas of the brain, as many as half the neurons die a short time after birth because they have not been able to make the right connections.

Organizing billions of individual nerve cells into a single brain is somewhat like drawing a map on which all the households in the United States connect through the nation's streets and highways. First, the cartographer would draw a general outline of the nation's borders and superhighways and the major cities along the routes. With time, the mapmaker would add more and more detail—the roads within the cities, the streets, and even the addresses of individual households. Each house would be like an individual nerve cell body linked to the whole by a system of neurological roads, or axons. This later, more detailed stage requires that the neurons themselves play some role in sorting out their locations along the highway. As the linkages between axons and their target cells grow, they create the networks, or neural circuits, that gather and process information in the brain.

How the Axon Finds Its Way

The first crucial structure in the embryo's developing nervous system is the neural plate, a flat sheet that runs along the embryo's back. Early in development, the neural plate curls up to form two ridges. It then curls further until the edges meet, creating the neural tube, which will later give rise to the brain and spinal cord. Along the center of the neural tube's ventral region, a thin structure called the floor plate defines the animal's midline—an invisible line that separates the left and right sides of the spinal cord and later determines the identities of the adjacent nerve cells.

It chooses its direction by sniffing around like a bloodhound...

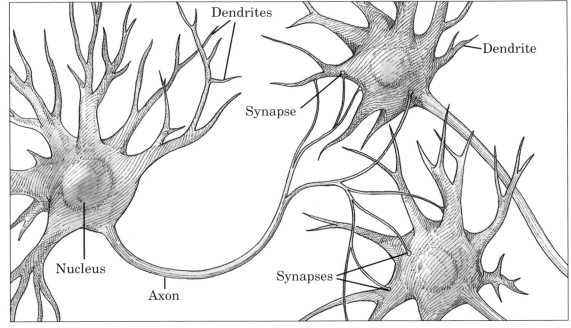

For a while the cells in the neural tube proliferate, especially around its central canal. Then they stop dividing and migrate outwards. Settling into their new environments, they mature and begin to form neural circuits: the axon of one cell connects to a target cell, which, in turn, sends out its axon to another cell, and so on. Information flows through these networks at lightning speed in the form of electrical impulses that travel down the axons. Each impulse triggers the release of a chemical neurotransmitter at the synapse, or junction, between an axon and another cell. The transmitter then diffuses across a small gap to reach receptors on the dendrites of the target cell. When enough receptors have been activated, an electrical impulse flashes down the axon of the target cell and starts the process all over again.

Axons are remarkable structures. They can stretch to nearly a meter (3.3 feet) in human beings, halfway from the brain to the toes. They may reach more than 10 meters in a whale. To form a particular circuit, the axon must tunnel its way through tissue, stretching its tip into a growth cone as it seeks its objectives. It chooses its direction by sniffing around like a bloodhound to detect the scent of the proper trail. As it grows, the axon navigates around physical obstructions, picking a path to its target.

Ever since Spanish neurobiologist Santiago Ramón y Cajal discovered axon growth in the 1890s, scientists have struggled to understand the systems that guide axons. "The axon grows along a general pathway, which has markers on it like street signs on a road," says Carla Shatz, a brain researcher at the University of California,

Berkeley. Scientists have just begun to identify these molecular street signs. Since the axon's growth cone may be relatively far away from the cell's nucleus and its genes, it must act in an independent fashion and make its own decisions at successive branch points, based on such cues. When it has journeyed far enough to reach its target, there must be some signal that says, "You've arrived; stop here." At this point, the growth cone transforms itself into a terminal—a structure that is able to form a synapse with the target cell and establish a neural circuit.

In studies that began in the 1940s, Roger Sperry of the California Institute of Technology showed that axons from the optic nerve in a frog's eye always connected to the same target region in the brain, even when he cut the growing axons and rotated the eye before letting the axons regenerate and reestablish their connections. From these and other experiments, Sperry concluded that each nerve cell must have its own chemical label, an identifying code like the street address of a house in a city, and that nerve cells have some way of recognizing each other—labels that match in a highly specific manner. Though he proved fundamentally right, his theory had some problems.

One neuron's axon connects with the dendrites of two other neurons to form part of a neural circuit.

"If you had to code each specific address genetically, the brain would require too many genes," says Carla Shatz. Individual chemical labels dictated by genes could not be the whole answer. Yet it took almost 20 years for alternative explanations to emerge in the early 1980s, as researchers began to identify the genes and proteins responsible for nerve cell differentiation and the growth and guidance of axons.

"You have to have multiple guidance cues for an axon to get from the cell body to its target," says Jessell. The axon "senses" these multiple, short-range chemical cues, which beckon it to move forward until it reaches the source; then it detects another cue that draws it in a new direction.

To work out the details of axonal growth in at least one type of neuron, Jessell's group picked a single class of spinal cord cells called the commissural neurons. These are the neurons that connect the right and left sides of the spinal cord. Born in the part of the neural tube that is close to the animal's tail, these cells put out axons that follow a complicated growth trajectory, moving down to the floor plate, crossing the midline, and then turning abruptly toward the brain, to end up in symmetrical areas on the left and right sides of the brain.

Jessell identified one set of diffusible factors released by the floor plate that attracts the axons to the midline, while another set helps the axons turn toward target cells in the brain.

For any cell—or part of a cell, such as an axon—to migrate, it must be able to cling to another surface. In the 1970s Gerald Edelman and his team at Rockefeller University discovered the first neural cell adhesion molecule, called N-CAM. Adhesion molecules, which are abundant on the surface of neurons, have a unique shape that allows them to bind to similar molecules on adjacent cells. That is, an N-CAM molecule on one neuron or axon can bind to an N-CAM on another neuron or axon, allowing the two cells to grow in contact with each other. Consequently, once pioneering neurons start producing axons in the right direction, they create a growth trail that

helps other axons of the same class find their target cells.

This may happen with Jessell's commissural neurons. Once their axons reach the floor plate and contact its sticky molecules, they turn toward target cells in the brain that they cannot yet detect. Then they set off in a long growth spurt next to the midline, where they appear to follow the pathway of related, trail-blazing axons that are already growing toward the brain.

The mechanisms of neural growth that Jessell and other scientists have uncovered in mammals turn out to be very similar to those that other researchers have found in grasshoppers and fruit flies. At the University of California, Berkeley, for instance, Corey Goodman, an HHMI investigator, and his team have been working on fly embryos. They showed that certain cell adhesion molecules are expressed on specific subsets of developing axons and act as labels, coding some developmental pathways much like the color codes that help commuters pick the right subway line.

"The whole process is dynamic," Goodman says. "The cells change over time what labels they are expressing. They have the ability to switch at choice points." This allows axons to respond to different growth signals or to follow a different set of trail-blazing axons.

Once again, as in mice, a good example of this guidance can be seen in the mechanisms that direct certain nerve cells to cross the midline. About twenty pathways eventually form on each side of the fly embryo's midline. Goodman studied three initial pathways that might be represented by colors and organized like this: black, brown, and blue; then, in a mirror image on the other side of the midline, blue, brown, and black. An axon born on one side and destined to follow the brown pathway on the other side would grow toward the midline, ignoring the brown pathway it encounters on its own side. Once it crosses the midline, however, a molecular switch changes the protein labels on its surface. When it next encounters a brown axon pathway—this time on the other side—it hops on like a subway rider whose train

They create a growth trail that helps other axons find their targets.

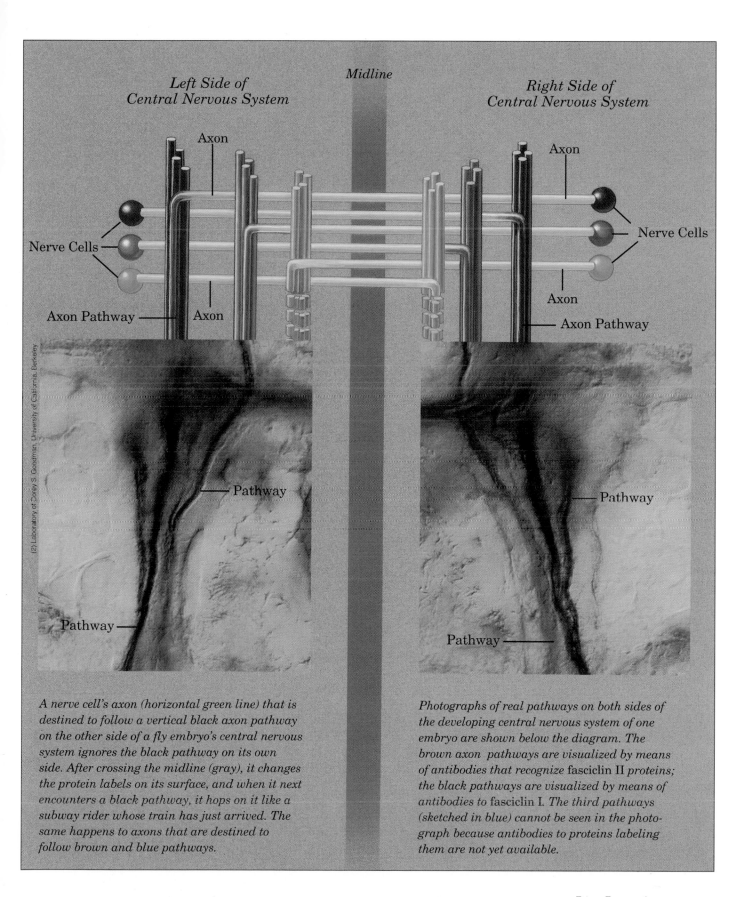

Left Side of
Central Nervous System

Midline

Right Side of
Central Nervous System

Axon

Nerve Cells

Axon Pathway ——— Axon

Axon

Nerve Cells

Axon

Axon Pathway

——— Pathway

——— Pathway

Pathway ———

Pathway ———

(2) Laboratory of Corey S. Goodman, University of California, Berkeley

A nerve cell's axon (horizontal green line) that is destined to follow a vertical black axon pathway on the other side of a fly embryo's central nervous system ignores the black pathway on its own side. After crossing the midline (gray), it changes the protein labels on its surface, and when it next encounters a black pathway, it hops on it like a subway rider whose train has just arrived. The same happens to axons that are destined to follow brown and blue pathways.

Photographs of real pathways on both sides of the developing central nervous system of one embryo are shown below the diagram. The brown axon pathways are visualized by means of antibodies that recognize fasciclin II proteins; the black pathways are visualized by means of antibodies to fasciclin I. The third pathways (sketched in blue) cannot be seen in the photograph because antibodies to proteins labeling them are not yet available.

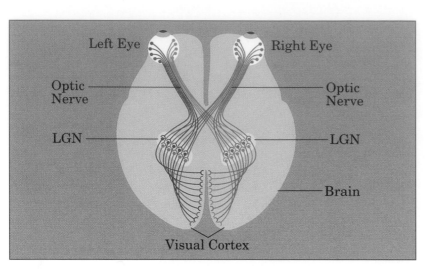

Left Eye • • Right Eye

Optic Nerve — — Optic Nerve

LGN — — LGN

— Brain

Visual Cortex

has just arrived.

The pathways are intricately labeled. "It is not simply one molecule, one pathway," Goodman explains. "The twenty individual pathways that run up and down the fly's nerve cord are marked by different combinations and concentrations of recognition molecules." His group has identified five of these molecules. Knocking out a gene that encodes one of them, *fasciclin II,* stalls growth cones that would normally follow pioneer axons labeled in part with this molecule. Goodman believes that about a hundred different recognition molecules may be required to build a complex central nervous system such as the fly's.

Shaping the Brain Through Activity

Genes establish the major pathways between different parts of the brain. More precise neural connections, such as those between the cells within a given part, depend on the brain's own activity.

"The brain's connections are malleable," says Goodman. "It is like building a computer and then, throughout life, letting the activity of the programs rewire the computer's connections. Your brain is continually fine-tuning and changing its connections."

Some clues to how these hookups are altered have come from recent studies of how the visual system assembles itself. In each eye, the retinal cells join together to form the optic nerve, a bundle of axons

that carries visual information to the rest of the brain. These axons project into two peanut-sized clusters of nerve cells buried within the left and right sides of the brain called the lateral geniculate nuclei (LGN)—relay stations which, in turn, send their fibers to the visual cortex, the cells that interpret vision at the back of the brain.

From the late 1950s on, Nobel Prize-winners David Hubel at Harvard University and Torsten Wiesel, now at Rockefeller University, used microelectrodes to monitor the activity of individual neurons in the visual cortex of anesthetized cats. They discovered that the cortex was organized into many parallel groups of cells, called ocular-dominance columns, and that the neurons in one column responded to input from one eye, while those in the adjoining column responded to the opposite eye. Blocking the cat's vision in one eye at birth could change these patterns irreversibly, making the cat functionally blind in that eye.

This means, according to Carla Shatz, that in the central visual system "growing axons divide up their target regions into alternate parts for the left and right eyes." Shatz decided to sort out the first half of the system: the connections between the retinal cells, whose axons form the optic nerve, and the LGN, which relays signals to the visual cortex.

She began by staining the retinal neurons with a yellow tracer dye that she injected into the cat's eye. The dye is absorbed by the neurons and their entire axons. Early in development, she saw, the retinal axons formed a dense network of cables that seemed to plug into target cells at random. The random connections were transient and were later replaced by orderly layers.

According to Shatz, several lines of evidence show that axons from neighboring retinal cells are supposed to connect only to LGN targets that are next to their neighbors' targets there. "It is like stringing phone wires from a whole series of homes on a city street to an identical neighborhood in a distant city," Shatz says. The phone lines from the corner house would always connect to the corner house in the new neighborhood; phone lines from the second house would connect to the second house in the new neighborhood, and so on.

Shatz then proposed a clever idea to explain how a retinal axon "knows" that it has found a target next to its neighbor: maybe the immature retinal cells are testing the connections of their axons by sending electrical signals, like a kind of test phone call, over their axons' "wires" to the LGN target cells. Using its connections in the LGN to listen for the ringing of the neighbor's phone there, a nerve cell might then be able to tell if it had connected to its proper target. "If they hear the phone ringing in the neighbor's house, then they know they are in the right place," Shatz says. She points out that the phoning home process would have to be generated by the immature nerve cells themselves, since this stage of brain development occurs before light ever enters the eye.

Shatz and her colleagues tested this idea with the aid of a device that eavesdrops on some 200 nerve cells simultaneously. It uses 60 microelectrodes at once, compared with Hubel and Wiesel's single microelectrode, and it allows them to determine which neurons are placing phone calls and which are not.

In this way they showed that neighboring retinal cells tend to send out nerve signals as a group, and that this induces many responses in the receiving neurons. A target neuron that receives a big enough impulse from several axons firing together may release chemicals that nourish those axons and encourage them to maintain their connections. The other axons do not get nourished, and their connections die off. In the developing visual circuits, this activity-dependent pruning eliminates most axonal connections in the LGN.

But why would evolution maintain such a seemingly wasteful system—establishing connections and then eliminating them? "The guidance cues for the growing axons operate with fidelity, yet errors in connection are made," Jessell suggests. "If even 5 percent of the axon projections were wrong, then you would get a whole set of inappropriate connections early in development. And if they were allowed to persist, many aspects of brain function would be impaired."

There is growing evidence that the same biological processes that drive the brain's development in the embryo—especially the nourishing feedback system Shatz described in the visual system—continue long after birth. Cats, for example, cannot see at birth because the final organization of their brain's visual system does not occur until their eyes begin taking in light. Humans can see at birth, but they cannot talk. A baby must hear spoken language for many months before it develops enough to babble and then speak.

Many of the mechanisms of embryonic brain development that are now being uncovered appear similar to mechanisms of learning and memory, according to Charles Stevens, of the HHMI unit at the Salk Institute in San Diego, California. "After development is done," he says, "these are the same mechanisms that we use throughout the rest of our life."

In brain development, the level of activity between axons and target neurons determines whether their connections will persist. Similarly, a phenomenon called long-term potentiation, which plays a key role in memory, occurs when several axons fire together and stimulate a nerve cell to have an increasingly strong response.

"If people could figure out how to strengthen the systems that tell axons where to maintain their connections during development or where to strengthen their activity in learning and memory, it would be like 'weight-lifting for the mind,'" Jessell says. We might then use some of the findings from studies of early brain development in flies, worms, and mice to keep our brains in top shape throughout our lives. We might be able to prevent some birth defects, repair some forms of damage, and avoid some of the ravages of old age. ●

◆

The phoning home process must be generated by the immature cells themselves...

◆

Some Striking
SIMILA

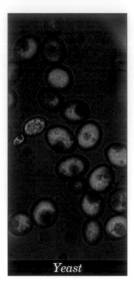

Yeast

Sea urchin

"Several years ago, I had a big surprise," says Ira Herskowitz, a professor of biochemistry at the University of California, San Francisco (UCSF), who is one of the world's experts on the genetics of yeast.

Herskowitz had been studying the "mating factors" that lead some of these single-celled creatures to fuse together. What shook him up was the discovery that the molecular machinery used by yeast to respond to these mating factors looked a lot like the machinery that visual cells use to respond to light.

It also looked like the machinery that brain cells use to respond to epinephrine (adrenaline), a hormone that stimulates the nervous system. In all three systems, a receptor threads its way through the cell membrane seven times and a companion protein, termed a G protein, acts on other molecules inside the cell.

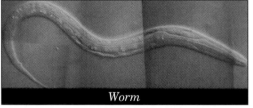

Worm

"I realized that the whole pathway which was known from studies of vertebrates exists in yeast," Herskowitz says. "Since then, it has become abundantly clear to me that one can discover all kinds of molecular machinery in yeast, flies, nematodes, or mammals, and then find the same machinery in the other organisms as well."

Many other scientists have been astounded by similar discoveries recently. They have found previously unthinkable similarities among their own experimental subjects—frogs, flies, worms, mice, chicks, sea urchins, fish, or yeast—and also between these organisms and human beings. The similarities extend to the order and organization of genes that control major pathways of development.

Each experimental organism still has its partisans, who can recite its special qualities at the drop of a hat. The fly people, who make up by far the largest and most established contingent of researchers on development, are very proud of how much has been learned through their work. Their closest rivals are the up-and-coming worm people. The sea urchin and frog people, who have made fundamental contributions to the field, feel unjustly neglected at the moment. The newly fashionable zebrafish people hope that their small vertebrate will bridge the gap between tiny, more easily studied insects or worms and the larger vertebrates. The mouse people are full of excitement about the new gene-targeting techniques that make it possible for them to knock out individual genes in mice and do experiments that were previously possible only in yeast.

The scientists who focus on different animals hold separate meetings and publish separate newsletters. Nevertheless, the mounting evidence of similarities between their subjects' developmental mechanisms has led some researchers to move beyond their own turf.

Zebrafish

RITIES

Frog

Christiane Nüsslein-Volhard, who pioneered the genetic study of the fly embryo's body plan, has shifted half of her famous lab in Tübingen, Germany, to research on zebrafish. And in the HHMI unit at UCSF, Yuh Nung Jan, who has identified genes that control the development of the fly embryo's nervous system, is starting to study corresponding genes in the brains of mice. "We want to extend our work to mammals," he says. "Nature regulates development with a very small number of shared mechanisms and molecules, and some general themes are now emerging from various directions."

Fruit Fly

Increasingly, biologists realize that each organism can contribute some pieces to the giant jigsaw puzzle of development. As CalTech's Paul Sternberg puts it, "If you run into a roadblock in one organism, you can get a clue from another organism because there are these universal pathways. The nicest thing that's coming out now is that some common mechanisms are used hundreds of times in nature, like little logic elements. So they can be studied at the molecular level in a variety of organisms."

Mouse

For example, *unc-86,* a gene that was found to control some aspects of development in the nervous system of worms, is so similar to *Pit-1,* a gene that affects the pituitary gland in the human brain, and also to *OCT-1* and *-2,* which were originally identified in the human immune system, that the whole group of genes is now called the POU group.

In fact, Sternberg says, known developmental genes are "a gold mine" that may lead to understanding—and perhaps preventing—a large number of disorders produced by errors in universal pathways. These include birth defects, as well as cancer and the degenerative diseases of the older brain.

The more similarities are found across the animal kingdom, the more promising this approach appears. "Developmental biologists are sort of whipped into a frenzy these days," Sternberg says, because of the striking similarities that have already emerged. ●

Humans

Two sets of experimental findings that once seemed impossible are now rocking the foundations of research on development.

One set promises an unlimited supply of the earliest, "totipotent" human stem cells—the cells that later differentiate into blood, bone, or other specialized tissue, or else form human organs. In work reported in 1998, teams from the University of Wisconsin-Madison and the Johns Hopkins University succeeded for the first time in growing these valuable cells in laboratory dishes. The scientists envision the use of such stem cells in medical care as well as in research.

Medically, the cells could replace sick or dying pancreatic cells in patients with juvenile diabetes, for instance, or brain cells in people with Parkinson's disease. Scientifically, the advance would finally allow researchers to study human cells as they mature into distinct kinds of tissues. This might reveal how cells "choose" what to become, possibly enabling researchers to guide such choices.

Another set of experiments suggests that not all of the cells' choices are as irreversible as scientists once thought. The researchers who cloned Dolly the sheep and other animals managed to wind back the developmental clock of certain specialized cells and make them stop dividing. Then they implanted the cells' nuclei into eggs whose own nuclei had been removed, and the eggs somehow reprogrammed the new nuclei to make the hybrid cells grow.

Both sets of findings go against long-held beliefs about early development. The unexpected ability to grow human stem cells in culture is particularly exciting to scientists. If these cells— the raw material for virtually every kind of human tissue—become readily available, medical investigators can devise completely new strategies for treating human diseases.

"People are just realizing it might be possible," says HHMI investigator Allan Spradling of the Carnegie Institution. "If so, it would have a huge impact on medicine: One could replace cells that are bad, especially in aging, and develop new therapies. It's a very promising and interesting area, but a lot of obstacles need to be overcome."

Ten Signaling Pathways

A current focus of interest, says Spradling, is how cells communicate with one another throughout the course of development. The cells use various kinds of chemical signals that can be classified into ten or more distinct "signaling pathways." These pathways link the cells much as multiple means of communication link people. Various combinations of pathways coordinate the development of specific tissues and organs.

So instead of looking for single morphogens or focusing on a single pathway, many scientists are now studying how different pathways interact with one another. It turns out that activating a particular pathway does not always guarantee a particular response, Spradling says, since the same pathways are often used in many different contexts.

The Worm's Surprises

Work on the microscopic roundworm *C. elegans* is becoming increasingly important because of the availability of the worm's genome, which was completed in 1998. Among the first surprises: It contains more genes (19,000) than previously believed (15,000). And as many as 70 percent of the genes so far identified in humans now appear to have counterparts in the worm.

Even before the worm genome was completed, research on *C. elegans* had produced some major achievements, particularly in the area of programmed cell death, or apoptosis.

Programmed Cell Death

Programmed cell death is now seen as a fundamental process that can cause serious diseases if it goes awry. It's a bit like a see-saw: Too much cell death can lead to severe loss of function, such as blindness in retinitis pigmentosa, and probably some neurodegenerative problems such as Alzheimer's, Parkinson's, or stroke. Too little cell death, on the other hand, leads to a dangerous proliferation of cells and can result in cancer.

"We can generate too many cells, as in cancer, not only by too much cell division, but also by too little cell loss," explains Robert Horvitz. A dramatic demonstration of the connection between cancer and apoptosis came when Stanley Korsmeyer, who was then an HHMI investigator at the Washington University School of Medicine in St. Louis, Missouri (he is now at the Dana Farber Cancer Institute in Boston, Massauchetts), and others identified *BCL2*, a human gene related to follicular lymphoma. They showed that when *BCL2* is deregulated, it "extends the survival of cells normally destined to die," Korsmeyer says.

Around the same time, Horvitz's team identified *ced9*, a roundworm gene that "protects cells from dying by programmed cell death." These two genes—*BCL2* and *ced9*—are "so similar that the human gene can work in worms to protect against worm cell death and to substitute for the worm gene," Horvitz reports.

By now a whole pathway that regulates apoptosis has been studied in worms, flies, and mammals. In the worm, it consists of at least 15 genes. "Mammalian counterparts have been found for 6 of the 15 genes," says Horvitz. He points out that once the entire pathway is known, there will be a wide choice of possible targets for drugs.

Vive La Différence
Scientists once thought the Y chromosome—the genetic key to maleness—held just one or two genes, says MIT's David Page. But investigators have now found 23 genes or gene families on the Y, and there may be more. Page expects the total to be 30 or 35. Many of these look like so-called housekeeping genes, he says, noting that they have counterparts on the X chromosome and are expressed in all tissues. But several Y genes are expressed only in the testes, and some are required for the production of sperm.

"We have come to understand that some fraction of infertility in healthy people is due to genetic defects," says Page. "Ten percent of American couples are infertile, half for male-related reasons. And the most common cause of male infertility that has been defined at the molecular level is a deletion of genetic matter on the Y chromosome— not a deletion of the *SRY* gene, but of other parts."

Embryos with the *SRY* gene develop testes, he explains—but if they have deletions of some of the Y chromosome's fertility genes, the adult males won't produce a significant number of sperm. "They will grow up infertile, though otherwise as perfectly healthy men," Page says. While the *SRY* gene is located on the short arm of the Y chromosome (which is a very short arm indeed), the genes for the fertility factors that have been identified so far are on the long arm of the Y.

Page's lab is now moving ahead in time, developmentally speaking—shifting its focus from the first stages of sexual development to slightly later stages. "We have started addressing the question, 'Do you make eggs or sperm?' That," he says, "is the most fundamental biological distinction between males and females."

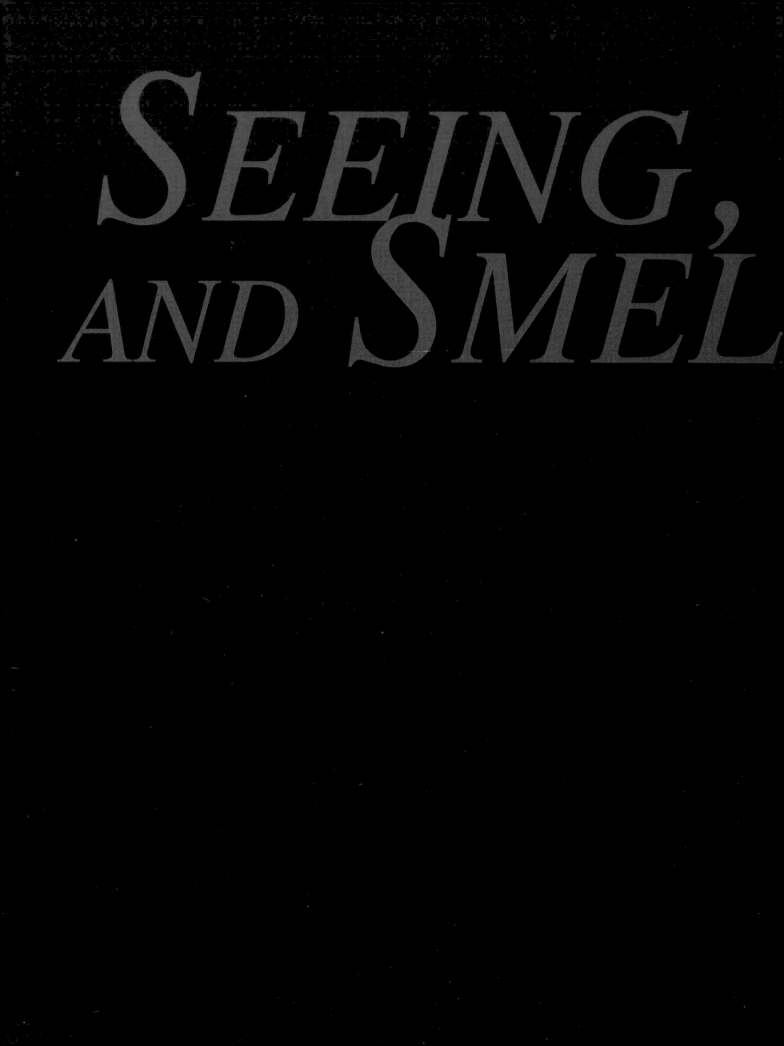

SEEING,
AND SMEL

Seeing, Hearing, and Smelling
AND Smelling
THE World

Howard Hughes Medical Institute
6000 Jones Bridge Road
Chevy Chase, Maryland 20815-6789
(301) 215-8500

A REPORT FROM THE HOWARD HUGHES MEDICAL INSTITUTE

SEEING,
HEARING,
AND
SMELLING THE WORLD

**NEW FINDINGS
HELP SCIENTISTS
MAKE SENSE OF
OUR SENSES**

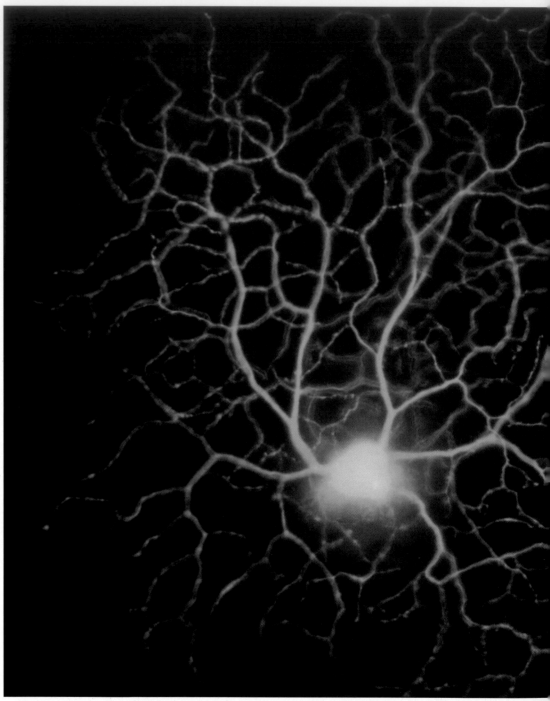

A nerve cell that can detect in what direction an object is moving branches out to make contact with many other cells in a rabbit's visual system. The cell glows yellow because it was injected with fluorescent dye.

SEEING, HEARING, AND SMELLING THE WORLD

New Findings Help Scientists Make Sense of Our Senses

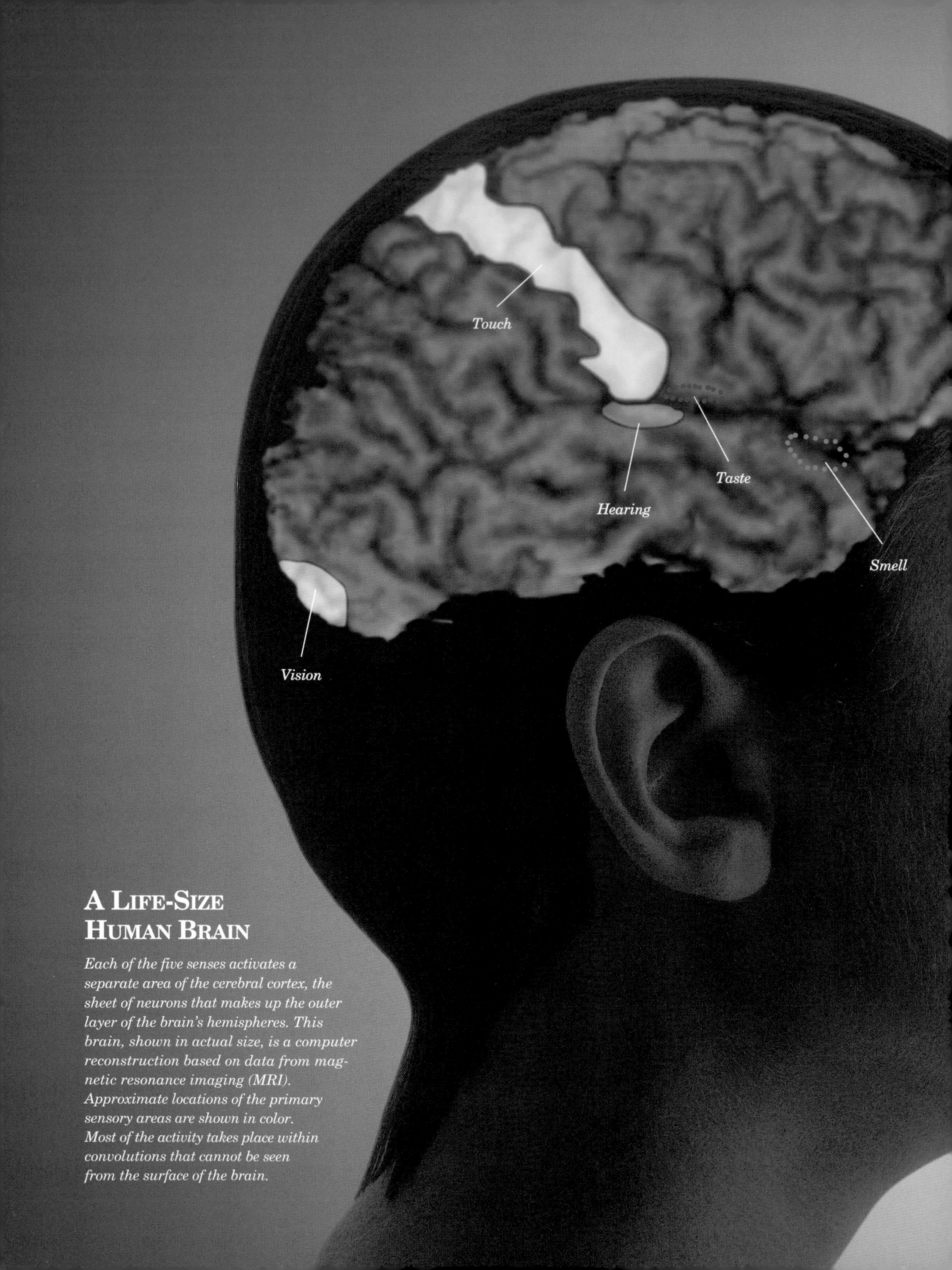

Touch

Hearing

Taste

Smell

Vision

A Life-Size Human Brain

Each of the five senses activates a separate area of the cerebral cortex, the sheet of neurons that makes up the outer layer of the brain's hemispheres. This brain, shown in actual size, is a computer reconstruction based on data from magnetic resonance imaging (MRI). Approximate locations of the primary sensory areas are shown in color. Most of the activity takes place within convolutions that cannot be seen from the surface of the brain.

W̲e can recognize a friend instantly—full-face, in profile, or even by the back of his head. We can distinguish hundreds of colors and possibly as many as 10,000 smells. We can feel a feather as it brushes our skin, hear the faint rustle of a leaf. It all seems so effortless: we open our eyes or ears and let the world stream in.

OUR COMMON SENSES

Yet anything we see, hear, feel, smell, or taste requires billions of nerve cells to flash urgent messages along linked pathways and feedback loops in our brains, performing intricate calculations that scientists have only begun to decipher.

"You can think of sensory systems as little scientists that generate hypotheses about the world," says Anthony Movshon, an HHMI investigator at New York University. Where did that sound come from? What color is this, really? The brain makes an educated guess, based on the information at hand and on some simple assumptions.

When you look at the illustration below, for instance, you see an X made of spheres surrounded by cavities. But if you turn the page upside down, all the cavities become spheres, and vice versa. In each case, the shapes seem real because "your brain assumes there is a single light source—and that this light comes from above," says Vilayanur Ramachandran, a professor of neuroscience at the University of California, San Diego. As he points out, this is a good rule of thumb in our sunlit world.

To resolve ambiguities and make sense of the world, the brain also creates shapes from incomplete data, Ramachandran says. He likes to show an apparent triangle that was developed by the Italian psy-

times "hear things" that are not really there. But suppose a leopard approached, half-hidden in the jungle—then our ability to make patterns out of incomplete sights, sounds, or smells could save our lives.

Everything we know about the world comes to us through our senses. Traditionally, we were thought to have just five of them—vision, hearing, touch, smell, and taste. Scientists now recognize that we have several additional kinds of sensations, such as pain, pressure, temperature, joint position, muscle sense, and movement, but these are generally included under "touch." (The brain areas involved are called the "somatosensory" areas.)

Although we pay little attention to them,

ILLUSIONS REVEAL SOME OF THE BRAIN'S ASSUMPTIONS

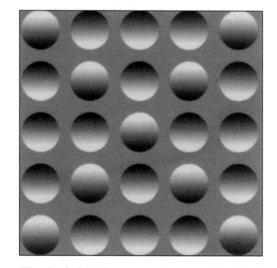

The shaded circles seem to form an X made of spheres. But if you turn the page upside down, the same circles form an X made of cavities, since the brain assumes that light comes from above.

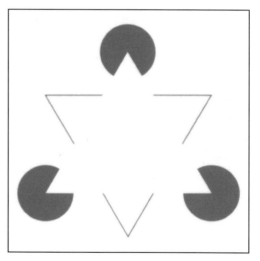

Are these triangles real? They appear to be, because the brain automatically fills in lines that are missing. But if you block out parts of the picture, the triangles vanish.

chologist Gaetano Kanizsa. If you hide part of this picture, depriving the brain of certain clues it uses to form conclusions, the large white triangle disappears.

We construct such images unconsciously and very rapidly. Our brains are just as fertile when we use our other senses. In moments of anxiety, for instance, we some-

each of these senses is precious and almost irreplaceable—as we discover, to our sorrow, if we lose one. People usually fear blindness above all other disabilities. Yet deafness can be an even more severe handicap, especially in early life, when children learn language. This is why Helen Keller's achievements were so extraordinary. As a result of an acute illness at the age of 19 months, she lost both vision and hearing and sank into a totally dark, silent universe. She was rescued from this terrible isolation by her teacher, Anne Sullivan, who managed to explain, by tapping signs into the little girl's palm, that things have names, that letters make up words, and that these can be used to express wants or ideas. Helen Keller later grew into a writer (her autobiography, *The Story of My Life,* was published while she was still an undergraduate at Radcliffe College) and a well-known advocate for the handicapped. Her remarkable development owed a great deal to her determination, her teacher, and her family. But it also showed that when a sense (or two, in Helen Keller's

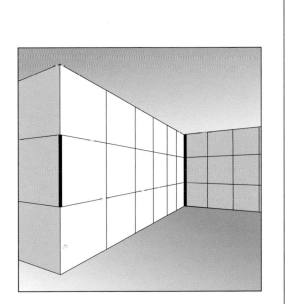

The black line in the back seems much longer than the one in the front because your brain assumes it is seeing the effects of perspective. Take a ruler to find out for yourself.

case) is missing, another sense (in her case, touch) may be trained to make up for the loss, at least in part.

What we perceive through our senses is quite different from the physical characteristics of the stimuli around us. We cannot see light in the ultraviolet range, though bees can, and we cannot detect light in the infrared range, though rattlesnakes can. Our nervous system reacts only to a selected range of wavelengths, vibrations, or other properties. It is limited by our genes, as well as our previous experience and our current state of attention.

What draws our attention, in many cases, is change. Our senses are finely attuned to change. Stationary or unchanging objects become part of the scenery and are mostly unseen. Customary sounds become background noise, mostly unheard. The feel of a sweater against our skin is soon ignored. Our touch receptors, "so alert at first, so hungry for novelty, after a while say the electrical equivalent of 'Oh, that again,' and begin to doze, so we can get on with life," writes Diane Ackerman in *A Natural History of the Senses.*

If something in the environment changes, we need to take notice because it might mean danger—or opportunity. Suppose an insect lands on your leg. Instantly the touch receptors on the affected leg fire a message that travels through your spinal column and up to your brain. There it crosses into the opposite hemisphere (the right hemisphere of the brain receives signals from the left side of the body, and vice versa) to alert brain cells at a particular spot on a sensory map of the body.

This map extends vertically along a strip of cerebral cortex near the center of the skull. The cortex—a deeply wrinkled sheet of neurons, or nerve cells, that covers the two hemispheres of the brain—governs all our sensations, movements, and thoughts.

The sensory map in humans was originally charted by the Canadian neurosurgeon Wilder Penfield in the 1930s. Before operating on patients who suffered from epilepsy, Penfield stimulated different parts of their brains with electrodes to locate the cells that set off their attacks. He could do this while the patients were awake, since the brain does not feel what is happening to it. In this way, Penfield soon learned exactly where each part of the body that was touched or moved was represented in the brain, as he showed in his famous "homunculus" cartoons of the somatosensory and motor areas.

...the patients were awake, since the brain does not feel what is happening to it.

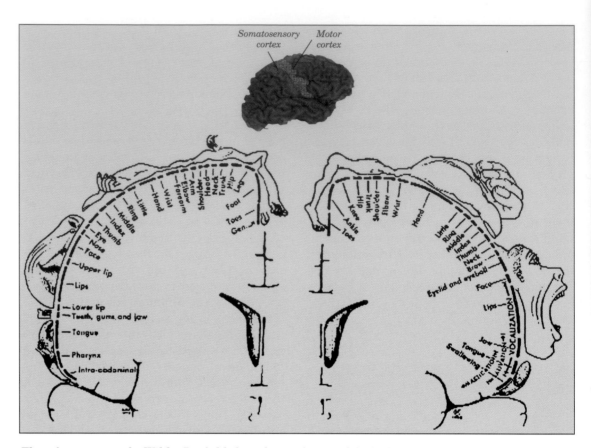

These famous maps by Wilder Penfield show that each part of the body is represented on two strips of the brain's cerebral cortex, the somatosensory cortex (left), which receives sensations of touch, and the motor cortex (right), which controls movements. Fingers, mouth, and other sensitive areas take up most space on both maps. Penfield called these cross sections the "sensory homunculus" and the "motor homunculus."

Surprisingly, these maps do not accurately reflect the size of body parts but rather, their sensitivity. Arms and legs take up very little space, despite their length. The face and hands, which have greater sensitivity, are given more space—especially the tips of the fingers. Nevertheless, the signal that a mosquito has landed on the back of your left leg comes through loud and clear. In a fraction of a second, through a decision process that is not yet understood, this signal leads you to swat the insect at just the right place.

Ever since humans have wondered about where their thoughts came from, they have tried to understand the senses. Much was learned from observing the results of head injuries and tumors, as well as by dissecting postmortem human brains and the brains of animals. In the 1930s and 1940s,

scientists applied electrodes to the surface of the brain or placed them on the skull of humans to study "evoked responses," the changing rhythms of electrical signals in the brain in response to specific stimuli such as light or sound. Unfortunately, these signals from billions of brain cells proved almost impossible to unscramble.

When extremely thin microelectrodes became available in the late 1950s, researchers implanted them into the brains of living animals to spy on the activity of individual cells. Sharp popping sounds could be heard as specific neurons fired, and the scientists tried to find out what provoked these electrical discharges.

This is how David Hubel and Torsten Wiesel, who were then at Johns Hopkins University, began the groundbreaking experiments on the visual cortex of cats and monkeys, for which they later won a Nobel Prize. They discovered that one neuron in the primary visual cortex at the back of a cat's brain might fire only when the animal's eye was exposed to a line at a particular location and angle, while another next to it would fire only in response to a line at a slightly different angle. No one had suspected that these neurons would dissect a scene—and respond to particular elements of it—with such amazing specificity. Hubel and Wiesel's success led to a general focus on the abilities of single neurons, especially in the visual system.

The past decade has seen an explosion of research on all the senses, partly because of the new tools supplied by molecular biology. Scientists can now analyze sensory neurons far more precisely, down to the level of specific genes and proteins within these neurons. This publication describes some recent research on three of our senses—vision, hearing, and smell—in which there have been particularly interesting developments.

The visual system, which involves roughly a quarter of the human cerebral cortex, has attracted more research than all the other sensory systems combined. It is also the most accessible of our senses. The retina, a sheet of neurons at the back of the eye that any physician can see through an ophthalmoscope, is the only part of the brain that is visible from outside the skull. Research on the visual system has taught scientists much of what they know about the brain, and it remains at the forefront of progress in the neurosciences.

Research on hearing is also gathering momentum. One group of scientists recently discovered how receptor neurons in the ear— the so-called "hair cells"—respond to sounds. Another group explored how animals use sounds to compute an object's location in space. This may be a model of similar operations in the auditory system of humans.

The olfactory system, which was almost a total mystery until a few years ago, has become the source of much excitement. The receptor proteins that make the first contact with odorant molecules appear to have been identified with the help of molecular genetics, and researchers are beginning to examine how information about smells is coded in the brain.

The use of molecular biology has enabled scientists to discover just how receptor neurons respond to light, to vibrations in the air, to odorant molecules, or to other stimuli. The receptor neurons in each sensory system deal with different kinds of energy—electromagnetic, mechanical, or chemical. The receptor cells look different from one another, and they exhibit different receptor proteins. But they all do the same job: converting a stimulus from the environment into an electrochemical nerve impulse, which is the common language of the brain (see p. 123). Researchers have uncovered many of the genes and proteins involved in this process of sensory transduction.

From their understanding of this first step on the sensory pathway, researchers have edged up to analyzing how messages about a sensory stimulus travel through the brain to the cerebral cortex and how these messages are coded.

They know that nearly all sensory signals go first to a relay station in the thalamus, a central structure in the brain. The messages then travel to primary sensory areas in the cortex (a different area for each sense), where they are modified and sent on to "higher" regions of the brain. Somewhere along the way, the brain figures out what the messages mean.

Many factors enter into this interpretation, including what signals are coming in from other parts of the brain, prior learning, overall goals, and general state of arousal. Going in the opposite direction, signals from a sensory area may help other parts of the brain maintain arousal, form an image of where the body is in space, or regulate movement.

These interactions are so complex that focusing on the activity of single neurons—or even single pathways—is clearly not enough. Researchers are now asking what the central nervous system does with all the information it gets from its various pathways.

In more authoritarian times, scientists believed that the brain had a strictly hierarchical organization. Each relay station was supposed to send increasingly complex information to a higher level until it reached the very top, where everything would somehow be put together. But now "we are witnessing

Sharp popping sounds could be heard as specific neurons fired...

a paradigm shift," says Terrence Sejnowski, an HHMI investigator who directs the Computational Neurobiology Laboratory at the Salk Institute in La Jolla, California. Instead of viewing the cortex as "a rigid machine," scientists see it as "a dynamic pattern-processor and categorizer" that recognizes which categories go together with a particular stimulus, as best it can, every step of the way. "There is no 'grandmother cell' at the top that responds specifically to an image of Grandma," Sejnowski emphasizes. "We recognize a face by how its features are put together in relation to one another."

Sejnowski, a leader in the new field of computational neuroscience, studies neural networks in which the interaction of many neurons produces surprisingly complex behavior. He has designed a computer model of how such a network might learn to "see" the three-dimensional shape of objects just from their shading, without any other information about where the light came from. After being "trained" by being shown many examples of shaded shapes, the network made its own generalizations and found a way to determine the objects' curvature.

Vision and the other senses evolved "to help animals solve vital problems—for example, knowing where to flee," says Sejnowski. Large populations of sensory neurons shift and work together in the brain to make this possible. They enable us to see the world in a unified way. They link up with the motor systems that control our actions. These neurons produce an output "that is more than the sum of its parts," Sejnowski says. Just how they do it is a question for the next century. ●

Maya Pines, Editor

Senses evolved "to help animals solve vital problems..."

SPECIAL RECEPTOR CELLS FOR EACH OF THE SENSES

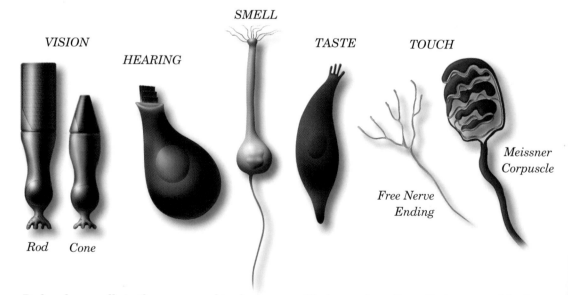

VISION

HEARING

SMELL

TASTE

TOUCH

Meissner Corpuscle

Free Nerve Ending

Rod *Cone*

Rod and cone cells in the eye respond to electromagnetic radiation—light.

The ear's receptor neurons are topped by hair bundles that move in response to vibrations—sound.

Olfactory neurons at the back of the nose respond—and bind—to odorant chemicals.

Taste receptor cells on the tongue and back of the mouth respond—and bind—to chemical substances.

Meissner corpuscles are specialized for rapid response to touch, while free nerve endings bring sensations of pain.

A Language the Brain Can Understand

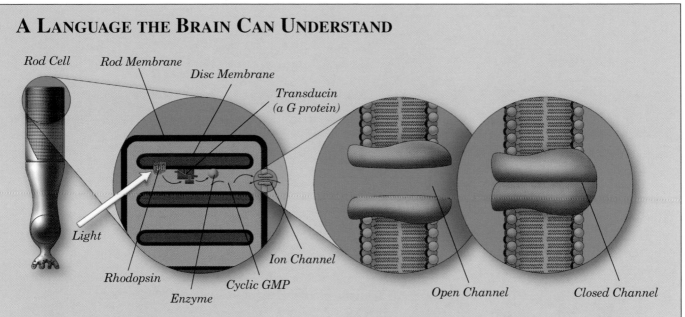

Rod Cell — Rod Membrane — Disc Membrane — Transducin (a G protein) — Light — Rhodopsin — Enzyme — Cyclic GMP — Ion Channel — Open Channel — Closed Channel

Almost at the very instant that light hits a cell in the retina, or a sound wave nudges the tip of a receptor cell in the ear, the receptor cell converts this stimulus into an electrical signal—the language of the brain.

This conversion, or transduction, is swift and precise. But it is also surprisingly intricate—so intricate that the process is not yet fully understood for most of the senses. In the past decade, however, it has been worked out quite thoroughly for vision.

It begins when a photon of light meets one of the photoreceptor cells of the retina (either a rod or a cone cell). A photon that strikes a rod cell is immediately absorbed by one of the 100 million molecules of a receptor protein—rhodopsin—that are embedded in the membranes of a stack of disks in the top part, or "outer segment," of each cell. These rhodopsin molecules have a snakelike shape, crisscrossing the membrane seven times, and contain retinal (a form of vitamin A), which actually absorbs the light. In the dark, the retinal fits snugly into a binding pocket in rhodopsin.

But on exposure to light, it straightens out. This alters the three-dimensional structure of the entire rhodopsin molecule, activating it and triggering a biochemical cascade.

The activated rhodopsin then stimulates transducin, a protein that belongs to the large family of so-called G proteins. This in turn activates an enzyme that breaks down cyclic GMP, a "second messenger," dramatically lowering its level. Cyclic GMP carries signals from the disks, where light is absorbed, to the cell's surface membrane, which contains a large number of channels that control the flow of ions (charged atoms) into the cell. As ions move into the cell, they alter its electrical potential.

"In the dark, the channels are constantly open because of a high level of cyclic GMP. This allows sodium and calcium ions, which carry positive charges, to flow into the cell," explains King-Wai Yau, an HHMI investigator at the Johns Hopkins University School of Medicine, who played an important role in deciphering the transduction process. "But in the light, the channels close. Then the electrical potential inside the cell becomes more negative. This reduces the amount of neurotransmitter that is released from the base of the cell to act on other cells"—and thus alerts neurons in the next layer of retinal cells that a photon of light has arrived.

This complex cascade of transduction events is repeated in a remarkably similar way in olfactory receptor cells, which respond to odors, says Yau. But the receptor cells that respond to sound use a very different system: their channels open and close as a direct response to a mechanical force—either tension or relaxation.

Whatever the means, the end result of transduction is the same: the cell generates an electrical signal that flashes through a dense thicket of nerve cell connections in the brain, bringing news from the outside world in a Morse code–like language the brain can understand.

ode of COLOR

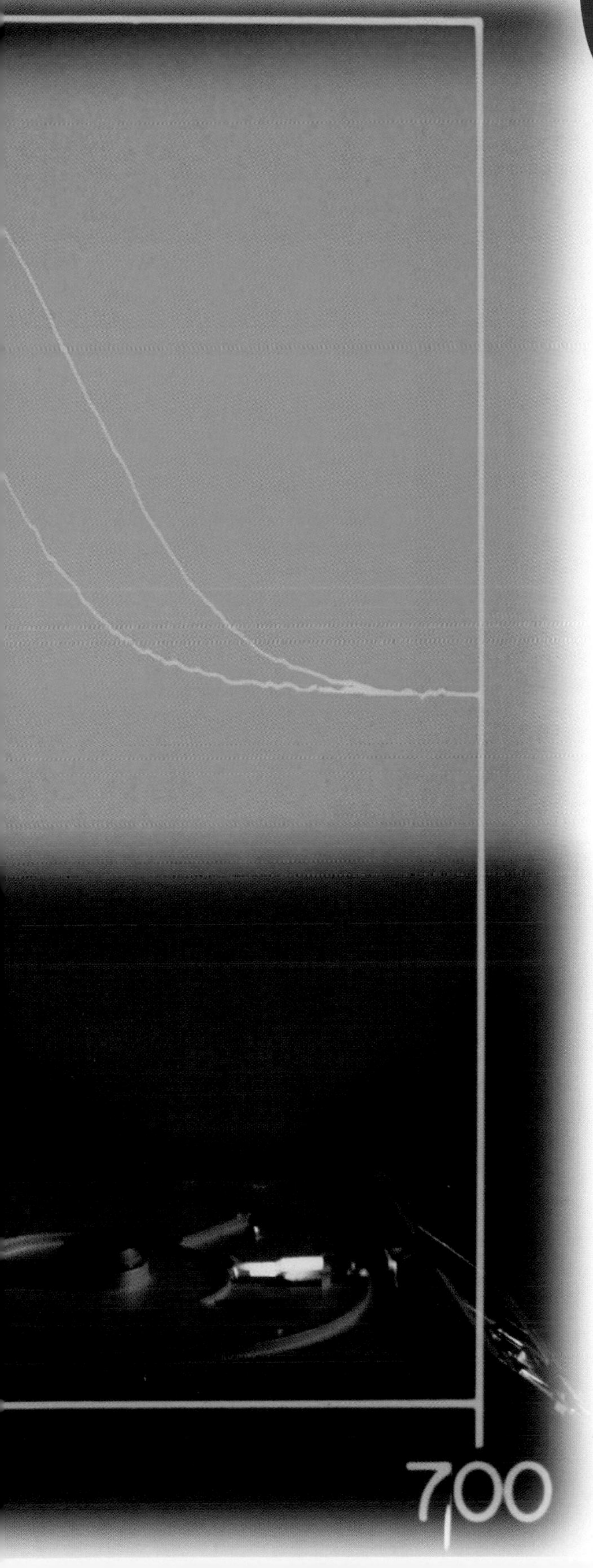

A bright red beach ball comes whirling toward you. You see its color, shape, and motion all at once—but your brain deals with each of these characteristics separately.

"We need parallel processing because neurons are relatively slow computing machines," says Jeremy Nathans, an HHMI investigator at the Johns Hopkins University School of Medicine. "They take several milliseconds to go from input to output. Yet you see things in a fraction of a second—time for no more than 100 serial steps. So the system has to have a massively parallel architecture."

Nathans adjusts a slide projector to show the colors that are detected by receptor proteins in red and green cone cells. The proteins were made from human DNA in his lab. The peaks in the graph indicate the wavelengths (in nanometers) of light best absorbed by each protein.

by Geoffrey Montgomery

700

The First Glimmer of Color

Nathans became interested in how we see in color the day he heard of new discoveries about how we see in black and white. It was 1980, and he was a student at Stanford Medical School, he recalls, when Lubert Stryer and Denis Baylor, both of Stanford, described their remarkable findings about the workings of rod cells. These cells—one of two kinds of photoreceptor cells in the retina—enable us to see in dim light, even by the muted starlight of a hazy night.

"Baylor showed that rod cells achieve the ultimate in light sensitivity—that they can respond to a single photon, or particle of light," says Nathans. "It was a beautiful experiment." (Baylor's work was done in collaboration with Trevor Lamb and King-Wai Yau.)

Then Stryer explained how rhodopsin, the light-sensitive receptor protein in the disk membranes of rod cells, announces the arrival of this tiny pulse of light to the signaling machinery inside the cell. Stryer had found that rhodopsin could do this only with the help of an intermediary, called a G protein, which belonged to a family of proteins that was already known to biochemists from their study of how cells respond to hormones and growth factors.

Nathans immediately realized this meant that the structure of rhodopsin itself might be similar to that of receptors for hormones. His mind began racing with possibilities. "And I ran—literally ran—to the library and started reading about vision," he says.

Until then, Nathans had been studying the genetics of fruit flies. But as he read a paper by Harvard University biologist George Wald—a transcript of Wald's 1967 Nobel Prize lecture on "The Molecular Basis of Visual Excitation"—Nathans set off on a different course. He determined to do what Wald himself had wished to do 40 years earlier: find the receptor proteins in the retina that respond to color.

Rod cells function only in dim light and

The intricate layers and connections of nerve cells in the retina were drawn by the famed Spanish anatomist Santiago Ramón y Cajal around 1900. Rod and cone cells are at the top. Optic nerve fibers leading to the brain may be seen at bottom right.

are blind to color. "Get up on a dark moonlit night and look around," suggests David Hubel of Harvard Medical School, a winner of the Nobel Prize for his research on vision. "Although you can see shapes fairly well, colors are completely absent. It is remarkable how few people realize that they do without color vision in dim light."

But the human retina also contains another kind of photoreceptor cell: the cones, which operate in bright light and are responsible for high-acuity vision, as well as color.

Rods and cones form an uneven mosaic within the retina, with rods generally outnumbering cones more than 10 to 1—except in the retina's center, or fovea. The cones are highly concentrated in the fovea, an area that Nathans calls "the most valuable square millimeter of tissue in the body."

Even though the fovea is essential for fine vision, it is less sensitive to light than the surrounding retina. Thus, if we wish to detect a faint star at night, we must gaze slightly to the side of the star in order to project its image onto the more sensitive rods, as the star casts insufficient light to trigger a cone into action.

In bright light, then, when the cones are active, how do we perceive colors? This question has attracted some of the finest minds in science. As early as 1672, by experimenting with prisms, Isaac Newton made the fundamental discovery that ordinary "white" light is really a mixture of lights of many different wavelengths, as seen in a rainbow. Objects appear to be a particular color because they reflect some wavelengths more than others. A red apple is red because it reflects rays from the red end of the visible spectrum and absorbs rays from the blue end. A blueberry, on the other hand, reflects the blue end of the spectrum and absorbs the red.

Thinking about Newton's discovery in 1802, the physician Thomas Young, who later helped decipher the hieroglyphics of the Rosetta Stone, concluded that the reti-

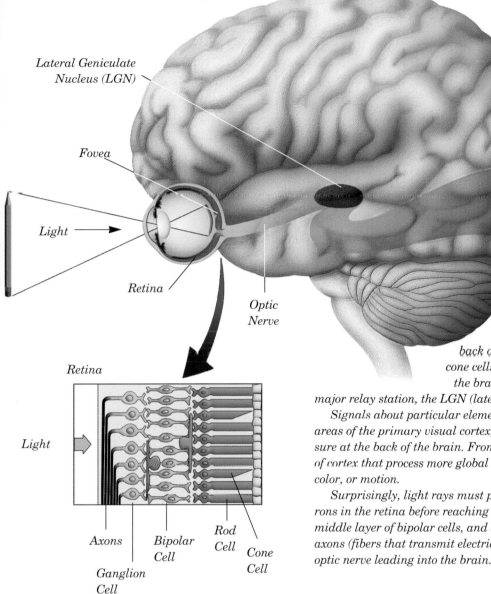

Lateral Geniculate
Nucleus (LGN)

Fovea

Light →

Retina

Optic
Nerve

Primary
Visual
Cortex (V1)

The visual pathway:
*Light rays reflected by an
object—for example, a pencil—
enter the eye and pass through its
lens. The lens projects an inverted
image of the pencil onto the retina at the
back of the eye. Signals produced by rod and
cone cells in the retina then start on their way into
the brain through the optic nerve and reach a
major relay station, the LGN (lateral geniculate nucleus).*

*Signals about particular elements of the pencil then travel to selected
areas of the primary visual cortex, or V1, which curves around a deep fis-
sure at the back of the brain. From there, signals fan out to "higher" areas
of cortex that process more global aspects of the pencil such as its shape,
color, or motion.*

*Surprisingly, light rays must penetrate two transparent layers of neu-
rons in the retina before reaching the precious rods and cones at the back: a
middle layer of bipolar cells, and a front layer of ganglion cells whose long
axons (fibers that transmit electrical impulses to other neurons) form the
optic nerve leading into the brain.*

Retina

Light

Axons Bipolar
Cell

Ganglion
Cell

Rod
Cell Cone
Cell

na could not possibly have a different
receptor for each of these wavelengths,
which span the entire continuum of colors
from violet to red. Instead, he proposed
that colors were perceived by a three-color
code. As artists knew well, any color of the
spectrum (except white) could be matched
by judicious mixing of just three colors of
paint. Young suggested that this was not
an intrinsic property of light, but arose
from the combined activity of three different "particles" in the retina, each sensitive
to different wavelengths.

We now know that color vision actually

depends on the interaction of three types of
cones—one especially sensitive to red light,
another to green light, and a third to blue
light. In 1964, George Wald and Paul
Brown at Harvard and Edward MacNichol
and William Marks at Johns Hopkins
showed that each human cone cell absorbs
light in only one of these three sectors of
the spectrum.

Wald went on to propose that the receptor
proteins in all these cones were built on the
same plan as rhodopsin. Each protein uses
retinal, a derivative of vitamin A, to absorb
light; and each tunes the retinal to absorb a

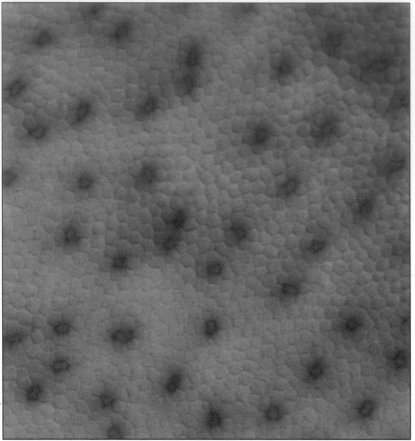

Laboratory of Richard Masland, Massachusetts General Hospital

The many rods (small circles) and fewer cones (dark outlines) in most of the retina form a mottled pattern, as shown in this photomicrograph. The central retina, or fovea, has only cones.

different range of wavelengths. Wald believed that the three receptor proteins in cones probably evolved from the same primordial gene—and so did rhodopsin. They were all "variations on a central theme," Wald wrote in his Nobel lecture.

This evolutionary message was music to Nathans' ears. It meant that if the gene encoding only one receptor protein could be located, the genes encoding the other receptor proteins could be found by the similarity of the sequence of bases in their DNA.

"I realized while reading Wald's lecture," says Nathans, "that Wald had laid out the whole problem of the genetic basis of color vision, and that this problem was now solvable, completely solvable, by molecular genetic methods." Wald had taken the problem as far as he could, Nathans pointed out. "But lacking these molecular methods, he couldn't go any further." Nathans' ambitious plan to isolate the genes that coded for the three color receptor proteins depended on Wald's view that the genes all evolved from the same pri-

mordial ancestor. The only visual receptor protein that had been studied with any intensity at that time was bovine rhodopsin—from the rod cells of cows' eyes. Scientists had purified bovine rhodopsin and deduced the sequence of a fragment of the DNA that coded for it. Nathans used this information to construct a lure—a single strand of DNA—with which he fished out the complete gene for bovine rhodopsin from a sea of bovine DNA.

Next he used part of this bovine gene as a lure to catch the gene for human rhodopsin from the jumble of DNA in a human cell. This took less than a year "because the genes for human and bovine rhodopsin are virtually identical, despite an evolutionary distance of 200 million years between cattle and humans," Nathans says.

Finding the human genes for the color receptors proved more challenging, however, since these genes are less closely related to the gene for rhodopsin. Nathans began to sift through DNA from his own cells. "I figured I'd be an unlimited source of DNA as long as I kept eating," he says. Eventually he fished out some pieces of DNA that belonged to three different genes, each of them clearly related to the rhodopsin gene. "This coincidence—three genes, three types of cones—didn't escape our notice," he said. Furthermore, two of these genes were on the X chromosome—"exactly what one would expect," says Nathans, "since it has long been known that defects in red and green color vision are X-linked."

Some 10 million American men—fully 7 percent of the male population—either cannot distinguish red from green, or see red and green differently from most people. This is the commonest form of color blindness, but it affects only 0.4 percent of women. The fact that color blindness is so much more prevalent among men implies that, like hemophilia, it is carried on the X chromosome, of which men have only one copy. (As in hemophilia, women are protected because they have two X chromosomes; a normal gene on one chromosome can often make up for a defective gene on the other.)

Wald and others had found that in color-blind men, the green or red cones worked improperly or not at all. Wald suggested

128 • EXPLORING THE BIOMEDICAL REVOLUTION

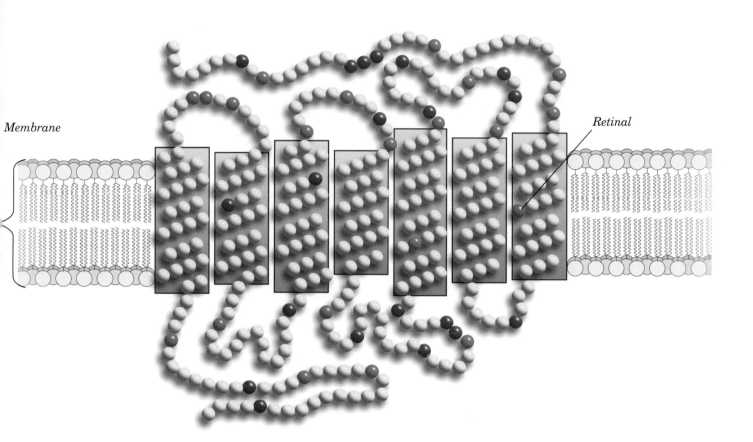

Membrane

Retinal

Rhodopsin, the receptor protein in rod cells, crosses the disk membrane seven times; its odd shape is shared by the three receptor proteins in cone cells. Retinal (which absorbs light) is shown in purple. The other colored balls represent amino acids that make up the rhodopsin structure.

that the genes for the red and green receptors were altered in these men. He also thought that these genes must lie near each other on the X chromosome. This tandem arrangement—which Nathans confirmed—probably results from the duplication of a DNA fragment in primates that occurred some 40 million years ago. The New World monkeys of South America, which broke from the continent of Africa at about that time, possess only a single functional copy of a red or green gene, much like color-blind men. But in Old World primates—the monkeys and apes of Africa and the ancestors of humans—a primordial red-green gene must have duplicated and then diverged slightly in sequence, leading to separate receptors of the red and green type. In keeping with this picture, Nathans found that the DNA sequences of the genes for red and green receptors differ by only 2 percent—evidence of their common origin and recent divergence.

Nathans himself is not color-blind. Before using his own DNA, he thoroughly tested his color vision to ensure that it was normal. Nevertheless, one of his initial findings presented a puzzle: Lying head to tail along his X chromosome were not just the two genes for the red and green receptors, but also an extra copy of the green receptor gene.

Here was the explanation for the prevalence of color blindness, he realized. Because the DNA sequences of the red and green receptor genes are so similar, and because they lie head to tail, it is easy for mistakes to occur during the development of egg and sperm, as genetic material is replicated and exchanged between chromosomes. One X chromosome—like Nathans'—may receive an extra green receptor gene, for instance, or maybe even two. This does no harm. But then the other chromosome with which it is exchanging bits of genetic information is

A NARROW TUNNEL OF LIGHT

Inability to see well in the dark may be an ominous sign in a child. Though it could just signal a need to eat more carrots or take vitamin A supplements, for more than a million people around the world (one out of every 4,000), it is the first symptom of retinitis pigmentosa (RP), a genetic disorder that may leave them totally blind by the age of 40.

"Sometime between their teens and their thirties, depending on the family, their retinas begin to degenerate," says Jeremy Nathans, who has been studying the genetic errors that cause the disease. First, the rod cells die at the retina's periphery. Then these zones of cell death slowly expand, leaving only a small patch of functioning retinal cells near the center of vision. The patients' visible world contracts to a narrow tunnel of light. Finally, the dying tissue may take everything with it, including the precious cones in the central retina, which are responsible for high-acuity vision.

"The retina doesn't regenerate," explains Nathans. "If any part of it goes, you won't get it back. And so far, there is no effective therapy for RP."

Until 1990, no one even knew the cause of RP. Thinking the disease might be related to a defect in rhodopsin, the receptor protein of rod cells (which are responsible for night vision), Nathans began to collect blood samples from patients so he could study their DNA. In 1989, Peter Humphries at Trinity College in Dublin, Ireland, found the location of a gene defect in a very large Irish family that had a dominant form of the disease (in which the inheritance of a mutated gene from just one parent causes the disease). Remarkably, Humphries mapped the defect to the same region of chromosome 3 in which Nathans had located the gene for rhodopsin.

Since then, two teams of scientists—Thaddeus Dryja at the Massachusetts Eye and Ear Infirmary, and Nathans and HHMI associate Ching-Hwa Sung at Johns Hopkins—have shown that about one-fourth of patients with the dominant form of the disease have mutations in their gene for rhodopsin. Other forms of RP result from mutations in different genes. Most of the errors in the rhodopsin gene cause the protein to be unstable,

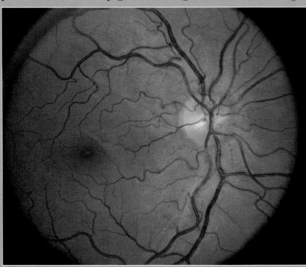

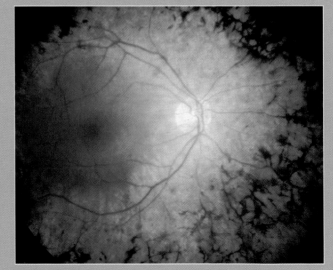

A healthy retina (below), as seen through an ophthalmoscope, has a firm, regular structure. In the retina of a person with retinitis pigmentosa (right), cells die, starting at the periphery. Cells laden with the black pigment melatonin invade the dead retinal tissue, producing black deposits that are characteristic of the disease.

Laboratory of Samuel Jacobson, University of Miami (2)

Nathans says. "Either it doesn't fold correctly to start with, or once it folds, it falls apart." There seems to be a correlation between the kind of mutation and the severity of the disease.

To find out how these mutations damage the retina and what drugs might be designed to prevent this process, several groups of researchers recently inserted defective genes for rhodopsin into mice, where these mutant genes cause an RP-like disease. They hope to use this mouse model to develop new treatments.

Nathans points out that the retina normally consumes more energy per gram than any other tissue in the body. A rod cell that needs to dispose of mutant rhodopsin must expend further energy still, which may "push the cell over the edge, so that it runs out of energy and dies," taking along the adjoining cones. Nathans speculates that perhaps a drug that reduced energy consumption in the rod cells might minimize or delay the retina's degeneration—and thereby save the patient's cones. "RP is a slow disease," says Nathans. "It may take 30 years to develop, so if we can delay its progression by another 30 years, that's virtually a cure."

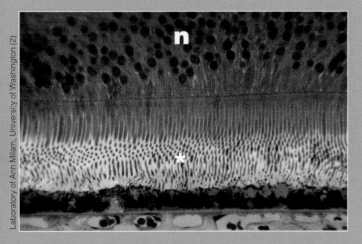

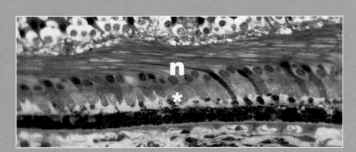

Cones are tightly packed in the fovea, which is specialized for high-acuity vision. In a healthy retina (top), cones appear tall and straight. In the retinas of people with advanced retinitis pigmentosa, cones lose their light-sensitive outer segment (shown with an *) and then die.

So many cones have died in the retina seen in the bottom picture that only one layer of cones remains (n indicates the cones' nuclei) and the whole area has shrunk. The two pictures were taken under a microscope at the same magnification.

left with only a red receptor gene. The man who inherits this slightly truncated chromosome will be color-blind, bereft of the genetic information needed to make a green receptor.

More than 95 percent of all variations in human color vision involve the red and green receptors in men's eyes. It is very rare for anyone—male or female—to be "blind" to the blue end of the spectrum. Nathans provided a genetic explanation for this phenomenon. He showed that the gene coding for the blue receptor lies on chromosome 7, which is shared equally by men and women, and that this gene does not have any neighbor whose DNA sequence is similar. Blue color blindness is caused by a simple mutation in this gene.

What Color Is It?

Seeing a color involves making comparisons. "All that a single cone can do is capture light and tell you something about its intensity," Nathans points out; "it tells you nothing about color." To see any color, the brain must compare the input from different kinds of cone cells—and then make many other comparisons as well.

The lightning-fast work of judging a color begins in the retina, which has three layers of cells. Signals from the red and green cones in the first layer, for instance, are compared by specialized red-green "opponent" cells in the second layer. These opponent cells compute the balance between red and green light coming from a particular part of the visual field. Other opponent cells then compare signals from blue cones with the combined signals from red and green cones.

On a broader scale, comparisons of neighboring portions of an image lead to our amazing ability to see colors as constants in an ever-changing world. Nathans vividly remembers demonstrations of this "color constancy" by the late Edwin Land, the inventor of instant photography and founder of the Polaroid Corporation. Land and his colleagues had made a large collage of multicolored geometric shapes, called a "Mondrian" after its resemblance to the works of the Dutch painter Piet Mondrian. They used

■

But the eye is not a camera.

■

three projectors that beamed light matching the wavelength sensitivity of the three human cone types. With these projectors, the wavelength composition reflected from any given patch on the Mondrian could be exactly controlled.

"Land pointed out a patch on the Mondrian that looked orange in the context of the surrounding colors," Nathans recalls. "Then he gave me a tube, like the tube inside a paper towel roll, and had me look at this patch in isolation. And it wasn't orange anymore. It was a perfect red."

The patch was in fact painted orange, but Land had beamed a high-intensity long-wave light from the red end of the spectrum on it so that it reflected a high proportion of red light. Under normal viewing conditions, however—when the patch was surrounded by other Mondrian colors—Nathans still saw the orange figure by its true color. Somehow, by comparing a patch of color with the surrounding colored region, the brain is able to discount the wavelength of the illuminating light and reconstruct the patch's color as it would be seen in daylight.

"Color constancy is the most important property of the color system," declares neurobiologist Semir Zeki of University College, London. Color would be a poor way of labeling objects if the perceived colors kept shifting under different conditions, he points out. But the eye is not a camera. Instead, the eye-brain pathway constitutes a kind of computer—vastly more complex and powerful than any that human engineers have built—designed to construct a stable visual representation of reality.

The key to color constancy is that we do not determine the color of an object in isolation; rather, the object's color derives from a comparison of the wavelengths reflected from the object and its surroundings. In the rosy light of dawn, for instance, a yellow lemon will reflect more long-wave light and therefore may appear orange; but its surrounding leaves also reflect more long-wave light. The brain compares the two and cancels out the increases.

Land's "Retinex" theory of color vision—a mathematical model of this comparison process—left open the question of where in the pathway between retina and cortex color constancy was achieved. This issue could only be addressed by studying the brain itself.

Working with anesthetized monkeys in the 1960s, David Hubel and Torsten Wiesel of Harvard Medical School had shown that the primary visual cortex or area V1, a credit card–sized region at the back of the brain, possesses a highly organized system of neurons for analyzing the orientation of an object's outlines. But in their early studies they found relatively few color-sensitive cells. Then in 1973, Semir Zeki identified a separate area that he called V4, which was full of cells that crackled with activity when the monkeys' visual field was exposed to different colors.

A few years later, Edwin Land paid Zeki a visit in London. "He showed me his demonstration, and I was much taken by that," Zeki says. "I was converted, in fact. So I used his Mondrian display to study the single cells in area V4."

In this way, Zeki discovered that some of the cells in area V4 consistently respond to the actual surface color of a Mondrian patch, regardless of the lighting conditions. He believes these are the cells that perform color constancy. More recently, with the aid of PET scans, he found an area similar in location to the monkeys' V4 that is specifically activated in humans when they look at Mondrian color displays. The color displays also stimulate the primary visual area and an area that is adjacent to it, V2.

Much controversy exists about all aspects of the color pathway beyond the retina, however. Researchers disagree about the exact role of cells in human V1 and V2, about the importance of V4, about the similarities between monkey and human brains.

To resolve such issues, scientists await the results of further experiments on humans. The new, noninvasive imaging techniques that can show the brain at work (see p. 142) may supply key answers. Within a few years, researchers hope, these techniques will reveal the precise paths of the neural messages that make it possible for us to see the wealth of colors around us. ●

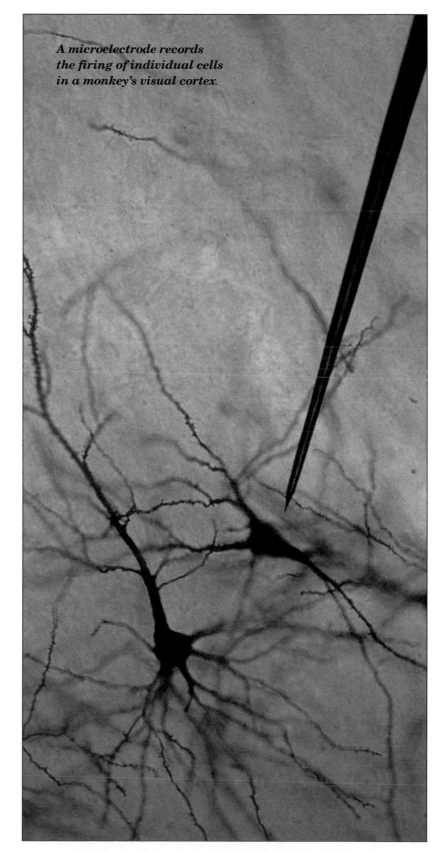

A microelectrode records the firing of individual cells in a monkey's visual cortex.

HOW WE SEE THINGS THAT MOVE

BY GEOFFREY MONTGOMERY

The patient had great difficulty pouring coffee into a cup. She could clearly see the cup's shape, color, and position on the table, she told her doctor. She was able to pour the coffee from the pot. But the column of fluid flowing from the spout appeared frozen, like a waterfall turned to ice. She could not see its motion. So the coffee would rise in the cup and spill over the sides.

More dangerous problems arose when she went outdoors. She could not cross a street, for instance, because the motion of cars was invisible to her: a car was up the street and then upon her, without ever seeming to occupy the intervening space.

Even people milling through a room made her feel very uneasy, she complained to Josef Zihl, a neuropsychologist who saw her at the Max Planck Institute for Psychiatry in Munich, Germany, in 1980, because "the people were suddenly here or there but I did not see them moving."

The woman's rare motion blindness resulted from a stroke that damaged selected areas of her brain. What she lost—the ability to see objects move through space—is a key aspect of vision. In animals, this ability is crucial to survival: Both predators and their prey depend upon being able to detect motion rapidly.

In fact, frogs and some other simple vertebrates may not even see an object unless it is moving. If a dead fly on a string is dangled motionlessly in front of a starving frog, the frog cannot sense this winged meal. The "bug-detecting" cells in its retina are wired to respond only to movement. The frog might starve to death, tongue firmly folded in its mouth, unaware that salvation lies suspended on a string in front of its eyes.

Unable to see motion, Gisela Leibold feels anxious as she rides down an escalator in Munich.

While the retina of frogs can detect movement, the retina of humans and other primates cannot. "The dumber the animal, the smarter its retina," observes Denis Baylor of Stanford Medical School. The large, versatile brain of humans takes over the job, analyzing motion through a highly specialized pathway of neural connections.

This is the pathway that was damaged in the motion-blind patient from Munich. Compared with the complex ensemble of regions in the visual cortex that are devoted to perceiving color and form, this motion-perception pathway seems relatively streamlined and simple. More than any other part of the cortex, it has yielded to efforts to unveil "the precise relationship between perception and the activity of a sensory neuron somewhere in the brain," says Anthony Movshon, an HHMI investigator at New York University. By studying the reactions of humans and monkeys to different moving stimuli and probing the parts of the visual cortex that are aroused at such times, researchers have begun to build a bridge between the objective world of electrically signaling neurons that can be observed in a laboratory and the subjective world of perception accessible only to an individual's own consciousness.

One way to visualize the key challenges for the motion-perception system, suggests Thomas Albright of the Salk Institute, is to consider what happens when we watch a movie. Each of the 24 frames projected per second on the theater screen is a still photograph; nothing in a movie truly moves except the film. The illusion of movement is created by the motion-processing system in our brains, which automatically fuses, for instance, the images of legs that shift position slightly from frame to frame into the appearance of a walking actor. The Munich patient is unable to perform this fusion. In life or in the movie theater, she sees the world as a series of stills.

"The motion system must match up image elements from frame to frame, over space and time," says Albright. "It has to detect which direction a hand is moving in, for instance, and not confuse that hand with a head when it waves in front of someone's face."

Researchers have now traced the path of neural connections that make up the motion pathway and tested the responses of cells at different steps along this path. This has revealed the basic stages by which the motion system senses which way a hand is waving.

Starting in the retina, large ganglion cells called magnocellular neurons, or M cells, are triggered into action when part of the image of a moving hand sweeps across their receptive field—the small area of the visual field to which each cell is sensitive. The M cells' impulses travel along the optic nerve to a relay station in the thalamus, near the middle of the brain, called the lateral geniculate nucleus. Then they flash to the middle layer of neurons in the primary visual cortex. There, by pooling together the inputs from many M cells, certain neurons gain a new property: they become sensitive to the direction in which the hand is moving across their window of vision.

Such direction-sensitive cells were first discovered in the mammalian visual cortex by David Hubel and Torsten Wiesel, who projected moving bars of light across the receptive fields of cells in the primary visual cortex of anesthetized cats and monkeys. Electrodes very close to these cells picked up their response to different moving lines, and the pattern of activity could be heard as a crackling "pop-pop-pop" when the signals were amplified and fed into a loudspeaker.

"Listening to a strongly direction-selective cell responding," Hubel has written, "the feeling you get is that the line moving in one direction grabs the cell and pulls it along and that the line moving in the other direction fails utterly to engage it, something like the feeling you get with a ratchet, in winding a watch."

The keystone of the motion pathway is an area of the cortex that lies just beyond the primary and secondary visual areas (V1 and V2)—a largely unexplored wilderness that used to be known as the "sensory association cortex." "It was thought that somewhere in this mishmash of association cortex visual forms were recognized and associated with information from other senses," says John Allman of Caltech. But studies in

Nothing in a movie truly moves...

the owl monkey by Allman and Jon Kaas (who is now at Vanderbilt) and in the rhesus monkey by Semir Zeki revealed that the area was not a mishmash at all. Instead, much of it was made up of separate visual maps, each containing a distinct representation of the visual field. In 1971, Zeki showed that one of these visual maps was remarkably specialized. Though its cells did not respond to color or form, over 90 percent of them responded to movement in a particular direction. American scientists usually call this map MT (middle temporal area), but Zeki called it V5. He also nicknamed it "the motion area."

"This very striking finding of this little hot spot, this little pocket, in which almost all the cells are sensitive for the direction of movement," says Anthony Movshon, was the impetus for many vision researchers to turn their attention to motion. Nowhere else in the visual cortex was there an area that seemed so functionally specialized.

The cells of this motion area, MT, are directly connected to the layer of direction-sensitive cells in the primary visual area, V1. And the two areas have a remarkably similar architecture. Hubel and Wiesel had discovered that V1 is organized into a series of columns. The cells in one column may fire only when shown lines oriented like an hour hand pointing to one o'clock, for instance, while the cells in the next column fire most readily to lines oriented at two o'clock, and so on around the dial. Amazingly, MT has the same kind of orientation system as V1, but in addition the cells in its columns respond preferentially to the direction of movement.

"When you see that an area, like V1 or MT, has this highly organized columnar structure," says Wiesel, "you get a sense of uncovering something fundamental about the way the cells in the visual area work."

In perceiving motion, as in determining color, the brain constructs a view of the world from pieces of information that can themselves be mistaken or ambiguous. Suppose you paint an X on a piece of paper and then move that paper up and down in front of someone's eyes. Direction-selective cells in the motion-pathway layer of V1—

each of which sees only a small part of the scene—will respond to the diagonal orientation of each of the lines making up the X but will not register the movement of the X as a whole. How, then, is this overall movement sensed?

There must be two stages of motion analysis in the cortex, suggested Movshon and Edward Adelson, who was then a postdoctoral fellow at New York University (he is now a professor at the Massachusetts Institute of Technology). At the second stage, certain cells must integrate the signals regarding the orientation of moving lines and produce an overall signal about the motion of the whole object.

When Movshon presented this idea at an annual meeting of vision researchers in 1981, William Newsome, then a postdoctoral fellow at the National Institutes of Health (he is now a professor of neurobiology at Stanford University School of Medicine), approached him. A lively three-hour dinner ensued and the two men resolved to collaborate. Together with Adelson, they would search for such cells in the motion area.

The researchers soon found that one-third of MT's cells could, in fact, signal the direction in which a hand waves through space. Later on, Albright's research group showed that MT cells can detect "transparent" motion, such as a shadow sweeping across the ground.

Then Allman and his colleagues discovered that many MT cells are able to integrate motion information from a large swath of the scene. "Even though an MT cell may respond directly to just one spot in the visual field," says Allman, "the cells have knowledge of what's going on in the region surrounding them." Using a computer display with a background texture that looks vaguely like a leafy forest, Allman showed that some MT cells will fire particularly furiously if the leafy background moves in a direction opposite to a moving object—the sort of visual pattern a cheetah would see when chasing an antelope along a stand of trees. If, however, the background moved in the same direction as the moving object, the cell's firing was sup-

The Urgent Need To Use Both Eyes

When you look at yourself in the mirror, "you are looking into a predator's eyes," writes Diane Acker-man in *A Natural History of the Senses*. Predators generally have eyes set right on the front of their heads so they can use precise, binocular vision to track their prey, she explains, whereas prey have eyes at the sides of their heads so they can be aware of predators sneaking up on them.

Binocular vision lets us see much more sharply—but only if our two eyes work smoothly together, starting early in life. Most of the time there is no problem. Each year, however, at least 30,000 babies in the United States develop strabis-mus, which means that their left and right eyes fail to align properly in the first few months after birth.

Until the 1970s, doctors did not realize the urgency of doing something about this condition. Treatment was generally delayed until the children were 4 or older—too late to do much good.

The need for earlier intervention became clear as a result of David Hubel and Torsten Wiesel's experi-ments with kittens. They showed that there is a crit-ical period, shortly after birth, during which the visual cortex requires normal signals from both eyes in order to develop properly. In kittens, the critical period lasts for about a month or six weeks. In humans, it continues until the age of 5 or 6.

A curious feature of cells in the visual cortex is that those responding to information from the left and right eyes form separate "ocular dominance columns," one for each eye. Normally these columns are arranged in a series of alternating bands that can be labelled by injecting a marker into one eye. This produces a pattern that resembles the black-and-white stripes of a zebra. But the columns are not fully wired at birth; they take shape during the first months of life, in response to visual experience.

If vision through one eye is blocked during the critical period, the ocular dominance columns responding to the open eye expand in the cortex, while the columns that would normally respond to the blocked eye progressively shrink. An adult who loses his vision because of a cataract (a clouding of the lens of the eye) will generally see normally again if the opaque lens is removed and replaced with a clear artificial lens. But a child whose cataract is not removed until the age of 7 will be blind in the eye that was blocked by the cataract,

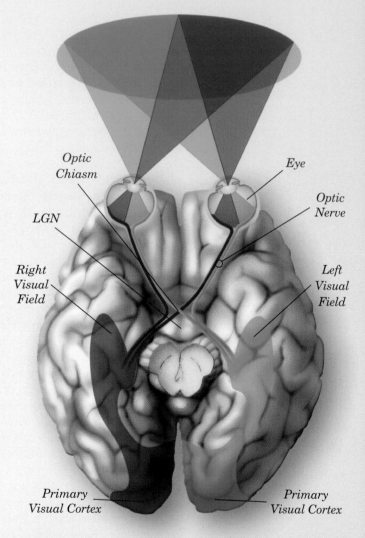

The two eyes provide slightly different views of the same scene. Information from the left visual field goes to the right side of the retina in both eyes. At the optic chiasm, half the nerve fibers from the left eye cross over to the right hemisphere and the rest stay uncrossed, so that all the information from the left visual field ends up in the right hemisphere. In this way, a given hemisphere gets information from the opposite half of the visual world—but each hemisphere gets input from both eyes.

even though the cataract is gone and the retina of the eye is able to function normally.

The same kind of amblyopia—a loss of vision without any apparent defect of the eye—occurs in children whose eyes are misaligned, as well as in those whose eyes focus at different distances. To avoid double vision, such children generally favor one eye and stop using the other. The brain then suppresses the signals coming from the unfavored or "lazy" eye. The neurons in the ocular dominance columns that should receive signals from this eye become wired incorrectly, and the child loses his ability to see with the neglected eye. After a few years, neither surgery nor exercises nor a patch over the favored eye can restore the lost vision.

Armed with this information, ophthalmologists now treat infants who have visual defects as early as possible, with either spectacles or surgery, since normal vision can be restored if treatment begins before the age of 3 or 4.

Anthony Movshon and other researchers have studied monkeys with artificially produced amblyopia. They found that the cells in these monkeys' retinas and lateral geniculate nuclei (LGN), the visual system's relay stations in the center of the brain, are all normal. But in the case of the neglected eye, "the signals that go from the LGN to the primary visual cortex don't make it," Movshon says. The loss occurs because the connections between cells in the LGN and cells in the corresponding eye dominance columns fail to develop or be maintained.

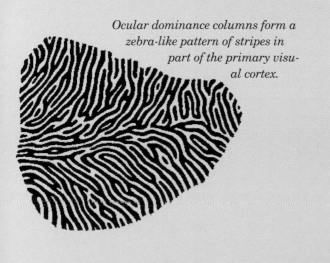

Ocular dominance columns form a zebra-like pattern of stripes in part of the primary visual cortex.

pressed. The cell acted as a large-scale detector of motion contrast, performing exactly the sort of operation an animal would need to sense a figure moving through the camouflage of the forest.

While MT cells do not respond to static forms and colors, Albright has found that they will detect a moving object much more easily if its form or color strongly contrasts with its background.

"Imagine you're looking down the concourse in Grand Central Station and you're supposed to find the woman in the red dress," says Albright. "There are hundreds of surrounding people moving in different directions. Yet there's no problem at all in detecting the woman in the red dress walking along. Your visual system uses the dress's color to filter out all the irrelevant noise around it and homes in on the moving object of interest."

Suppose scientists could record from the MT cells in a laboratory monkey that looked at the woman in the red dress crossing Grand Central Station. They could determine that a particular cell fired when the woman in the red dress passed through its receptive field. But how would they know that the firing of this specific MT cell—and not a network of thousands of other cells in the brain, of which this cell is only one node—actually causes the monkey to perceive the direction of the woman's movement? How could they ever get inside the monkey's mind and determine what it perceives?

Since Hubel and Wiesel's pioneering studies in the visual cortex, most visual scientists have assumed that the perception of form, color, depth, and motion corresponds to the firing of cells specialized to detect these visual qualities. In a spectacular series of experiments conducted since the mid-1980s, Newsome and his colleagues at Stanford have been directly testing this link between perception and the activity of specific neurons.

They use a device that was developed in Movshon's laboratory at NYU: a blizzard of white dots moving on a computer monitor. When all the white dots are moving randomly, the display looks like a TV tuned to a nonbroadcasting channel. However, the experimenters can gradually increase the percentage of dots moving in the same direction. When 10 percent of the dots move coherently together, their motion becomes apparent. By 25 percent, it is unmistakable.

Movshon had found that whenever a human subject could detect the dots' motion at all, he or she could also tell the direction in which the dots were moving. "This means that the part of the visual pathway carrying the information used for motion detection is also carrying a label that says what direction is being detected," says Movshon. This is precisely how one would expect MT, with its columns of direction-selective cells, to encode a moving target.

Next, Newsome began to teach rhesus monkeys to "tell" him what they saw on the computer screen. When they saw dots moving downward, for instance, the monkeys were supposed to move their eyes to a downward point on the screen. Correct responses were rewarded with fruit juice. Soon the monkeys could signal with eye movements that they saw the dots move in any of six directions around the clock. And after much training on low-percentage moving dot displays, the monkeys were able to perform nearly as well as Movshon's human subjects.

Everything was in place. Newsome, Movshon, and their colleagues were ready to study the relationship between the monkeys' perception of motion and the activity of cells in particular columns of MT.

"We found, very much to our surprise," says Newsome, "that the average MT cell was as sensitive to the direction of motion as the monkey was." As more dots moved together and the monkey's ability to recognize their direction increased, so did the firing of the MT neuron surveying the dots.

If the monkeys were actually "listening" to the cells in a single MT column as they made their decision about the direction of movement of the dots on the screen, could the decision be altered by stimulating a different MT column, the researchers wondered. So they electrically stimulated an MT "up" column while the monkeys looked at a downward-moving display. This radically changed the monkeys' reports of what they saw.

"The tenth experiment was an unforgettable experience," remembers Newsome. "We got the first of what became known in the lab as 'whoppers'—when the effects of microstimulation were just massive. Fifty percent of the dots would move down, and yet if we'd stimulate an 'up' column, the monkey would signal 'up' with its eyes. That was really a remarkable day."

The monkeys' perceptual responses no longer seemed to be driven by the direction of dots on the screen. Instead, the animals' perceptual responses were being controlled by an electric stimulus applied to specific cells in the brain by an experimenter. "Intellectually," says Newsome, "it's like putting a novel gene into a bacterium and seeing a novel protein come out. We're putting a signal into this motion circuitry, and we're seeing a predictable behavior come out that corresponds to the signal we put in."

These experiments, says Movshon, "close a loop between what the cells are doing and what the monkey's doing." Allman calls the finding "the most direct link that's yet been established between visual perception and the behavior of neurons in the visual cortex."

It is still possible, however, that when the dots are moving down and the experimenters stimulate an MT "up" column, the stimulation changes what the monkey decides without actually changing what it sees.

"This is a key question," says Newsome. "We now know a lot about the first and last stages of this process. But we are almost totally ignorant about the decision process out there in the middle—the mechanism that links sensory input to the appropriate motor output. How does the decision get made?"

It is a burning question not only for research on the visual system, but for all of cognitive neuroscience, Newsome believes. The answer would provide a bridge from the study of the senses, where so much progress has been made, to the much more difficult study of human thought. At long last, Newsome says, "we're now poised to approach this question." ●

BRAIN SCANS THAT SPY

ON THE SENSES

For centuries, scientists dreamed of being able to peer into a human brain as it performs various activities—for example, while a person is seeing, hearing, smelling, tasting, or touching something. Now several imaging techniques such as PET (positron emission tomography) and the newer fMRI (functional magnetic resonance imaging) make it possible to observe human brains at work.

The PET scan on the left shows two areas of the brain (red and yellow) that become particularly active when volunteers read words on a video screen: the primary visual cortex and an additional part of the visual system, both in the left hemisphere.

Other brain regions become especially active when subjects hear words through earphones, as seen in the PET scan on the right.

To create these images, researchers gave volunteers injections of radioactive water and then placed them, head first, into a doughnut-shaped PET scanner. Since brain activity involves an increase in blood flow, more blood—and radioactive water—streamed into the areas of the volunteers' brains that were most active while they saw or heard words. The radiation counts on the PET scanner went up accordingly. This enabled the scientists to build electronic images of brain activity along any desired "slice" of the subjects' brains. The images below were produced by averaging the results of tests on nine different volunteers.

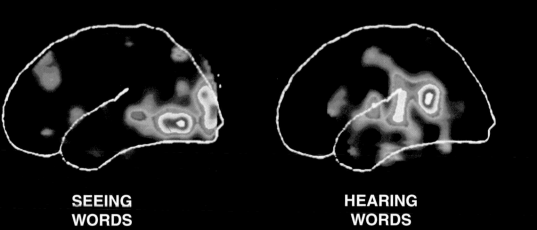

SEEING WORDS

HEARING WORDS

Laboratory of Marcus Raichle, Washington University

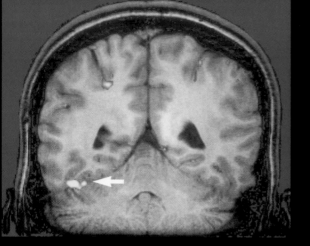

Section on Funsitional Brain Imaging, NIMH (2)

The volunteer's brain is particularly active in an area of her right hemisphere called the fusiform gyrus (arrow) as she matches one of the two faces at the bottom of the display with the face at the top. This "slice" of her brain is seen as though looking through her face.

A GIANT MAGNET REVEALS THE BRAIN'S ACTIVITY

Much excitement surrounds a newer technique, fMRI, that needs no radioactive materials and produces images at a higher resolution than PET. In this system, a giant magnet surrounds the subject's head. Changes in the direction of the magnetic field induce hydrogen atoms in the brain to emit radio signals. These signals increase when the level of blood oxygen goes up, indicating which parts of the brain are most active.

Since the method is non-invasive, researchers can do hundreds of scans on the same person and obtain very detailed information about a particular brain's activity, as well as its structure. They no longer need to average the results from tests on different subjects, whose brains are as individual as fingerprints.

Here a normal volunteer prepares for a fMRI study of face recognition. She will have to match one of the two faces at the bottom of the display with the face at the top. James Haxby, chief of the section on functional brain imaging at the National Institute of Mental Health in Bethesda, Maryland, adjusts the mirror that will allow her to see the display from inside the magnet.

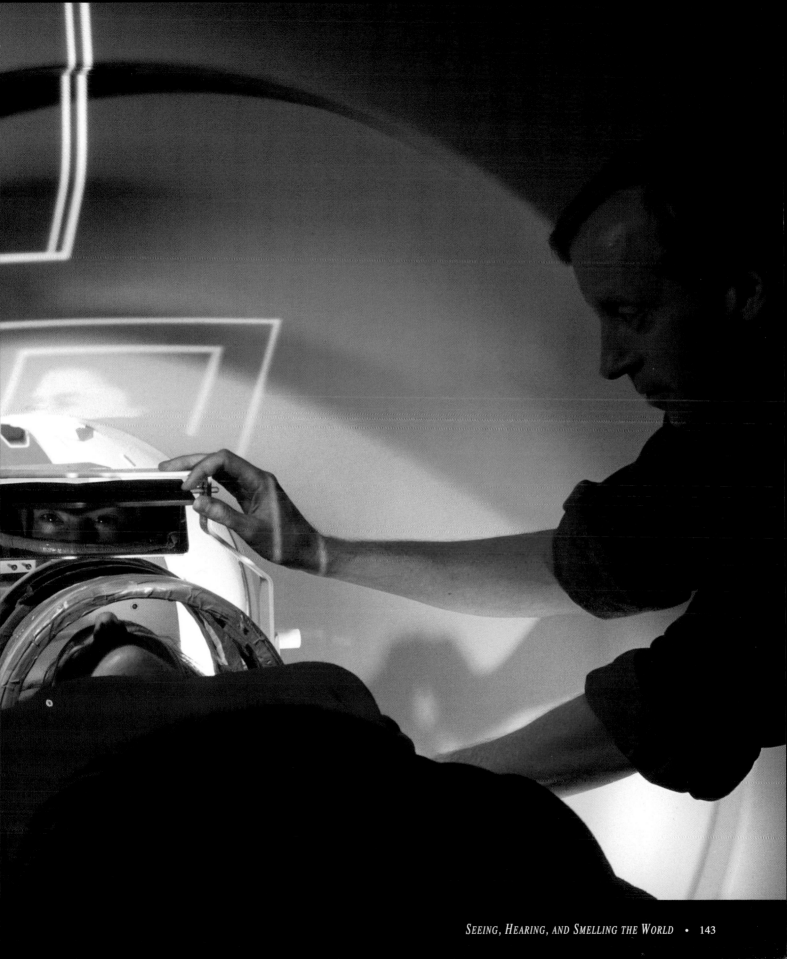

David Corey and James Hudspeth, who played major roles in discovering how the ear's hair cells respond to sound, discuss the cells' conversion of vibrations into nerve signals. In the background, a slide shows hair cells in the inner ear.

THE
QUIVERING

BUNDLES
THAT LET US
HEAR

BY
JEFF GOLDBERG

An unusual dance recital was videotaped in David Corey's lab at Massachusetts General Hospital recently. The star of the performance, magnified many times under a high-powered microscope, was a sound-receptor cell from the ear of a bullfrog, called a hair cell because of the distinctive tuft of fine bristles sprouting from its top. The music ranged from the opening bars of Beethoven's Fifth Symphony and Richard Strauss's "Thus Spake Zarathustra" to David Byrne and the Beatles.

As the music rose and fell, an electronic amplifier translated it into vibrations of a tiny glass probe that stimulated the hair cell, mimicking its normal stimulation in the ear. The bristly bundle of "stereocilia" at the top of the cell quivered to the high-pitched tones of violins, swayed to the rumblings of kettle drums, and bowed and recoiled, like tiny trees in a hurricane, to the blasts of rock-and-roll.

The dance of the hair cell's cilia plays a vital role in hearing, Corey explains. Now an HHMI investigator at MGH and Harvard Medical School, Corey was a graduate student at the California Institute of Technology when he began working with James Hudspeth, a leading authority on hair cells. Together, the two researchers have helped discover how movements of the cilia, which quiver with the mechanical vibrations of sound waves, cause the cell to produce a series of brief electrical signals that are conveyed to the brain as a burst of acoustic information.

In humans and other mammals, hair cell bundles are arranged in four long, parallel columns on a gauzy strip of tissue called the basilar membrane. This membrane, just over an inch long, coils within the cochlea, a bony, snail-shaped structure about the size of a pea that is located deep inside the inner ear.

Sound waves generated by mechanical forces, such as a bow being drawn across a string, water splashing on a hard surface, or air being expelled across the larynx, cause the eardrum—and, in turn, the three tiny bones of the middle ear—to vibrate. The last of these three bones (the stapes, or "stirrup") jiggles a flexible layer of tissue at the base of the cochlea. This pressure sends waves rippling along the basilar membrane, stimulating some of its hair cells. These cells then send out a rapid-fire code of electrical signals about the frequency, intensity, and duration of a sound. The messages travel through auditory nerve fibers that run from the base of the hair cells to the center of the cochlea, and from there to the brain. After several relays within the brain, the messages finally reach the auditory areas of the cerebral cortex, which processes and interprets these signals as a musical phrase, a dripping faucet, a human voice, or any of the myriad sounds in the world around us at any particular moment.

"The mechanics of the hair cell are fascinating—the fact that simply pushing a little bundle of cilia magically allows us to hear. And the cells are beautiful. I never get tired of looking at them," says Corey.

Corey and Hudspeth have explored the

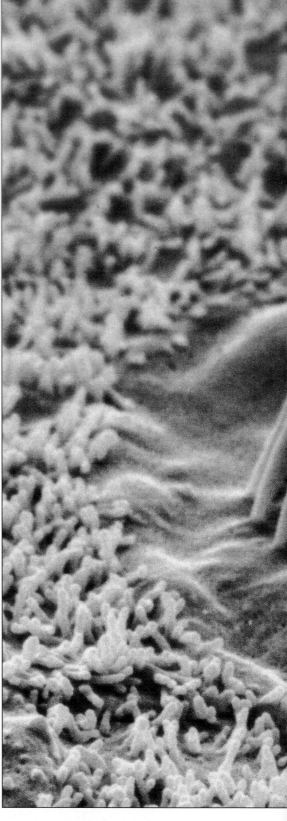

When this bundle of 50 to 60 cilia at the top of a hair cell vibrates in response to sound, the hair cell (from a bullfrog's inner ear) produces an electrical signal. Tiny tip links can be seen joining the tops of shorter cilia to the sides of taller ones (arrow).

Laboratory of David Corey, Massachusetts General Hospital

microscopic inner workings of hair cells in finer and finer detail over the past 20 years, gaining a solid understanding of how the cells work. Some pieces of the puzzle have fallen into place recently with the discovery of a unique mechanism that endows hair cells with their two most distinctive properties—extreme sensitivity and extreme speed.

This success has attracted a large number of scientists to the study of the auditory sys-

been suspected of playing an important role in hearing. This view was bolstered by clinical evidence that the majority of hearing impairments—which affect some 30 million Americans—involve damage to hair cells.

There are only 16,000 hair cells in a human cochlea, compared to 100 million photoreceptors in the retina of the eye, and they are extremely vulnerable. Life in a high-decibel society of pounding jackhammers, screeching subway cars, and heavy metal rock music can take a devastating toll on them. But whatever the cause—overexposure to loud noises, disease, heredity, or aging—people tend to lose 40 percent of their hair cells by the age of 65. And once destroyed, these cells do not regenerate.

Hudspeth's investigation of these cells was initially a solitary, frustrating effort. "I was struck by the fact that so little was known about them," recalls Hudspeth, who is now an HHMI investigator at the University of Texas Southwestern Medical Center. "So I decided to apply myself to solving this one basic problem." He wanted to see whether movements of the cilia bundle on top of the cell could convert mechanical vibrations into electrical signals to the brain, a process known as transduction.

Together with Corey, who joined him in 1975, Hudspeth began a series of experiments that focused on transduction in hair cells. Such experiments, now routine in their labs, are tricky. Protected deep inside the skull, hair cells cannot easily be studied in living creatures—and once removed from laboratory animals, these cells quickly die. Even now, Corey acknowledges, "a good experiment would be to study three or four cells for maybe 15 minutes each."

The measurements are so delicate that they are usually carried out on a table mounted on air-cushioned legs, to reduce any external movements or vibrations; otherwise, the building's own vibration would deafen a hair cell in seconds. Hudspeth found that an unused swimming pool built on bedrock in a basement at the University of California, San Francisco, where he worked previously, made the perfect laboratory for hair cell experiments—especially after he had it filled with 30 truckloads of

tem. But until the early 1970s, when Hudspeth set out to determine precisely what hair cells did and how they did it, research into the basic biology of the auditory system lagged so far behind the exciting advances being made in vision that it was dubbed the "Cinderella sense" by some researchers.

Because of the hair cell bundles' uncanny resemblance to little antennae and their location in the inner ear, the cells had long

concrete for more stability.

Hair cells from bullfrogs were exposed by removing the sacculus, a part of the inner ear, and pinning the pinhead-sized tissue to a microscope slide. Working under a microscope, Hudspeth and Corey were then able to manipulate an individual hair cell's bundle of cilia with a thin glass tube. They slipped the tube over the bundle's 50 to 60 stereocilia, which are arranged like a tepee on the top of each hair cell, and moved the tube back and forth, deflecting the bundle less than a ten-thousandth of an inch. The hair cell's response was detected by a microelectrode inserted through the cell membrane.

Corey and Hudspeth found that the bundle of stereocilia operated like a light switch. When the bundle was prodded in one direction—from the shortest cilia to the tallest—it turned the cell on; when the bundle moved in the opposite direction, it turned the cell off.

Based on data from thousands of experiments in which they wiggled the bundle back and forth, the researchers calculated that hair cells are so sensitive that deflecting the tip of a bundle by the width of an atom is enough to make the cell respond. This infinitesimal movement, which might be caused by a very low, quiet sound at the threshold of hearing, is equivalent to displacing the top of the Eiffel Tower by only half an inch.

At the same time, the investigators reasoned that the hair cells' response had to be amazingly rapid. "In order to be able to process sounds at the highest frequency range of human hearing, hair cells must be able to turn current on and off 20,000 times per second. They are capable of even more astonishing speeds in bats and whales, which can distinguish sounds at frequencies as high as 200,000 cycles per second," says Hudspeth.

Photoreceptors in the eye are much slower, he points out. "The visual system is so slow that when you look at a movie at 24 frames per second, it seems continuous, without any flicker. Contrast 24 frames per second with 20,000 cycles per second. The auditory system is a thousand times faster."

ON THE TRAIL OF A "DEAFNESS" GENE

Being able to hear speech is taken for granted—"Is it possible for a hearing person to comprehend the enormity of its absence in someone else?" asks Hannah Merker in her poignant book *Listening*. "The silence around me is invisible... ."

Most of the 28 million deaf or hearing-impaired people in the United States were born with normal hearing, as was Merker, who became deaf after a skiing accident in her twenties. Deafness generally results from overexposure to loud noise, disease, or old age. But genetic factors are also an important cause of hearing loss, especially in children.

It has been estimated that 1 in every 1,000 newborns is profoundly deaf, while nearly 1 in 20 has significant hearing impairment. In more than half of these cases, the cause is genetic.

Large families in which a single type of deafness is clearly inherited are rare, however. Geoffrey Duyk, until recently an HHMI investigator at Harvard Medical School, often spends hours on the telephone with health care workers and geneticists in the United States and abroad, trying to track down leads on families with similar hearing disorders so he can search for the genetic error leading to their deafness. He has found an unusually large family in Worcester, Massachusetts, whose DNA he can analyze, as well as an entire tribe of Bedouin Arabs in Northern Israel.

Hearing loss has taken an extremely high toll among the members of both families. Thirteen of fifty members of the Worcester family who were examined by a team of specialists coordinated by Duyk were going through the same disastrous sequence of events: although they could hear well at birth, they started to lose their hearing in their teens; by their early forties they were profoundly deaf. The Bedouin tribe suffered from an even more damaging kind of hereditary disorder: about a quarter of their children were born deaf.

Duyk took samples of blood from both families and set out to find mutations in their DNA that could account for their hearing loss. The task was formidable. "Deafness is associated with over 100 different genetic disorders, and there are upwards of 30 forms of hereditary hearing loss alone, each caused by a different mutation," Duyk says. Yet he

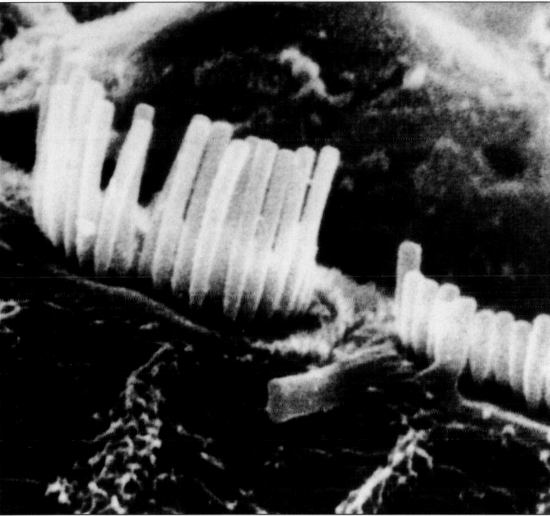

A two-hour exposure to loud noise—such as that of loud rock bands—is enough to seriously damage cilia bundles on the hair cells of a cat's inner ear. Normal mammalian hair cell bundles have two or three parallel rows of cilia, one taller than the next. The tall cilia are most vulnerable to noise. After exposure to loud noise, all the tall cilia on the right of this picture have disappeared or fused together and fallen over.

is hot on the trail of a genetic error that appears to be responsible for the Bedouin tribe's deafness.

Meanwhile, David Corey and his colleague Xandra Breakefield at Massachusetts General Hospital are examining the gene that is defective in Norrie disease, a different disorder that causes not only a progressive loss of hearing similar to that of the Worcester family but, in addition, blindness at birth. The scientists are analyzing the protein made by normal copies of this gene and trying to understand its function, which might lead to ways of preventing the disorder.

For a deeper understanding of such disorders,

scientists need to work with animal models. Duyk's research group is studying two strains of mice, called "jerkers" and "shakers," that have been found to suffer from inherited forms of progressive hearing loss (as well as the peculiar movement disorders that give them their names.) The researchers are looking for fragments of DNA from these mice that might be similar to pieces of DNA from families with genetic deafness.

"We would like to develop new kinds of treatment for hearing loss," Duyk explains. "But first we need to identify the proteins and genes that are essential to hearing."

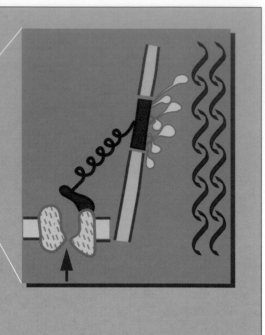

A TIP LINK PULLS UP THE GATE OF A CHANNEL.

In this sketch, James Hudspeth suggests how the movement of a hair cell's cilia bundle (top) opens ion channels at the tips of the cilia. When the bundle tilts to the right, tip links from higher cilia pull up the gates of ion channels on adjoining, shorter cilia.

A close-up shows how a tip link between two cilia opens an ion channel on the shorter cilium.

Even more highly magnified (right), the open channel allows ions into the cell. A cluster of myosin molecules in the taller cilium is shown in green and some actin filaments are shown in blue.

How do hair cells do this?

Unlike other types of sensory receptor cells, hair cells do not rely on a cascade of chemical reactions to generate a signal. Photoreceptor cells in the eye, for instance, require a series of intricate interactions with a G protein and a second messenger before their ion channels close, sending a signal to the brain. This process would be much too slow to deal with sounds. Hair cells have to possess a mechanism that allows their ion channels to open and close more rapidly than those of any other sensory receptor cells.

The answer is that hair cells use something very much like a spring to open their channels when the cilia bend, without the need for a time-consuming chemical exchange.

Corey and Hudspeth first theorized that such a "gating spring" mechanism existed in the early 1980s. They proposed that hair cells had a previously unknown type of ion channel—a channel directly activated by mechanical force. They also developed a biophysical theory to account for the hair cells' rapid response. But their

theory didn't tell them where the channels were or what the spring was.

By painstakingly measuring the electrical field around the cilia with an electrode, Hudspeth detected a tiny drop in voltage at the cilia's tips, as if the current were being sucked into a minute whirlpool. This led him to conclude that the channels through which charged particles move into the cell, changing its electrical potential, were located at the cilia's tips. He then reasoned that the gating springs that opened these channels should be there as well.

The springs themselves were first observed in 1984, in electron microscope images taken by James Pickles and his colleagues in England. Called tip links, these minute filaments join each stereocilium to its tallest neighbor. Pickles pointed out that the geometry of the cilia bundle would cause the bundle to stretch the links when it was deflected in one direction and relax them when it was moved in the other. If the tip links were the hypothetical gating springs, it would explain everything.

"This was a completely new kind of mechanism, unlike anything ever observed before," says Corey, who provided compelling evidence that the tip links pull on the channels. By "cutting" the tip links with a chemical, Corey could stop the cell's response cold. "Within less than a second, as the tip links became unstable, the whole mechanical sensitivity of the cell was destroyed," Corey observed.

Recently, both he and Hudspeth have been independently investigating another property of hair cells: their ability to adapt to being deflected. At first, when a hair cell bundle is deflected, the ion channels open. But if the bundle remains deflected for a tenth of a second, the channels close spontaneously. It appears from electron microscope images and physiological evidence that the channels close when the tip links relax. This is related to the activity of the tip links' attachment points, which can move up and down along the cilia to fine-tune the tension on the channels. When the attachment points move down, the tip links are relaxed and the ion channels close.

While the researchers are still trying to figure out what enables the attachment points to move, they strongly suspect that myosin plays a role. Myosin is the protein that gives muscle cells their ability to contract and relax, and Hudspeth's group has found evidence of myosin in cilia bundles. Both labs have now cloned and sequenced the gene for a myosin molecule in hair cells. A cluster of such molecules in each stereocilium could provide the force to move the attachment point up or down.

Slight movements of the attachment points allow the hair cell to set just the right amount of tension on each channel so it is maximally sensitive. They also permit the cell to avoid being overloaded when it is barraged by sound.

A second type of hair cell in the highly specialized cochlea of mammals may enable us to distinguish the quietest sounds. These outer hair cells, which are shaped like tiny hot dogs, look distinctly different from inner hair cells. The outer hair cells also have a peculiar ability to become shorter or longer within microseconds when stimulated, doing so with a flamboyant, bouncy, up-and-down

❑

...our ears not only receive sounds, but emit them as well.

❑

motion not found in any other cell type. They outnumber inner hair cells 3 to 1. However, the 4,000 inner hair cells are connected to most of the auditory nerve fibers leading to the brain and are clearly the main transmitters of sound.

The precise function of the outer hair cells is still unclear. Auditory researchers speculate that these cells may serve as an amplification mechanism for tuning up low-frequency sound waves, possibly by accelerating the motion of the basilar membrane.

Hudspeth is also intrigued by the possibility that outer hair cells may be responsible for something that has puzzled researchers for years: the fact that our ears not only receive sounds, but emit them as well. When sensitive microphones are placed in the ear and a tone is played, a faint echo can be detected resonating back out. Such otoacoustic emissions are considered normal; in fact, their presence in screening exams of newborn babies is thought to be indicative of healthy hearing. However, in certain cases, otoacoustic emissions can be spontaneous and so intense that they are audible without the aid of special equipment. "In some people, you can actually hear them. The loudest ones ever recorded were in a dog in Minnesota, whose owner noticed the sound coming out of the animal's ear and took the dog to a specialist, who did recordings and analysis," says Hudspeth.

"What may be happening is that the amplification system driven by the movements of outer hair cells is generating feedback, like a public address system that's turned up too high," he speculates, adding that such otoacoustic emissions gone awry may account for certain unusual forms of tinnitus, or ringing in the ear.

Hudspeth and Corey's research is providing such a detailed picture of the hair cell that it is now possible to begin to identify the individual proteins making up the tip links, ion channels, and motor mechanisms involved, as well as the genes that produce them. Malfunctions in those genes, resulting in defects in these important structures, may be the cause of inherited forms of deafness (see p. 148). ●

Locating a Mouse by Its Sound

by Jeff Goldberg

While some scientists investigate the mystery of how we hear from the bottom up, beginning with the ear's sound receptors, others search for answers from the top down, mapping networks of auditory neurons in the brain in an effort to understand how the brain processes sounds.

At Caltech in the mid-1970s, Masakazu ("Mark") Konishi began studying the auditory system of barn owls in an effort to resolve a seemingly simple question: Why do we have two ears?

While most sounds can be distinguished quite well with one ear alone, the task of pinpointing where sounds are coming from in space requires a complex process called binaural fusion, in which the brain must compare information received from each ear, then translate subtle differences into a unified perception of a single sound—say a dog's bark—coming from a particular location.

Konishi, a zoologist and expert on the nervous system of birds, chose to study this process in owls. The ability to identify where sounds are coming from based on auditory cues alone is common to all hearing creatures, but owls—especially barn owls—excel at the task. These birds exhibit such extraordinary sound localization abilities that they are able to hunt in total darkness.

In total darkness, a barn owl swoops down on a mouse.

Working with Eric Knudsen, who is now conducting his own research on owls at Stanford University, Konishi undertook a series of experiments on owls in 1977 to identify networks of neurons that could distinguish sounds coming from different locations. He used a technique pioneered by vision researchers, probing the brains of anesthetized owls with fine electrodes. With the electrodes in place, a remote-controlled sound speaker was moved to different locations around the owl's head along an imaginary sphere. As the speaker moved, imitating sounds the owl would hear in the wild, the investigators recorded the firing of neurons in the vicinity of the electrodes.

Over the course of several months, Konishi and Knudsen were able to identify an area in the midbrain of the birds containing cells called space-specific neurons—about 10,000

in all—which would fire only when sounds were presented in a particular location. Astonishingly, the cells were organized in a precise topographic array, similar to maps of cells in the visual cortex of the brain. Aggregates of space-specific neurons, corresponding to the precise vertical and horizontal coordinates of the speaker, fired when a tone was played at that location.

"Regardless of the level of the sound or the content of the sound, these cells always responded to the sources at the same place in space. Each group of cells across the circuit was sensitive to sound coming from a different place in space, so when the sound moved, the pattern of firing shifted across the map of cells," Knudsen recalls.

The discovery of auditory brain cells that could identify the location of sounds in space quickly produced a new mystery. "The lens

of the eye projects visual space onto receptors on a 2-dimensional sheet, the retina, and the optic nerve fibers project the same spatial relationships to the brain," says Konishi. "But in the auditory system, only the frequency of sound waves is mapped on the receptor layer, and the auditory nerve fibers project this map of frequency to the brain. How can the brain create a map of auditory space, based only on frequency cues?"

The answer, Konishi believes, may shed light on how the brain and the auditory system process all sounds.

To enable the brain to process efficiently the rapid stream of impulses emanating from the hair cells in the ear, the auditory system must first filter out simple, discrete aspects of complex sounds. Information about how high- or low-pitched a sound is, how loud it is, and how often it is heard is then channeled along separate nerve pathways to higher-order processing centers in the brain, where millions of auditory neurons can compute the raw data into a recognizable sound pattern.

This filtering process begins with the hair cells, which respond to different frequencies at different locations along the basilar membrane. Hair cells at the bottom of the basilar membrane respond more readily when they detect high-frequency sound waves, while those at the top are more sensitive to low-frequency sounds. David Corey compares the arrangement to the strings of a grand piano, with the high notes at the base of the cochlea, where the basilar membrane is narrow and stiff, and the bass notes at the apex, where the membrane is wider and more flexible.

Hair cells also convey basic information about the intensity and duration of sounds. The louder a sound is at any particular frequency, the more

After wearing prism spectacles for a few months, this owl began to miss auditory targets because the sound localization system in its brain tried to harmonize with the visual system, which received erroneous cues.

Laboratory of Eric Knudsen, Stanford University

vigorously hair cells tuned to that frequency respond, while their signaling pattern provides information about the timing and rhythm of a sound.

Konishi hypothesized that such timing and intensity information were vital for sound localization. So he placed microphones in the ears of owls to measure precisely what they were hearing as the portable loudspeaker rotated around their head. He then recorded the differences in time and intensity as sounds reached each of the owl's ears. The differences are very slight. A sound that originates at the extreme left of the animal will arrive at the left ear about 200 microseconds (millionths of a second) before it reaches the right ear. (In humans, whose sound localization abilities are keen but not on a par with those of owls, the difference between a similar sound's time of arrival in each ear would be about three times greater.)

As the sound source was moved toward the center of the owl's head, these interaural time differences diminished, Konishi observed. Differences in the intensity of sounds entering the two ears occurred as the speaker was moved up and down, mostly because the owl's ears are asymmetrical—the left ear is higher than eye level and points downward, while the right ear is lower and points upward.

Based on his findings, Konishi delivered signals separated by various time intervals and volume differences through tiny earphones inserted into the owls' ear canals. Then he observed how the animals responded. Because owls' eyes are fixed in their sockets and cannot rotate, the animals turn quickly in the direction of a sound, a characteristic movement. By electronically monitoring these head-turning movements, Konishi and his

CAN FUNCTIONAL MRI TELL WHETHER
A PERSON IS HEARING MUSIC OR JUST MEANINGLESS CLICKS?

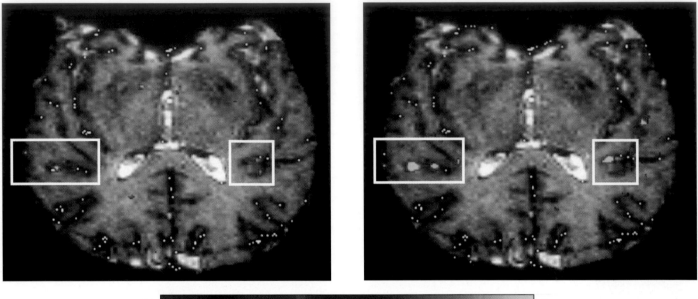

NMR Center, Massachusetts General Hospital

Parts of a volunteer's brain were activated (white box on left of first picture) when he heard a series of sharp but meaningless clicks while inside a fMRI magnet at Massachusetts General Hospital. Some of the same areas became much more active and several new areas were activated as well (square box on right of second picture) when he listened to instrumental music, reflecting the richer meaning of the sounds.

assistants showed that the owls would turn toward a precise location in space corresponding to the interaural time and intensity differences in the signals. This suggested that owls fuse the two sounds that are delivered to their two ears into an image of a single source—in this case, a phantom source.

"When the sound in one ear preceded that in the other ear, the head turned in the direction of the leading ear. The longer we delayed delivering the sound to the second ear, the further the head turned," Konishi recalls.

Next, Konishi tried the same experiment on anesthetized owls to learn how their brains carry out binaural fusion. Years earlier, he and Knudsen had identified space-specific neurons in the auditory area of the owl's midbrain that fire only in response to sounds coming from specific areas in space. Now Konishi and his associates found that these space-specific neurons react to specific combinations of signals, corresponding to

the exact direction in which the animal turned its head when phantom sounds were played. "Each neuron was set to a particular combination of interaural time and intensity difference," Konishi recalls.

Konishi then decided to trace the pathways of neurons that carry successively more refined information about the timing and intensity of sounds to the owl's midbrain. Such information is first processed in the cochlear nuclei, two bundles of neurons projecting from the inner ear. Working with Terry Takahashi, who is now at the University of Oregon, Konishi showed that one of the nuclei in this first way station signals only the timing of each frequency band, while the other records intensity. The signals are then transmitted to two higher-order processing stations before reaching the space-specific neurons in the owl's midbrain.

One more experiment proved conclusively that the timing and intensity of sounds are processed along separate pathways. When the researchers injected a minute amount of local anesthetic into one of the cochlear nuclei (the magnocellular nucleus), the space-specific neurons higher in the brain stopped responding to differences in interaural time, though their response to differences in intensity was unchanged. The converse occurred when neurons carrying intensity information were blocked.

"I think we are dealing with basic principles of how an auditory stimulus is processed and analyzed in the brain. Different features are processed along parallel, almost independent pathways to higher stations, which create more and more refined neural codes for the stimulus," says Konishi. "Our knowledge is not complete, but we know a great deal. We are very lucky. The problem with taking a top-down approach is that often you find nothing."

Konishi has been able to express the mechanical principles of the owl's sound localization process as a step-by-step sequence. He has collaborated with computer scientists at Caltech in developing an "owl chip" that harnesses the speed and accuracy of the owl's neural networks for possible use in computers.

At Stanford University, Eric Knudsen fitted owls with prism spectacles to determine whether distortions in their vision affect their sound localization abilities. Despite their exceptionally acute hearing, he found, the owls trust their vision even more. When they wear distorting prisms, their hunting skills deteriorate over a period of weeks as their auditory systems try to adapt to the optical displacement of the prisms. "The visual system has ultimate control and basically dictates how the brain will interpret auditory localization cues," Knudsen says.

He is also examining a particular network of neurons in the animals' brains where he believes auditory and visual system signals converge. "This network makes it possible for the owls to direct their eyes and attention to a sound once it's heard," Knudsen explains. His research is part of a new wave of studies that focus not just on single sensory pathways, but on how the brain combines information it receives from many different sources.

HELP FROM A BAT

Perhaps the finest achievement in sound processing is the ability to understand speech. Since this is a uniquely human trait, it would seem difficult to study in animals. Yet a researcher at Washington University in St. Louis believes it can be examined—by working with bats.

Bats navigate and locate prey by echolocation, a form of sonar in which they emit sound signals of their own and then analyze the reflected sounds. Nobuo Suga, who has spent nearly 20 years investigating the neural mechanisms used by bats to process the reflected signals, is convinced that such research can shed light on the understanding of human speech.

When Suga slowed down recordings of the high-frequency, short-duration sounds that bats hear, he found that the sounds' acoustic components were surprisingly similar to those of mammalian communication, including human speech. There were some constant frequencies and noise bursts, not unlike vowel and consonant sounds, as well as frequency-modulated components that were similar to those in combinations of phonemes such as "papa."

Each of these acoustic elements is processed along a distinct pathway to higher-order neurons, which combine and refine different aspects of the sonar pattern in much the same way that space-specific neurons combine the timing and intensity cues of sound signals.

Suga also identified maps of neurons in the bats' auditory cortex which register slight variations in these components of sound. Humans may use similar maps to process the basic acoustic patterns of speech, though speech requires additional, higher-level mechanisms, he points out.

"The ability to recognize variations in sound is what enables us to understand each other. No two people pronounce vowels and consonants in exactly the same way, but we are able to recognize the similarities," says Suga. He believes that neuronal maps may also play a role in human voice recognition—the ability to recognize who is speaking as well as what is being said.

The MYSTERY

After taking a mixture of mind-altering drugs one night, Stephen D., a 22-year-old medical student, dreamed that he had become a dog and was surrounded by extraordinarily rich, meaningful smells. The dream seemed to continue after he woke up—his world was suddenly filled with pungent odors.

Walking into the hospital clinic that morning, "I sniffed like a dog. And in that sniff I recognized, before seeing them, the twenty patients who were there," he later told neurologist Oliver Sacks. "Each had his own smell-face, far more vivid and evocative than any sight-face." He also recognized local streets and shops by their smell. Some smells gave him pleasure and others

disgusted him, but all were so compelling that he could hardly think about anything else.

The strange symptoms disappeared after a few weeks. Stephen D. was greatly relieved to be normal again, but he felt "a tremendous loss, too," Sacks reported in his book *The Man Who Mistook His Wife for a Hat and Other Clinical Tales*. Years later, as a successful physician, Stephen D. still remembered "that smell-world—so vivid, so real! It was like a visit to another world, a world of pure perception, rich, alive, self-sufficient, and full…I see now what we give up in being

Linda Buck (right) sniffs an odorant used to study the sense of smell. She and Richard Axel (left) discovered what appear to be the long-sought odorant receptor proteins.

# of SMELL

BY MAYA PINES

civilized and human."

Being civilized and human means, for one thing, that our lives are not ruled by smells. The social behavior of most animals is controlled by smells and other chemical signals. Dogs and mice rely on odors to locate food, recognize trails and territory, identify kin, find a receptive mate. Social insects such as ants send and receive intricate chemical signals that tell them precisely where to go and how to behave at all times of day. But humans "see" the world largely through eyes and ears. We neglect the sense of

smell—and often suppress our awareness of what our nose tells us. Many of us have been taught that there is something shameful about odors.

Yet mothers can recognize their babies by smell, and newborns recognize their mothers in the same way. The smells that surround us affect our well-being throughout our lives. Smells also retain an uncanny power to move us. A whiff of pipe tobacco, a particular perfume, or a long-forgotten scent can instantly conjure up scenes and emotions from the past. Many writers and artists have marveled at the haunting quality of such memories.

In *Remembrance of Things Past,* French novelist Marcel Proust described what happened to him after drinking a spoonful of tea in which he had soaked a piece of a madeleine, a type of cake: "No sooner had the warm liquid mixed with the crumbs touched my palate than a shudder ran through my whole body, and I stopped, intent upon the extraordinary thing that was happening to me," he wrote. "An exquisite pleasure had invaded my senses...with no suggestion of its origin....

"Suddenly the memory revealed itself. The taste was of a little piece of madeleine which on Sunday mornings...my Aunt Leonie used to give me, dipping it first in her own cup of tea....Immediately the old gray house on the street, where her room was, rose up like a stage set...and the entire town, with its people and houses, gardens, church, and surroundings, taking shape and solidity, sprang into being from my cup of tea."

Just seeing the madeleine had not brought back these memories, Proust noted. He needed to taste and smell it. "When nothing else subsists from the past," he wrote, "after the people are dead, after the things are broken and scattered...the smell and taste of things remain poised a long time, like souls...bearing resiliently, on impalpable droplets, the immense edifice of memory."

Proust referred to both taste and smell—and rightly so, because most of the flavor of food comes from its aroma, which wafts up the nostrils to sensory cells in the nose and also reaches these cells through a passageway in the back of the mouth. Our taste buds provide only four distinct sensations: sweet, salty, sour, and bitter. Other flavors come from smell, and when the nose is blocked, as by a cold, most foods seem bland or tasteless.

Both smell and taste require us to incorporate—to breathe in or swallow—chemical substances that attach themselves to receptors on our sensory cells. Early in evolution, the two senses had the same precursor, a common chemical sense that enabled bacteria and other single-celled organisms to locate food or be aware of harmful substances.

How we perceive such chemical substances as odors is a mystery that, until recently, defeated most attempts to solve it. Anatomical studies showed that signals from the olfactory cells in the nose reach the olfactory area of the cortex after only a single relay in the olfactory bulb. The olfactory cortex, in turn, connects directly with a key structure called the hypothalamus, which controls sexual and maternal behavior. When scientists tried to explore the details of this system, however, they hit a blank wall. None of the methods that had proved fruitful in the study of vision seemed to work.

To make matters worse, very little was known about the substances to which the olfactory system responds. The average human being, it is said, can recognize some 10,000 separate odors. We are surrounded by odorant molecules that emanate from trees, flowers, earth, animals, food, industrial activity, bacterial decomposition, other humans. Yet when we want to describe these myriad odors, we often resort to crude analogies: something smells like a rose, like sweat, or like ammonia. Our culture places such low value on olfaction that we have never developed a proper vocabulary for it. In *A Natural History of the Senses,* poet and essayist Diane Ackerman notes that it is almost impossible to explain how something smells to someone who hasn't smelled it. There are names for all the pastels in a hue, she writes—but none for the tones and tints of a smell.

Nor can odors be measured on the kind of linear scale that scientists use to measure the wavelength of light or the frequency of sounds. "It would be nice if one smell corresponded to a short wavelength and another to a long wavelength, such as rose versus skunk, and you could place every smell on this linear scale," says Randall Reed, an HHMI investigator at the Johns Hopkins University School of Medicine, who has long been interested in olfaction. "But there is no smell scale," since odorous molecules vary widely in chemical composition and three-dimensional shape.

To find out how these diverse odorants trigger our perception of smell, researchers needed to examine the olfactory cells and identify the receptor proteins that actually bind with the odorants. This task was made more difficult by the awkward location of the olfactory cells.

Axel. "It's because there are a large number of odorant receptors, and each was expressed only at a very low level."

Finally, Buck came up with what Axel calls "an extremely clever twist." She made three assumptions that drastically narrowed the field, allowing her to zero in on a group of genes that appear to code for the odorant receptor proteins, though the final proof is not yet in.

Her first assumption—based on bits of evidence from various labs—was that the odorant receptors look a lot like rhodopsin, the receptor protein in rod cells of the eye. Rhodopsin and at least 40 other receptor proteins criss-cross the cell surface seven times, which gives them a characteristic, snake-like shape. They also function in similar ways, by interacting with G proteins (see p. 123) to transmit signals to the cell's interior. Since many receptors of this type share certain DNA sequences, Buck designed probes that would recognize these sequences.

Next, she assumed that the odorant receptors are members of a large family of related proteins. So she looked for groups of genes that had certain similarities. Third, the genes had to be expressed only in a rat's olfactory epithelium.

"Had we employed only one of these criteria, we would have had to sort through thousands more genes," says Axel. "This saved several years of drudgery."

Buck recalls that "I had tried so many things and had been working so hard for three and a half years, with nothing to show for it. So when I finally found the genes, I couldn't believe it! None of them had ever been seen before. They were all different but all related to each other. That was very satisfying."

The discovery made it possible to study the sense of smell with the techniques of modern molecular and cell biology and to explore how the brain discriminates among odors. It also allowed researchers to "pull out" the genes for similar receptor proteins in other species by searching through libraries of DNA from these species. Odorant receptors of humans, mice, catfish, dogs, and salamanders have been identified in this way.

The team's most surprising finding was that there are so many olfactory receptors. The 100 different genes the researchers identified first are just the tip of the iceberg, according to Axel. He thinks there must be a total of "about 1,000 separate receptor proteins" on rat—and probably human—olfactory neurons.

"That's really a lot of genes," Axel says. "It's 1 percent of the genome! This means that, at least in the rat, 1 out of every 100 genes is likely to be engaged in the detection of odors." This staggering number of genes reflects the crucial importance of smell to animals.

Large as the number of receptors may be, however, it is probably smaller than the number of odors we can recognize. "Most likely, the number of odorants far exceeds the number of receptor proteins—by a ratio of at least 10 to 1," Axel says. "In that case, how does the brain know what the nose is smelling?"

The visual system needs only three kinds of receptors to distinguish among all the colors that we can perceive, he points out. These receptors all respond to the same thing—light. Light of different wavelengths makes the three kinds of receptors react with different intensity, and then the brain compares their signals to determine color. But the olfactory system must use a different strategy in dealing with the wide variety of molecules that produce odors.

To figure out this strategy, Axel began by asking how many kinds of receptor proteins are made by a single olfactory neuron. "If a single neuron expresses only one or a small number of receptors, then the problem of determining which receptors have been activated reduces to determining which neurons have been activated," he says.

He thought he would make more rapid progress by working with simpler organisms than rats. So he turned to fish, which respond to fewer odorants and were likely to have fewer receptors. From studies with catfish, whose odorant receptors proved very similar to those of rats, Axel and his associates soon concluded that a given olfactory neuron can make only one or, at most, a few odorant receptors. (Buck and

"When I finally found the genes, I couldn't believe it!"

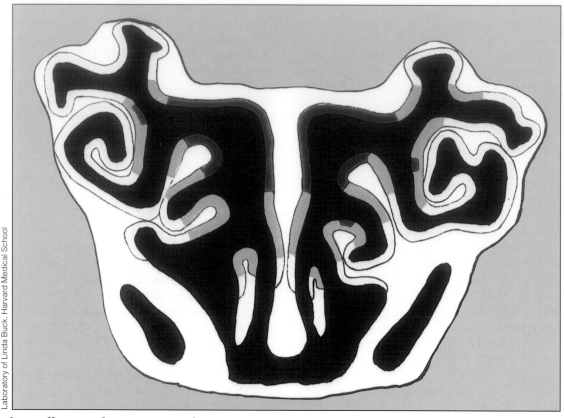

her colleagues have come to the same conclusion from their work with mice.)

The next step was to find out how these odorant receptors—and the neurons that make them—are distributed in the nose. Also, what parts of the brain do these neurons connect with? "We want to learn the nature of the olfactory code," Axel says. "Will neurons that respond to jasmine relay to a different station in the brain than those responding to basil?" If so, he suggests, the brain might rely on the position of activated neurons to define the quality of odors.

Each olfactory neuron in the nose has a long fiber, or axon, that pokes through a tiny opening in the bone above it, the cribriform plate, to make a connection, or synapse, with other neurons in the olfactory bulb, which is a part of the brain. A round, knob-like structure, the olfactory bulb is quite large in animals that have an acute sense of smell. It decreases in relative size as this ability wanes. Thus, bloodhounds, which can follow the scent of a person's tracks for long distances over varied terrain, have larger olfactory bulbs than

humans do—even though humans are more than twice the total size of these dogs and have brains that are several times as large.

In the olfactory epithelium of the nose, Axel's and Buck's groups found, neurons that make a given odorant receptor do not cluster together. Instead, these neurons are distributed randomly within certain broad regions of the nasal epithelium. Then their axons converge on the same place in the olfactory bulb, Axel believes.

"The brain is essentially saying something like, 'I'm seeing activity in positions 1, 15, and 54 of the olfactory bulb, which correspond to odorant receptors 1, 15, and 54, so that must be jasmine,'" Axel suggests. Most odors consist of mixtures of odorant molecules, so other odors would be identified by different combinations.

Buck, who has been trying to solve the same problem at Harvard, recently found that the olfactory epithelium of mice is divided into regions that she calls expression zones, each of which contains a different set of odorant receptors. These zones are symmetrical on the two sides of the animals' nasal cavities (see p. 164). "This suggests that there may be an initial broad organization of sensory information that occurs in the nose, even before the information is sent on to the brain," she declares.

Earlier researchers had traced the anatomical connections between neurons in the olfactory epithelium and the olfactory bulb, using radioactive labels. When Buck and her associates examined this older work recently, "we were amazed," she says, "because their patterns (of connections) looked just like our zones." Putting the two sets of findings together, she has produced a tentative map of the connections between expression zones in the olfactory epithelium of mice and certain parts of the olfactory bulb. She believes the initial organization of information about smells is maintained as it reaches the bulb. She has preliminary evidence that once the axons get to the bulb, they reassort themselves so that all those that express the same receptor converge at a specific site.

And so the first stages of olfaction are beginning to yield to researchers. But many mysteries remain.

One riddle is how we manage to remember smells despite the fact that each olfactory neuron in the epithelium only survives for about 60 days, and is then replaced by a new cell. "The olfactory neurons are the only neurons in the body that are continually replaced in adults," points out Randall Reed. Other neurons die without any successors, and it is thought that we lose increasing numbers of brain cells as we age. The olfactory neurons are far more exposed and vulnerable than other neurons, since they come into direct contact with the outside environment. But as they die, a layer of stem cells beneath them constantly generates new olfactory neurons to maintain a steady supply.

"Then how can we remember smells?" asks Buck. "How do we maintain perceptual fidelity when these neurons are constantly dying and being replaced, and new synapses are being formed? You'd have to recreate the same kind of connections between olfactory epithelium and bulb over and over again, throughout life, or you

wouldn't be able to remember smells in the same way."

An even deeper mystery is what happens to information about smells after it has made its way from the olfactory epithelium to the olfactory bulb. How is it processed there, as well as in the olfactory cortex? How does it go into long-term memory? How does it reach the higher brain centers, in which information about smells is linked to behavior?

Some researchers believe that such questions can best be answered by studying the salamander, in which the nasal cavity is a flattened sac. "You can open it up more or less like a book" to examine how its olfactory neurons respond to odors, says John Kauer, a neuroscientist at Tufts Medical School and New England Medical Center in Boston, Massachusetts, who has been working on olfaction since the mid-1970s.

Salamanders will make it possible to analyze the entire olfactory system—from odorant receptors to cells in the olfactory bulb, to higher levels of the brain, and even to behavior, Kauer thinks. His research group has already trained salamanders to change their skin potential—the type of behavioral response that is measured in lie detector tests—whenever they perceive a particular odor. To study the entire system non-invasively, Kauer uses arrays of photodetectors that record from many sites at once. He applies special dyes that reveal voltage changes in the membranes of cells. Then he turns on a videocamera that provides an image of activity in many parts of the system.

"We think this optical recording will give us a global view of what all the components do when they operate together," says Kauer. He hopes that "maybe 10 years from now, or 20 years from now, we'll be able to make a very careful description of each step in the process."

This would be amazing progress for a sensory system that was relatively neglected until a few years ago. Axel and Buck's discoveries have galvanized the study of olfaction, and scientists now flock to this field, aroused by the possibility of success, at last, in solving its mysteries. ●

"How can we remember smells… when these neurons are constantly dying and being replaced?"

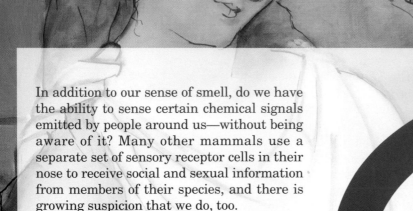

In addition to our sense of smell, do we have the ability to sense certain chemical signals emitted by people around us—without being aware of it? Many other mammals use a separate set of sensory receptor cells in their nose to receive social and sexual information from members of their species, and there is growing suspicion that we do, too.

A whiff of airborne chemicals from a female mouse, for instance, may spur a male mouse to mate immediately. Certain chemical messages from other males may make him aggressive. Other messages may produce changes in his physiology—as well as in that of the responding female.

The effects of such messages would be far less obvious in humans. If we do receive chemical signals from people in our vicinity, these signals must compete with many other factors that influence our behavior. Yet our physiology may be just as responsive to chemical messages as that of other mammals. It is known that certain chemical messages from other mice lead to the onset of puberty in young males, while a different set of signals brings young female mice into estrus. Similarly, there are some suggestions that women may alter their hormonal cycles when exposed to chemical signals from other people.

In the past five years, scientists have become extremely interested in these signals, as well as in the "accessory olfactory system" that responds to them in many ani-

mals. This system starts with nerve cells in a pair of tiny, cigar-shaped sacs called the vomeronasal organs (VNOs), where the signals are first picked up.

"The VNO appears to be a much more primitive structure that uses a different set of molecular machinery than the main olfactory system," says Richard Axel, who recently became intrigued with this system. "It seems to work in a different way—and we don't know how."

The VNOs are located just behind the nostrils, in the nose's dividing wall (they take their name from the vomer bone, where the nasal septum meets the hard palate). In rodents, at least, signals travel from the VNO to the accessory olfactory bulb (rather than the main olfactory bulb) and then, as Sally Winans of the University of Michigan showed in 1970, to parts of the brain that control reproduction and maternal behavior.

"It's an alternate route to the brain," explains Rochelle Small, who runs the chemical senses program at the National

Institute on Deafness and Other Communicative Disorders in Bethesda, Maryland. If the accessory olfactory system functions in humans as it does in rodents, bypassing the cerebral cortex, there is likely to be no conscious awareness of it at all.

This system is particularly important to animals that are inexperienced sexually. Experiments by Michael Meredith, a neuroscientist at Florida State University in Tallahassee, Charles Wysocki, of the Monell Chemical Senses Center in Philadelphia, and others have shown that the VNOs play a key role in triggering sexual behavior in naive hamsters, mice, and rats.

A virgin male hamster or mouse whose vomeronasal organs are removed generally will not mate with a receptive female, even if the male's main olfactory nerves are undamaged. Apparently, the VNOs are needed to start certain chains of behavior that are already programmed in the brain.

Losing the VNOs has a much less drastic effect on experienced animals, says Wysocki,

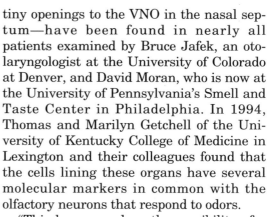

BY MAYA PINES

IN THE HUMAN NOSE?

who has been studying the VNOs for nearly 20 years. When male mice have begun to associate sexual activity with other cues from females, including smells, they become less dependent on the VNOs. Sexually experienced males whose VNOs are removed mate almost as frequently as intact males.

Do human beings have VNOs? In the early 1800s, L. Jacobson, a Danish physician, detected likely structures in a patient's nose, but he assumed they were non-sensory organs. Others thought that although VNOs exist in human embryos, they disappear during development or remain "vestigial"—imperfectly developed.

Yet both VNOs and vomeronasal pits—

tiny openings to the VNO in the nasal septum—have been found in nearly all patients examined by Bruce Jafek, an otolaryngologist at the University of Colorado at Denver, and David Moran, who is now at the University of Pennsylvania's Smell and Taste Center in Philadelphia. In 1994, Thomas and Marilyn Getchell of the University of Kentucky College of Medicine in Lexington and their colleagues found that the cells lining these organs have several molecular markers in common with the olfactory neurons that respond to odors.

"This has opened up the possibility of a new sensory system in humans," says Rochelle Small. "We were often told that the VNO does not exist in adults, so we have taken a big step just to show that the structure is there." She cautions that we still don't know whether this organ actually has connections to the brain, however. "The question now," she says, "is what its function might be."

Just what do the VNOs of rodents—or, perhaps, humans—respond to? Probably pheromones, a kind of chemical signal originally studied in insects. The first pheromone ever identified (in 1956) was a powerful sex attractant for silkworm moths. A team of German researchers worked 20 years to isolate it. After removing certain glands at the tip of the abdomen of 500,000 female moths, they extracted a curious compound. The minutest amount of it made male moths beat their wings madly in a "flutter dance." This clear sign that the males had sensed the attractant enabled the scientists to purify the pheromone. Step by step, they removed extraneous matter and sharply reduced the amount of attractant needed to provoke the flutter dance.

When at last they obtained a chemically pure pheromone, they named it "bombykol" for the silkworm moth, *Bombyx mori,* from which it was extracted. It signaled, "come to me!" from great distances. "It has been soberly calculated that if a single female moth were to release all the bombykol in her sac in a single spray, all at once, she could theoretically attract a trillion males in the instant," wrote Lewis Thomas in *The Lives of a Cell.*

In dealing with mammals, however, scientists faced an entirely different problem. Compared to insects, whose behavior is stereotyped and highly predictable, mammals are independent, ornery, complex creatures. Their behavior varies greatly, and its meaning is not always clear.

What scientists need is "a behavioral assay that is really specific, that leaves no doubt," explains Alan Singer of the Monell Chemical Senses Center. A few years ago, Singer and Foteos Macrides of the Worcester Foundation for Experimental Biology in Massachusetts did find an assay that worked with hamsters—but the experiment would be hard to repeat with larger mammals. It went as follows: First the researchers anesthetized a male golden hamster and placed it in a cage. Then they let a normal male hamster into the same cage. The normal hamster either ignored the anesthetized stranger or bit its ears and dragged it around the cage. Next the researchers repeated the procedure with an anesthetized male hamster on which they had rubbed some vaginal secretions from a female hamster. This time the normal male hamster's reaction was quite different: instead of rejecting the anesthetized male, the hamster tried to mate with it.

Eventually Singer isolated the protein that triggered this clear-cut response. "Aphrodisin," as the researchers called it, appears to be a carrier protein for a smaller molecule that is tightly bound to it and may be the real pheromone. The substance seems to work through the VNO, since male hamsters do not respond to it when their VNOs have been removed.

Many other substances have powerful effects on lower mammals, but the pheromones involved have not been precisely identified and it is not clear whether they activate the VNO or the main olfactory system, or both.

Humans are "the hardest of all" mammals to work with, Singer says. Yet some studies suggest that humans may also respond to some chemical signals from other people. In 1971, Martha McClintock, a researcher who is now at the University of Chicago (she was then at Harvard University), noted that college women who

...some evidence of real, measurable sexual chemistry.

The opening of an adult woman's VNO is seen as a small pit (arrow) in the picture below, which was taken with an angled telescope. The VNOs are narrow sacs, only a few millimeters long. They lie on either side of the nasal septum, quite far from the olfactory epithelium.

Pit

lived in the same dormitory and spent a lot of time together gradually developed closer menstrual cycles. Though the women's cycles were randomly scattered when they arrived, after a while their timing became more synchronized.

McClintock is now doing a new study of women's menstrual cycles, based on her findings from an experiment with rats. When she exposed a group of female rats—let's call them the "A" rats—to airborne "chemosignals" taken from various phases of other rats' estrous cycles, she discovered that one set of signals significantly shortened the A rats' cycles, while another set lengthened them. Now she wants to know whether the same is true for humans—whether there are two opposing pheromones that can either delay or advance women's cycles. In this study, she is focusing on the exact time of ovulation rather than on synchrony.

The most direct scientific route to understanding pheromones and the VNO may, once again, be through genetics. Many researchers, including Axel, Buck, and Reed, are now racing to find the genes for the receptor proteins that actually bind to pheromones in the VNOs of rodents. These genes would lead them to the first receptors for pheromones ever identified in mammals—a prize tool for studying the mechanism and function of the VNO.

Once the genes for such receptors are identified, it should be easy to find out whether equivalent genes exist in humans. Scientists could then determine, once and for all, whether such genes are expressed in the human nose. If they are, the receptors may provide a new scientific clue to the mystery of attraction between men and women— some evidence of real, measurable sexual chemistry.

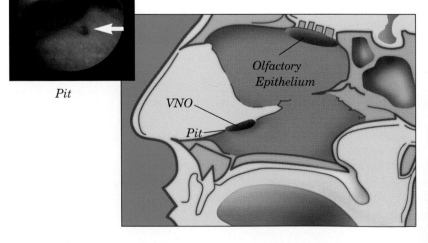

Olfactory Epithelium

VNO

Pit

Messages from the senses travel so swiftly through the brain that imaging machines such as PET and fMRI cannot keep up with them. To track these messages in real time, scientists now use faster methods—electrical recording techniques such as MEG (magnetoencephalography) or EEG (electroencephalography). These techniques rely on large arrays of sensors or electrodes that are placed harmlessly on the scalp to record the firing of brain cells almost instantaneously. Their data are then combined with anatomical information obtained by structural MRI scans.

One of the first experiments in which structural MRI was used jointly with MEG produced a three-dimensional map of the areas of the brain that are activated by touching the five fingers of one hand (below). A New York University research team headed by Rodolfo Llinás found this map to be distorted in the brain of a patient who had two webbed fingers since birth. A few weeks after the man's fingers were separated by surgery, however, parts of his brain reorganized and the map became almost normal.

Laboratory of Rodolfo Llinas, New York University

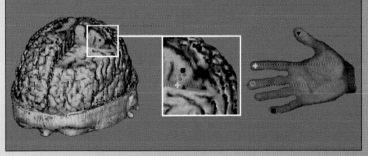

Each of the color-coded areas in this combined MRI/MEG image of the brain responds to the touch of a different finger of the right hand.

THE NEXT GENERATION

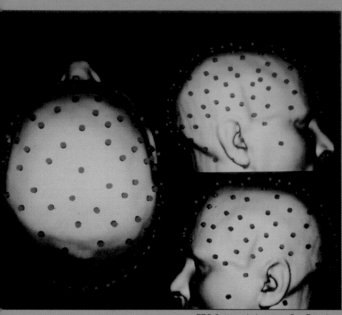

In this high-tech version of EEG, the positions of 124 recording electrodes (attached to a soft helmet) are carefully plotted on an MRI model of the head.

EEG Systems Laboratory, San Francisco

The rapidly-shifting patterns of activity in the six images below reflect what goes on in the brain of a woman who is looking at a letter on a screen during a test at the EEG Systems Laboratory, a private research center directed by Alan Gevins in San Francisco. The woman's task is to decide whether the letter is located in the same place as a letter she has seen before.

In the "low load" test she compares the new letter's location to a previous one. In the "high load" test she compares the new location to three previous ones, and the brighter colors reflect a higher degree of brain activation.

The images are based on data from 124 recording electrodes positioned in a soft helmet that covered the woman's head. The scientists used an MRI-derived model

MATCHING A LOCATION

Comparing (High load)

Comparing (Low load)

Updating (High load)

EEG Systems Laboratory, San Francisco

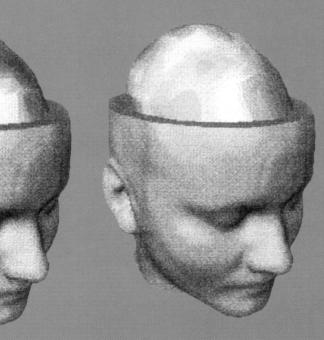

These computer-generated images recreate the electrical signals that flash across the brain of a volunteer during the matching test. A strong electrical signal (first image) sweeps across the frontal cortex of her right hemisphere 320 milliseconds after a new letter has appeared on the screen, as she compares the letter's location to three locations that she has seen before. The same areas of her brain are activated—but less intensively—in the second image, as she compares a new letter's location to only one location that she has seen before.

Only 140 milliseconds later, a different set of electrical signals is recorded from the volunteer's brain and recreated in these images. This time the frontal cortex of her left

of her head to correct for any distortions in electrical transmission that might be caused by variations in the thickness of her skull.

The resulting images clearly show that various areas of the woman's brain are activated in turn. However, these images are limited to the brain's surface.

The next generation of imaging technology will use functional MRI in various combinations with MEG and EEG, predicts John Belliveau, director of cognitive neuroimaging at the Massachusetts General Hospital in Cambridge. Functional MRI shows activity deep in the brain with high spatial resolution, but is relatively slow since it is based on the blood-flow response, which takes about 450 milliseconds. "If you do a visual stimulation experiment, four to five differ-

ent areas may have turned on within that time," Belliveau says. "We know where those areas are, but we don't know which one turned on first." By contrast, EEG's spatial resolution is relatively poor, but because of its speed it may reveal the sequence of events. His group has already done some EEG recordings right inside the magnet of an fMRI machine, to get simultaneous measurements.

Together, such techniques will offer scientists a glimpse of how information from the senses is processed in different parts of the brain. Building on the studies shown here, the new hybrids may then begin to tackle neural networks. They may help researchers examine how various parts of the brain exchange information and—most intriguing—how sensory information leads to thought. ●

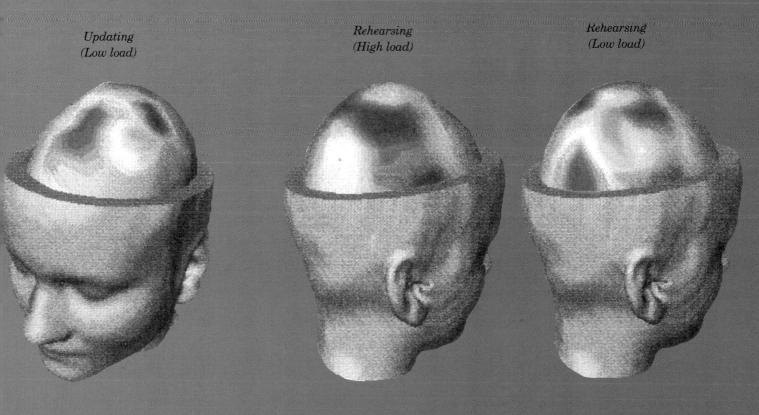

Updating
(Low load)

Rehearsing
(High load)

Rehearsing
(Low load)

hemisphere is activated as she enters the location of the new letter into her working memory. The signals are more intense in the high load than in the low load condition.

After the screen goes blank, the volunteer rehearses the new memory. As the next two images show, this activity produces yet another electrical signal over her right hemisphere. The signal is stronger in the high load than in the low load condition, but in both cases it is maintained until a new letter appears on the screen.

One by one, the very earliest stages in our perception of sights, sounds, smells, and taste are giving up their secrets through molecular genetics.

A recent entry: two genes that encode what appear to be taste receptor proteins, newly identified by HHMI investigator Charles Zuker of the University of California, San Diego, and Nicholas Ryba of the National Institute of Dental and Craniofacial Research. The researchers think that the TR1 receptor, which was isolated from the taste buds of rats and mice, may recognize sweet (which usually means nutritious), while the TR2 receptor recognizes bitter. Both receptors seem distantly related to the receptors for pheromones.

But how do signals from sense receptors reach other parts of the brain? And how does the brain interpret these signals and respond to them? Progress has been much slower in this area. Some of the most interesting findings about brain connections that lead to perception come from studies of smell—particularly from the work of Richard Axel and Linda Buck, who have solved long-standing problems in this field.

"One riddle was, how can we remember smells over long periods of time when the olfactory neurons in the epithelium survive for only about 60 days, to be replaced by new cells which have to form new synapses?" says Buck. "Now we know the answer: Memories survive because the axons of neurons that express the same receptor always go to the same location in the brain."

How We Recognize Odors

In March 1999, Buck proved that mammals recognize and process odors through a code based on varying combinations of receptors. She likens olfactory receptors to letters of the alphabet, which can be used over and over again to compose a vast vocabulary.

At the Life Electronics Research Center in Amagasaki, Japan, Buck and her colleagues wafted 30 different odorants over some 600 olfactory nerve cells they had taken from the noses of mice. A special dye inside these cells lit up whenever an odorant receptor had been stimulated. Then, at Harvard, the scientists analyzed each responding cell's RNA to identify the olfactory protein it produced. In this way, they found out which receptors had been triggered by which odorants. They concluded that mammals use different combinations of receptors to recognize smells and to distinguish, for instance, between the odors of roses and of goats.

Linking Odors and Behavior

That same month, Axel reported that he had discovered odor-detecting receptors in the fruit fly *Drosophila*—a finding that could open the way to linking odor perception to behavior. Fruit flies are tractable experimental subjects and have sophisticated scent-sniffing organs, which they use to recognize a large repertoire of aromas. Axel's team identified 11 genes that encode *Drosophila* odor receptors. He estimates they belong to a family of between 100 and 200 genes and plans to use these genes to study how specific odors influence the flies' behavior.

If his group succeeds in identifying the receptors that are activated by odors that induce mating, for instance, the researchers may be able to map the neural circuitry of the mating response. This could lead to a simple way of preventing crop-eating insects from reproducing. Going further, Axel hopes to link certain olfactory connections in the flies' brains to specific kinds of learning and memory.

The Erotic Nose

Exploring the vomeronasal organ, or VNO, which some scientists now call "the erotic nose," two teams of researchers announced in April 1999 that they had mapped out how sensory neurons in the VNOs of mice connect to specific areas of the accessory olfactory bulbs.

One team was led by Peter Mombaerts of the Rockefeller University; the other was led by two HHMI investigators, Catherine Dulac of Harvard University and Richard Axel. Both groups suggest, but have not yet proved, that pheromones bind to special receptors on sensory neurons in the VNOs of mice. This is difficult to demonstrate, they point out, because very few mammalian pheromones have been identified.

Do Humans Sense Pheromones?

The best evidence that humans communicate through pheromones comes from Martha McClintock, who in 1998 completed a study in which she manipulated the timing of women's menstrual cycles. Every day for two months, she and Kathleen Stern of the University of Chicago collected cotton pads from the armpits of nine women in various phases of their ovulatory cycles and then wiped these pads just under the noses of 20 other women, who were asked not to wash their faces for the next 6 hours. The recipients did not know the source of the compounds and could smell only the alcohol, which served both as a control and as a carrier of the compound.

Women who had been exposed to pads from women in the follicular phase (before ovulation) ovulated earlier, shortening their menstrual cycles. However, pads taken from the same donors during their time of ovulation had the opposite effect, delaying the recipients' ovulation and lengthening their menstrual cycles. "This study provides definitive evidence of human pheromones," the researchers say. "Well-controlled studies of humans are now needed to determine whether there are other types of pheromones, with effects that are as far-reaching in humans as they are in other species."

Researchers still do not know how humans sense pheromones, however. If—unlike the VNOs of mice—the human VNOs turn out to be nonfunctional, humans may sense pheromones through ordinary odor receptors in the nose, after all.

BLOOD: BEARER AND

A REPORT FROM THE HOWARD HUGHES MEDICAL INSTITUTE

BEARER OF LIFE AND DEATH

New Ways to Fight
Diseases Caused by
Faults in the
Bloodstream

OF LIFE AND DEATH

DEATH

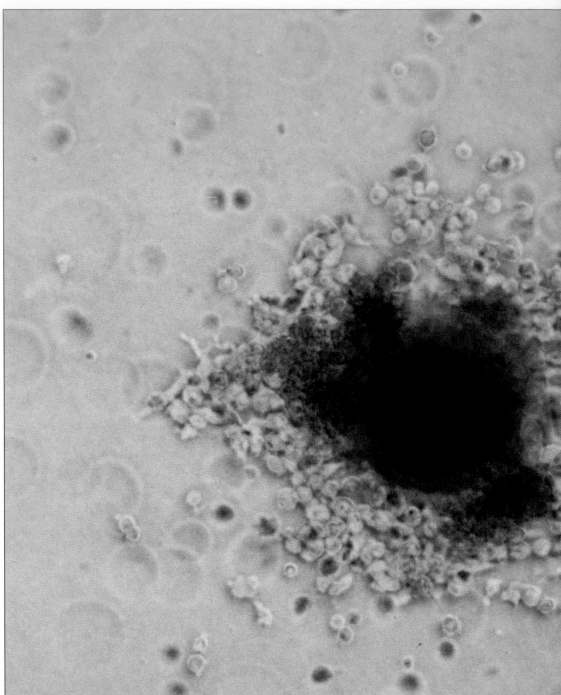

. . . .

Clumps of red blood cells arise from mouse embryonic cells—but only when a recently identified gene, GATA-1, is active, showing that the gene is essential for red cell development.

. . . .

BLOOD: BEARER OF LIFE AND DEATH

New Ways to Fight Diseases Caused by Faults in the Bloodstream

INTRODUCTION

I n his book *Fantastic Voyage*, Isaac Asimov imagined that a scientist who held a secret vital to the world's survival was in a coma, with a potentially fatal blood clot in his brain. Asimov fantasized that a team of doctors and technicians in a submarine—miniaturized sufficiently to enter the bloodstream of the comatose scientist—dove in to save his life by destroying the clot. As they began their journey through his blood vessels, they were amazed at what they saw.

"It was a vast, exotic aquarium they faced," wrote Asimov 30 years ago, "one in which not fish but far stranger objects filled the vision." He described red cells, "large rubber tires, the centers depressed but not pierced through," and a white cell that threatened to swallow up the ship, "huge, milky, and pulsating...a frightening object." The circulatory system of a single human being, he pointed out, is 100,000 miles long. Rearranged in a single line, "it would go four times around the earth or, if you prefer, nearly halfway to the moon."

Many researchers who study blood disorders today share a similar awe of the mighty human bloodstream and a similar urge to save people in distress—AIDS victims whose white cells are decimated, cancer patients unable to make new blood cells after chemotherapy, people who suffer painful crises from sickle cell disease or hemophilia, or who are incapacitated by blood clots. But the greatest spur for these scientists is their growing optimism—a feeling that, thanks to the latest tools of molecular biology, they are on their way to a wealth of discoveries about the red river that courses through our bodies.

The new biochemical techniques that were developed after World War II were immediately applied to blood disorders. In 1949, sickle cell disease became known as "the first molecular disease" because it was traced to a single faulty molecule, hemoglobin. And in 1978, while seeking the genetic basis of this disease, HHMI investigator Yuet Wai Kan devised the first technique for studying inherited variations in DNA. This launched the whole field of genetic diagnosis of disease and started a revolution in medical research that is still gathering steam.

The cause of sickle cell disease is now clear, and future cases of the disease can be prevented through genetic counseling. However, doctors still lack a fully effective treatment for it.

By contrast, hemophilia, a bleeding disorder, can now be treated successfully with transfusions of blood-clotting factor VIII. The story of a young man who lived through the last two decades of research on hemophilia may be found on page 209. The "golden age" of hemophilia treatment, when it seemed that the use of concentrated forms of factor VIII would solve all the major problems of hemophiliacs, came to an abrupt halt in the late 1970s as the AIDS virus invaded the blood supply and infected thousands of children and adults. Today's factor VIII concentrates are safe again—either because any virus that might have lurked in them was inactivated, or because these factor VIII concentrates were produced by means of recombinant DNA technology. But people with hemophilia find themselves split into two groups: those over age 13, three-fourths of whom are

HIV-positive; and the younger ones, who are generally free of the virus.

To maintain a person's health, red blood cells must develop in exactly the right shape and number, contain a sufficient amount of the right type of hemoglobin (which carries oxygen), and bring exactly the right amount of oxygen to different tissues at different times and under different conditions. How red cells achieve this is one of the miracles of life. Key stages in the birth and busy life of a red blood cell are highlighted in a foldout that begins on page 201.

Far more people die from the effects of blood clots than from bleeding disorders, however. The body has developed elaborate systems to plug holes and repair blood vessels after injuries, since tissues deprived of oxygen for more than a few minutes would lose critical functions. But these remedies can sometimes lead to problems of their own. Scientists have only recently uncovered the checks and balances at work in normal clotting, through the discovery of the genes and proteins involved. Each gene or molecule that is found to cause deficiencies or excesses of any blood component offers a clue to the systems involved, thereby suggesting new targets for drugs.

Many blood disorders could be cured by transplanting healthy blood cells into the bone marrow, where blood cells are formed. Perhaps the most exciting prospect is that of isolating the "pluripotent" hematopoietic stem cells for this purpose. These stem cells live in the bone marrow and give rise to all other blood cells. They would be particularly useful if scientists could make them produce specific types of blood cells on command, so as to treat a variety of ailments.

Maya Pines, *Editor*

WANTED

the MOTHER

of all blood cells

by Maya Pines

Lying face down on a hospital bed, the slim graduate student keeps her eyes averted. She volunteered to give researchers some of her bone marrow, but now she does not want to see what is being done to her.

David Williams, an HHMI investigator at the Indiana University School of Medicine, approaches her with a syringe. He plunges the needle into her hip bone and withdraws some soupy red fluid—three quarters of a teaspoon of bone marrow. Then he does it three more times.

The student waits until she hears the door close behind him. Relieved, she looks up and says she didn't feel much pain during the 10-minute procedure. "They numb you up with a local anesthetic," she explains. "It was an easy 100 bucks, something I can do while in school."

Meanwhile Williams strides up to his laboratory on another floor with four syringes containing the rich, warm marrow he has just extracted, plus two larger syringes filled with blood that he drew at the same time. He needs fresh supplies of human marrow twice a week. It is the raw material for his study of some extraordinary cells called hematopoietic (blood-forming) cells, which live inside the marrow of our longest bones and produce the trillions of specialized cells that circulate in the bloodstream: the red blood cells that carry oxygen throughout the body; the six categories of white cells that defend us from infections; and the platelets that stop the loss of blood at sites of injury. He is particularly interested in the most powerful and rarest type of blood-forming cell—the pluripotent stem cell, the mother of them all.

Dozens of scientists around the world are racing to isolate this precious cell and make it multiply. Whoever wins will be able to rebuild an entire blood system with just a few cells. Most blood cells live only a few days or months and must constantly be replaced, but the pluripotent stem cells renew themselves—apparently throughout life—and keep replenishing the bloodstream with over 260 billion new cells every day.

Dozens of scientists around the world are racing to isolate this precious cell and make it multiply.

Each nodule on this mouse spleen consists of thousands of diverse blood cells generated by a single injected stem cell.

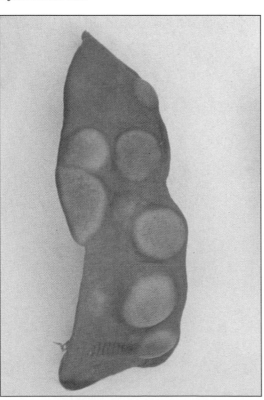

Courtesy of David A. Williams, Indiana University

If doctors could give injections of these pluripotent stem cells to people with certain blood disorders, they could save thousands of lives. The pluripotent cells could also rescue cancer patients whose own bone marrow has been destroyed by radiation or chemotherapy. Bone marrow transplants would be made easier and safer by these cells. AIDS patients' ability to fight infections might be greatly improved. And scientists might be enabled to do really effective gene therapy.

Unfortunately, pluripotent stem cells "are not very user-friendly," as one scientist puts it; "there are very few of them, and you can hardly tell them apart from the others." Some researchers have become so exasperated with their failure to isolate these cells that they even question whether such cells exist. Yet the pluripotent stem cells betray their presence in many ways, and after years of frustration, scientists are now closing in on them.

Human blood-forming cells first appear in the yolk sac of the embryo. These early cells then migrate to the liver as the fetus develops. Before birth, the blood-forming cells leave the liver and home in on the bone marrow, where they remain throughout life. The cells must then maintain their own numbers and also produce the hundreds of billions of new blood cells we need every day. In trying to understand how the blood-forming cells do this, scientists invented the concept of the stem cell: a cell that either makes more cells like itself or creates descendants that eventually mature into the various blood cells.

Much of the recent progress in the study of blood cells can be traced to a collaboration between hematologists (who study blood) and specialists on the effects of radiation. The collaboration started in the 1950s as a joint effort to save the lives of people who had been exposed to high doses of radiation from atomic fallout—or from cancer treatments. Radiation kills cells by damaging their DNA as it is duplicated during cell division, and since bone-marrow cells divide rapidly, they are particularly sensitive to radiation. People who have been exposed to whole-body radiation may die of bleeding (because they run out of platelets to clot their blood), of infections (because they have no white blood cells to defend them), or of anemia (because they lack red cells to provide oxygen to their tissues).

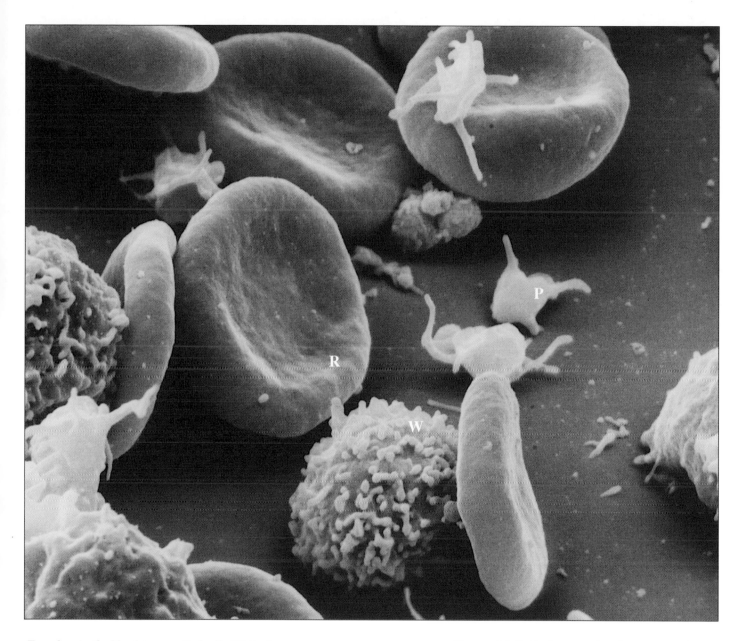

Travelers in the bloodstream: Red cells (R) that bring oxygen, white cells (W) that fight infections, and platelets (P) that start the clotting process when blood vessels are injured.

The researchers found they could rescue mice that had been exposed to lethal doses of radiation by injecting them with bone marrow cells from healthy, genetically compatible mice. This suggested that the injected cells included some stem cells.

The proof that stem cells exist came in 1961, when two Canadian researchers, James Till and Ernest McCulloch, examined some irradiated mice they had rescued with injections of healthy bone marrow cells. Unexpectedly, twelve days after the injection, the mice grew a series of nodules in their spleens. The researchers found that each nodule consisted of hundreds of thousands of blood cells of several different types. Most important, all the cells in each nodule had descended from a single injected cell. This finding—the best evidence of the existence of stem cells—also gave scientists a wonderful tool: a way of counting the number of stem cells injected into an irradiated mouse by counting the

A Single Stem Cell Gives Rise to Eight Different Kinds of Blood Cells

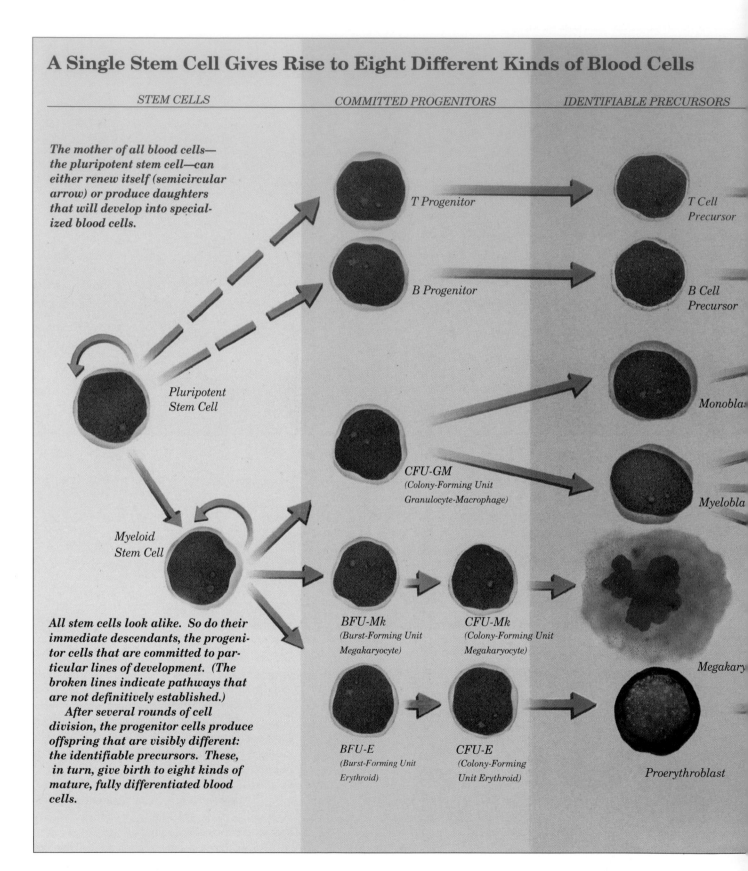

STEM CELLS *COMMITTED PROGENITORS* *IDENTIFIABLE PRECURSORS*

The mother of all blood cells— the pluripotent stem cell—can either renew itself (semicircular arrow) or produce daughters that will develop into specialized blood cells.

Pluripotent Stem Cell

Myeloid Stem Cell

All stem cells look alike. So do their immediate descendants, the progenitor cells that are committed to particular lines of development. (The broken lines indicate pathways that are not definitively established.)

After several rounds of cell division, the progenitor cells produce offspring that are visibly different: the identifiable precursors. These, in turn, give birth to eight kinds of mature, fully differentiated blood cells.

T Progenitor

B Progenitor

CFU-GM
(Colony-Forming Unit Granulocyte-Macrophage)

BFU-Mk
(Burst-Forming Unit Megakaryocyte)

CFU-Mk
(Colony-Forming Unit Megakaryocyte)

BFU-E
(Burst-Forming Unit Erythroid)

CFU-E
(Colony-Forming Unit Erythroid)

T Cell Precursor

B Cell Precursor

Monobla

Myelobla

Megakary

Proerythroblast

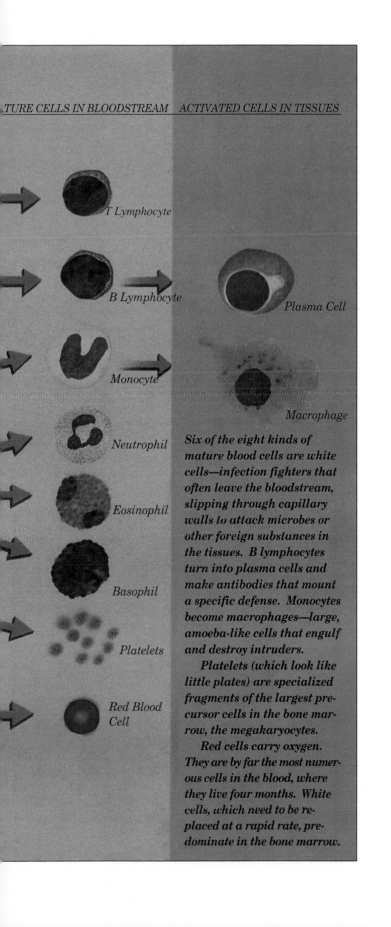

T Lymphocyte

B Lymphocyte

Plasma Cell

Monocyte

Macrophage

Neutrophil

Eosinophil

Basophil

Platelets

Red Blood Cell

Six of the eight kinds of mature blood cells are white cells—infection fighters that often leave the bloodstream, slipping through capillary walls to attack microbes or other foreign substances in the tissues. B lymphocytes turn into plasma cells and make antibodies that mount a specific defense. Monocytes become macrophages—large, amoeba-like cells that engulf and destroy intruders.

Platelets (which look like little plates) are specialized fragments of the largest precursor cells in the bone marrow, the megakaryocytes.

Red cells carry oxygen. They are by far the most numerous cells in the blood, where they live four months. White cells, which need to be replaced at a rapid rate, predominate in the bone marrow.

number of colonies in its spleen (see picture on p. 182).

The stem cells could be counted after they had done their job, but they couldn't be identified. Back at his desk after delivering the freshly extracted bone marrow to a technician in his laboratory, David Williams grabs a piece of paper and draws a diagram to illustrate the problem.

At the top of the page he sketches a single cell, placing a semicircular arrow around it—a sign, universally recognized among hematologists, that the cell can renew itself. This is the pluripotent hematopoietic stem cell, he explains; it can generate all the other types of blood cells, but it generally stays quiescent and doesn't divide. When it does divide, it either duplicates itself or gives birth to cells that have already moved one step down the path to differentiation and maturity. Though these stem cells have fewer options for future development, they can still renew themselves indefinitely.

Williams then draws a dotted line to separate the stem cells from their more numerous offspring, the committed "progenitor" cells. "The progenitors are restricted in their options—once they start along one pathway of development, they can't switch to another—and they are restricted in their ability to self-renew," he says. Most of the committed progenitors can differentiate into only one type of cell.

The next cycle of cell division gives rise to "precursor" cells, which are well on their way to producing differentiated, mature blood cells. The precursor cells cannot renew themselves. And, increasingly, they begin to look like the cells they will generate.

"When you take a sample of bone marrow cells and look at them under the microscope, you can see some cells that you know will become red blood cells or some other kind of blood cell," says Williams. "Those are the precursors. As a hematologist, I'm trained to recognize them, and I can tell what they will become. But these cells back here (he points to the progenitors and the stem cells), they're a black box—they all look the same. We're just beginning to tell them apart functionally."

If a bone marrow cell forms a colony of differentiated cells that look red, for instance, scientists know that the colony came from a red-cell progenitor. Progenitors can also be recognized by the fact that certain molecules will bind to particular proteins on the progenitor cell's surface.

Detecting the stem cells is much more difficult, however. Scientists have tried to identify them with a variety of tests based on the cells' density, sensitivity to certain chemicals, or surface proteins. They have used fluorescence-activated cell sorters, magnetic beads, and other techniques to pick out groups of cells with a high concentration of stem cells. But they could not be sure their tests were sufficiently discriminating to select only stem cells.

The problem is that the blood-forming stem cells—particularly the pluripotent stem cells—are so rare. Less than one tenth of 1 percent of the cells in the bone marrow qualify as pluripotent stem cells, based on the number of mixed colonies these cells produce in the spleens of irradiated mice and on other tests. Researchers who wish to rescue irradiated mice must inject at least 200,000 normal bone-marrow cells into each mouse to ensure that enough stem cells have been transferred to save 95 percent of the mice. To be safe—and because humans are so much larger—a typical bone-marrow transplant in human patients requires about 40 billion healthy bone-marrow cells. These are extracted from a donor and then injected into the patient's veins, from where they find their way to the patient's own bone marrow.

In 1988, a Stanford University team astounded its competitors with the announcement that it had succeeded in isolating the pluripotent stem cells of mice. Gerald Spangrude (who was then at Stanford but is now at the National Institutes of Health's Rocky Mountain Laboratory in Hamilton, Montana), Shelly Heimfeld, and Irving Weissman of Stanford had picked these cells out of ordinary bone marrow, subdividing the original marrow cells again and again with the help of antibodies until the team produced what it called "a virtually pure population" of primitive stem cells. The selected cells represented only 1 in 2,000 bone marrow cells.

When the researchers injected these elite cells into lethally irradiated mice, the mice survived for more than four months and made a full complement of blood cells. The next question to be addressed was how few of the selected cells were needed to do the job. Trying smaller and smaller numbers, the team found that 95 percent of the

mice survived after injections of only 100 selected cells (as opposed to 200,000 unselected marrow cells), and half the mice survived after injections of only 30 selected cells.

This mouse experiment provided what scientists call "the gold standard" in the hunt for the pluripotent stem cell. It opened the floodgates to attempts to isolate similar cells in human bone marrow.

A few years later, Charles Baum of Systemix Inc. in Palo Alto, California, Irving Weissman, and their associates reported that they had zeroed in on some "candidate" human blood-forming stem cells. They had picked out these cells with the aid of a series of antibodies in a highly complex process. Then, since human beings could not be used for experiments, they tested the cells' activity in a strain of mice known as *scid* (severe combined immunodeficiency) mice, which are born without an immune system of their own and readily accept transplanted human organs.

The candidate human stem cells were injected into cylindrical segments of human fetal bones that had been implanted in the mice to give the

A typical bone-marrow transplant requires about 40 billion healthy bone-marrow cells.

All blood cells of adults are made in the marrow of bones in the skull, upper arm, breastbone, ribs, spine, pelvis, and upper leg—the central part of the body.

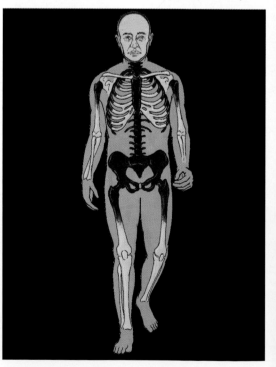

THE DANGER OF AIDS

Each of the small round bumps on the surface of an infected white blood cell in this picture is a dangerous, freshly made virus—a human immunodeficiency virus (HIV), which causes AIDS. As the new viruses break loose from the cell's surface, they spill out into the bloodstream to infect other cells.

Even a slight prick with a needle containing some drops of blood that is contaminated with HIV could transmit the deadly virus. When addicts shoot drugs directly into their veins with shared needles, they are at serious risk of contracting AIDS. The virus can also make its way into the bloodstream through sexual contact with an infected person—the most common form of transmission. Mothers can transmit it to infants during pregnancy or delivery or through their milk. AIDS was first reported in the United States in 1981. For a few years thereafter, the virus lurked in the blood supply because there was no way to detect its presence, and thousands of people were infected through blood transfusions. Since 1985, however, blood banks have been able to screen whole blood for antibodies to HIV and reject any blood that tests positive. Individual components of blood, such as various clotting factors, are sturdier than whole blood and can be treated with heat or chemicals to kill the virus.

The chances of acquiring HIV from blood transfusions have become very small in the United States—somewhere between 1 in 40,000 to 1 in 250,000 units of whole blood, according to Anthony Fauci, director of the National Institute of Allergy and Infectious Diseases in Bethesda, Maryland. Since the average transfusion requires five or more units of whole blood, a slight risk still exists. For this reason—and to avoid any danger of immune reactions—people who need elective surgery often prefer to donate their own blood in advance.

AIDS kills people when their blood has lost so many white cells that they can no longer fight off infections or cancer. Two kinds of white cells are the targets of HIV: T lymphocytes, also known as "helper T cells," the immune system's first line of defense, and monocytes, which nor-

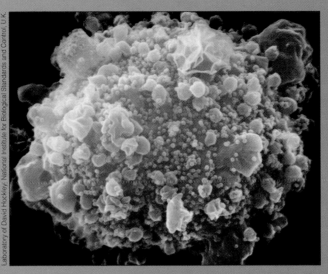

New AIDS viruses emerge from an infected human cell.

mally go out of the bloodstream into tissues, in the form of macrophages, to swallow up and destroy bacteria and other foreign invaders. The surfaces of both of these cells are studded with a protein called CD4. The virus sees this protein as a receptor, clamps onto it, and uses it to enter the cell.

Once inside, the virus takes over the cell's machinery. It forces the cell to produce thousands of new viral subunits, which eventually assemble into complete viruses and bud out of the cell. This destroys the cell. Meanwhile, the new copies of HIV travel in the bloodstream or inside white blood cells.

Antibodies to HIV begin to appear in blood about one to three months after infection, but it may take as long as six months for them to be generated in large enough quantities to show up in standard blood tests. This is a worrisome window of time during which people who have been infected with the virus still test HIV-negative, though in fact they are themselves infected and can infect others.

Most people remain healthy for several years after infection, with no symptoms of the disease. Yet their immune systems go through an ominous change—a relentless decline in the number of CD4-bearing cells in their blood. An uninfected person's blood will usually contain about 1,000 CD4 cells per cubic millimeter. As that count falls below 200 in people who are infected with HIV, their risk of developing various infections and cancers increases greatly. Some AIDS patients have no detectable CD4 cells left at all.

Much about HIV's activities inside white blood cells remains unclear, despite intensive study by hundreds of research teams around the world. Each stage in the virus's life cycle is now being analyzed, in the hope that it will provide a weak spot at which HIV can be attacked. Many different strategies are being tried. A new generation of drugs and experimental vaccines against AIDS is being tested in humans. The epidemic continues, however, with no end in sight.

cells a normal environment in which to grow. Two out of nine mice in which this experiment was tried produced human blood cells of various types (all except T cells) that could be shown to come from the injected human cells. When similar cells were injected into a human fetal thymus that was implanted in *scid* mice, they gave rise to human T cells.

Are these candidate cells the true, pluripotent, human stem cells? Weissman believes they are, and Sandoz Ltd, a large Swiss drug company, had enough confidence in the process to invest $392 million in Systemix in 1992. However, "the final test that they truly are stem cells is what happens when you put them into humans," Weissman emphasizes. The company hopes to develop the process to the point where the cells can be tried in cancer patients who have received chemotherapy. Meanwhile, other groups of researchers continue to seek the human stem cell by other means, and the race goes on.

The stem cells that are ultimately used to treat patients may not be taken from bone marrow, however. Instead of subjecting patients or donors to a hospital stay and about 100 bone punctures to withdraw enough marrow for a transplant (25 times the amount David Williams took from the volunteer), a growing number of physicians are using peripheral blood, which can be drawn from a vein as in an ordinary transfusion. But first they must "enrich" the blood in stem and progenitor cells. This involves subjecting the patients to chemotherapy (which compels their bone marrow to produce new stem cells to replace those killed by the treatment) and then giving them certain growth factors (which make the new stem and progenitor cells spill out into the bloodstream). After drawing about a pint of the patients' blood—the same amount as is collected during blood drives—the physicians use antibodies to select cells that have certain markers on their surface. The group of cells they end up with appears to be rich in stem and progenitor cells: patients in whom such cells are transplanted make a rapid recovery from radiation or chemotherapy.

Several groups of researchers are now vying to perfect this technique and develop ways of clearly identifying the stem cells in enriched peripheral blood. The cells would then be used as a focus of treatment.

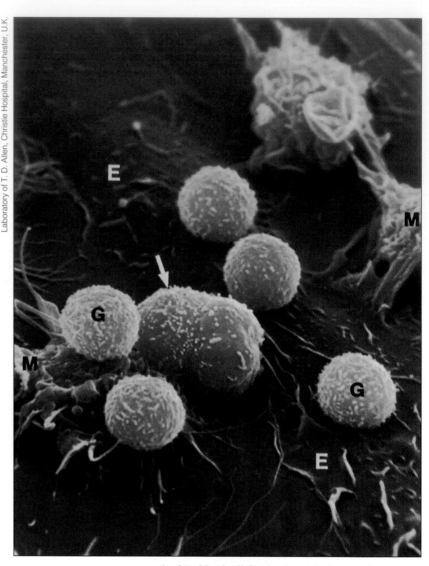

Laboratory of T. D. Allen, Christie Hospital, Manchester, U.K.

A white blood cell divides (arrow) after growing on a layer of sticky endothelial cells (E) that mimic the environment of normal bone marrow. Granulocytes (G)—white cells that contain granules—monocytes (M), and stem cells thrive for months in this kind of culture.

Another possible source of stem cells is an organ that is usually discarded: the human umbilical cord.

According to Hal Broxmeyer, a researcher at the Indiana University School of Medicine who is a world leader in the study of cord blood, the idea of salvaging the contents of the umbilical cord occurred to Edward Boyse, an immunologist who was then at the Memorial Sloan Kettering Cancer Center in New York City (as was Broxmeyer), in the 1980s. "Boyse thought it was a waste to throw away something that might be useful," recalls Broxmeyer. "He asked me what could be done with cord blood. I mentioned there was some evidence that there are stem cells and progenitor cells in cord blood, from the literature and from my own work. So he said, 'Well, then, why don't we transplant the cord blood, instead of bone marrow?'"

This suggestion started a lively national and then international collaboration that led in 1988 to the first successful transplant of cord blood.

The patient was a five-year-old boy with Fanconi's anemia, a fatal disease that can sometimes be cured by bone marrow transplants from a well-matched sibling. When his mother became pregnant again, doctors were on the alert. Prenatal tests showed that the fetus, a female, did not suffer from the same disorder and was immunologically well matched. At the infant's birth, the researchers collected blood from her umbilical cord and sent it to Broxmeyer, who froze it, keeping a small sample for analysis. The cord blood had about the same proportion of progenitor cells as did normal bone marrow, Broxmeyer's lab found. This prompted the researchers to try treating the boy with his sister's cord blood. They knew that if they failed, they could still use some of his newborn sister's bone marrow to save him.

The actual transplant was done in Paris, at a special bone-marrow transplant unit directed by Eliane Gluckman at the Hôpital Saint-Louis. First the child's own bone marrow had to be destroyed by chemotherapy and radiation. Then the cord blood was thawed out and dripped into his veins; from there, the stem cells found their way to his bone marrow. He responded well, adopting his sister's blood system as his own, and continues to make healthy blood cells to this day. This proves that he received a sufficient number of pluripotent stem cells from the cord blood.

At least 17 people in various parts of the world received cord blood transplants from sibling or unrelated donors. All of them were children; researchers are not yet sure that an umbilical cord would contain enough stem cells for an adult.

"I think that every person born should have his cord blood frozen and stored," Broxmeyer says. "It's an insurance policy. Hopefully, you'll never need it—and the odds are that you won't—but if you do, it's there."

No matter how—or where—human stem cells are finally isolated, scientists will need to know how to keep them alive. "Our goal is to completely define what's needed to keep the stem cells alive—and even better, multiplying—in the lab," David Williams declares. "If we could do that—if we could expand the stem-cell population in the lab—it would have implications for inserting genes into those cells as part of gene therapy and also for many aspects of bone marrow transplantation."

For a long time, such work was hampered by the fact that no bone marrow cells of any kind could be grown in a test tube. Then, in the mid-1960s, two groups of scientists—Dov Pluznik and Leo Sachs in Israel and Thomas Bradley and Donald Metcalf in Australia—independently figured out that bone marrow cells would grow if dispersed in a semisolid medium—something with the consistency of gelatin. The researchers also had to provide some nutrient-filled liquid ("conditioned medium") that had surrounded other growing cells and evidently contained proteins secreted by those cells. This technique allowed them to grow colonies of bone marrow cells that originated from a single cell. At the same time, it enabled scientists to compare the effects of various kinds of conditioned media.

"That is what led to the purification of growth factors," Williams explains. Different types of conditioned media contained different "colony-stimulating factors," or growth factors—proteins that were secreted by neighboring cells and stimulated different kinds of blood cells to proliferate or mature. Each kind of bone marrow cell had a different set of receptors on its surface, enabling it to respond to specific growth factors. So depending on the type of growth factor used, one could get colonies of particular blood cells.

A major drawback of the cell culture technique soon became clear, however: while most of the bone marrow cells lived a normal life span in such cultures, the stem cells did not. These either died or differentiated. They seemed to need a more natural environment to sustain them, something similar to the stroma, the tissue that surrounds these cells in real bone. In 1973 an English scientist, Michael Dexter, set

Depending on the type of growth factor used, one can get colonies of particular blood cells.

out to reproduce such an environment—in a flask. He started out with a layer of sticky cells derived from the stroma. When he covered this layer with fresh bone-marrow cells, the bone-marrow cells thrived for several months, which meant that some stem cells must have survived to replenish them.

By 1983, scientists had purified five different growth factors, but only in minute quantities. The development of recombinant-DNA techniques led to a big spurt in research. Very rapidly the genes that encoded these growth factors were identified, making it possible to mass-produce them in the laboratory.

Many more growth factors have been discovered and produced in the past few years. Some scientists are particularly interested in one that is called "Steel factor" (after a mutant mouse called "Steel") or "stem-cell factor." Derived from stromal cells, it contributes to the growth of stem cells. Williams himself is focusing on IL-11 (Interleukin-11), which was discovered and cloned by researchers from his laboratory in collaboration with Genetics Institute. IL-11 seems to control the speed with which human stem cells differentiate and proliferate. It is now being tested on cancer patients to see if it will help them withstand some of the damaging effects of chemotherapy.

Several growth factors are already being used to increase the production of blood cells that are present in low concentrations in certain diseases. For example, erythropoietin is now the preferred treatment for the anemia caused by chronic kidney disease. Two other growth factors, GM-CSF (granulocyte-macrophage colony-stimulating factor) and G-CSF (granulocyte colony-stimulating factor), have been tried in AIDS patients, as well as patients who have had chemotherapy or marrow transplants, to help them make more white cells.

Other growth factors are still so new that they have not yet entered clinical trials. And this is only the beginning. Scientists expect to find at least one growth factor that controls the mother of all blood cells, telling it whether to stay quiescent, duplicate itself, or proliferate. At that point—assuming that pluripotent stem cells can be clearly identified—scientists will be able to adjust the blood-forming system on demand and correct a wide range of disorders. ●

WHEN BLOOD CELLS GRO

The pluripotent stem cells usually do a superlative job of producing exactly the right number of white cells, red cells, and platelets at precisely the right time, despite wild fluctuations in need. During infections, for instance, stem cells supply additional white cells to fight the invaders. At high altitudes, they make more red cells to carry extra oxygen to the tissues.

"The blood-making system responds to stress," says Owen Witte, an HHMI investigator at the University of California, Los Angeles. "You produce an average number of red blood cells per day because you have an average turnover. But if you bleed, you can produce 10 times more red cells to compensate for the loss. The system is like a seesaw that maintains balance by going up and down. If you lose this balance—if you have too much of one type of cell, or not enough of another—you get sick."

Having too many blood cells of a particular type is a hallmark of leukemia, cancer of the blood. Normal stem cells and progenitors are delicately poised between resting and dividing. But in chronic myelogenous leukemia (CML), a form of leukemia that kills about 5,000 people in the United States every year, the bone marrow's stem cells tilt toward division. Even a subtle change of this sort can have dramatic consequences: as one type of blood cell—usually a white cell—grows out of control, it crowds out other types of cells.

The first clue to what causes this uncontrolled growth came when scientists in Philadelphia examined bone-marrow cells from patients with CML and noticed that a piece of one small chromosome was missing. This was in 1960, before techniques for a more precise analysis of the 23 pairs of human chromosomes became available, so the scientists were not even sure which of two small chromosomes was involved—number 21 or 22.

"The fate of the missing piece was unclear

as well," says Janet Rowley, a University of Chicago geneticist who resolved this question in 1972. Rowley discovered that the truncated chromosome—the so-called Philadelphia chromosome—was formed from chromosome 22 by a translocation (a transfer of chromosomal parts) rather than a simple deletion. Repeatedly in her patients, a chunk of chromosome 22 had broken off and come to rest at the tip of chromosome 9, a larger chromosome. Trading places with it, a piece of chromosome 9 migrated to the tip of the shortened chromosome 22. This was the first chromosomal translocation specifically associated with a cancer.

As she went on to discover several other translocations in other leukemias, Rowley realized that they were extremely consistent, always involving the same parts of two particular chromosomes for each type of leukemia. She proposed that the critical change leading to leukemia was "the abnormal juxtaposition of two very specific genes," at least one of which was probably involved in regulating growth.

Focusing on the chromosomal regions where these consistent breaks occurred, scientists were astonished to find a familiar gene in the Philadelphia chromosome: the *Abelson (ABL)* gene, the human equivalent of a viral gene known to cause leukemia in mice. This gene is an "oncogene"—a gene that can cause cancer when activated inappropriately. Though its normal location is on chromosome 9, it had moved to an abnormal position next to another gene at the breakpoint of chromosome 22. This produced a new "fusion" gene, part 22 and part 9, which changed the function of the *ABL* gene and led to leukemia.

Now researchers are studying all the abnormal junctions produced by translocations and seeking the genes that may be responsible for leukemia. "In the past few years, many new genes involved in translocations have been identified," says Janet Rowley. After such genes are found, scientists can analyze exactly how the abnormal proteins they produce trigger cancer; this may reveal how to counteract or prevent the process.

"Cancer is a genetic disease," says Rowley. "In the future, we will treat each subtype of leukemia differently because we will be able to define the genetic changes in a patient's malignant cells and tailor the therapy." This will be more effective, as well as less toxic, and will allow precise monitoring of the disease, she points out. In view of the rapid progress being made in mapping the human genome, Rowley expects most of the genes involved in leukemia to be identified very soon.

Broken Chromosome 9
Plus Chunk of Chromosome 22

Normal Chromosome 22

Normal Chromosome 9

Philadelphia Chromosome
(Broken Chromosome 22
Plus Chunk of Chromosome 9)

A chunk of chromosome 22 (green) breaks off and trades places with a piece of chromosome 9, a larger chromosome (red). An abnormal gene formed by this fusion at the tip of the Philadelphia chromosome (the altered chromosome 22) then activates an oncogene, causing leukemia. The chromosomes are identified by fluorescent probes.

TURNING BACK THE BIOLOGICAL CLOCK TO CURE SICKLE CELL DISEASE

Maya Pines

Five-year-old Maurice Harrell drinks an experimental drug against sickle cell disease, hoping it will reduce his pain. Susan Perrine, who developed the drug, observes him.

Ever since she spent a summer working in a Saudi Arabian health clinic as a freshly minted physician, Susan Perrine has been seeking a cure for sickle cell disease. As a pediatrician and researcher at Children's Hospital Oakland Research Institute in Oakland, California, Perrine completed a pilot study of a highly promising experimental treatment for this painful, life-threatening disorder.

As yet, there is no cure for sickle cell disease, nor even an accepted, specific treatment. The disease affects some 80,000 African-Americans and millions of people in Africa and other parts of the globe. It strikes children who have inherited two copies of a gene for abnormal hemoglobin, one from each parent (who may not even be aware of carrying this faulty gene and thus having the sickle cell trait). About 2.5 million African-Americans—1 in 12—are carriers.

The mutant genes contain a tiny mistake, a kind of typographical error involving just one nucleotide of DNA out of 3 billion in each human cell. Yet this is enough to change the chemical properties of hemoglobin, the protein that carries oxygen within red blood cells, causing these cells to become stiff and sometimes sickle shaped when they release their load of oxygen. The sickled cells tend to get stuck in narrow blood vessels, blocking the flow of blood. This produces symptoms ranging from mild to excruciating "crises," stroke, blindness, or damage to the lungs, kidneys, or heart.

In Saudi Arabia, in the 1970s, Susan Perrine saw that many of the Arab patients who came to the clinic had surprisingly mild cases of sickle cell disease. They complained of unrelated problems, but displayed no symptoms of the ailment. As her father, Richard Perrine, a physician for the Arabian American Oil Company, and his colleagues reported in a medical journal in 1972, their type of sickle cell disease was "benign." Most remarkable, their blood contained exceptionally high levels of fetal hemoglobin—the kind of hemoglobin that all humans produce before birth but generally turn off soon afterwards, replacing it with the adult kind. In the Arab patients, about 25 percent of the total hemoglobin was of the fetal type, rather than 5 percent, as in other ethnic groups, and it seemed to protect them.

"That was fascinating to me," Susan Perrine recalls. "When you have a good gene as a baby, why do you have to turn on a bad one? From then on, I looked for groups of people who didn't switch off their fetal hemoglobin as completely as others do."

Her first clue came in 1984, when a baby born of a diabetic mother was found to have a very high level of fetal hemoglobin and a very low level of the adult kind. The baby produced large amounts of insulin in response to the mother's high blood sugar, as well as abnormally large quantities of certain byproducts of sugar metabolism. This set Perrine thinking. She knew that sheep were being given large doses of insulin in a research project at the University of California, San Francisco (UCSF), not far from the Stanford University School of Medicine, where she was teaching pediatrics. So she decided to draw some blood from a fetal lamb in this project and test it.

When she read the results of the test and saw that the lamb, too, had an unusually high level of fetal hemoglobin for its gestational age, she was very excited. "I took the wet, soggy gel that showed this result to Dr. Kan, an expert on hemoglobin genes," she recalls. [Yuet Wai Kan, an HHMI investigator, is at UCSF.] "I dripped some of the gel all over his desk—I didn't have enough grant money for a gel dryer—and said, 'Dr. Kan, I think something stops the globin genes from switching!' "

Intrigued, Kan decided to help her. He got her more sheep to study and began to collaborate with her in a series of experiments designed to identify the mechanism involved.

Both fetal and adult hemoglobin molecules consist of four globin chains that act together (see p. 201). Each globin chain curves around a heme group that binds oxygen, and each is specified by a different gene. Fetal hemoglobin—which has a higher affinity for oxygen, so that

After releasing their load of oxygen, some of the red blood cells of sickle cell patients become stiff and sickled, as shown in this photo (amplified 1,500 times). The misshapen cells often get stuck in small blood vessels, causing extreme pain and damage.

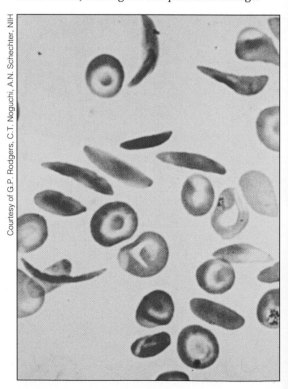

Courtesy of G.P. Rodgers, C.T. Noguchi, A.N. Schechter, NIH

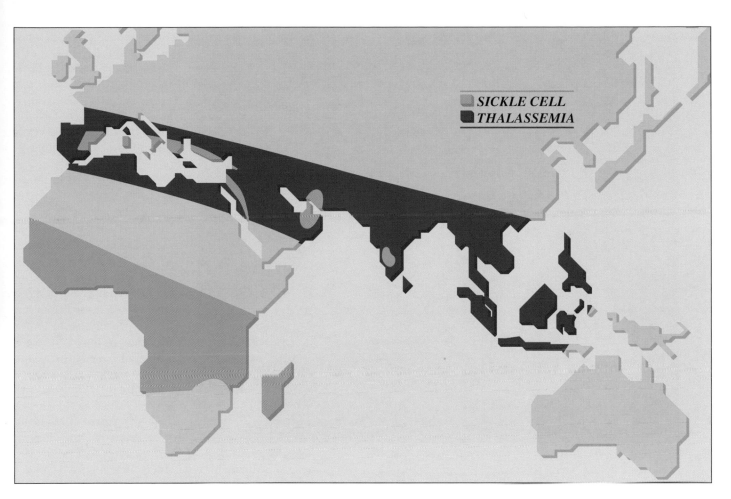

The sickle cell mutation originated in tropical parts of Africa and spread to Mediterranean areas where malaria was prevalent. It gave its carriers a selective advantage: some protection against malaria. As shown on this map, the gene never took hold around the Sahara desert, which is free of malaria-bearing mosquitoes.

The thalassemias, including a severe form known as Cooley's anemia, are caused by insufficient production of one or more globin chains (resulting from defects in the hemoglobin gene). Like the sickle cell gene, these mutations offer some protection against malaria to persons who have only one copy of the defective gene. The thalassemias are prevalent in Mediterranean areas and southeast Asia.

the fetus can pick up oxygen from the placenta—consists of two alpha and two gamma globins. The alpha chains do not change with age, but around the time of birth the gamma globins are turned off and replaced by beta globins in a still-mysterious process.

Kan wanted to find out why this process was delayed in the infants of diabetic mothers. He focused on certain chemicals that were abnormally abundant in the infants' blood. Testing each chemical in turn, the researchers discovered that one of them—butyric acid, a simple fatty acid that is widely used as a food additive—sharply increased the amount of fetal

hemoglobin produced in normal precursors of red blood cells. Furthermore, the chemical had the same effect on the cells of patients with sickle cell disease. The researchers then pumped some butyric acid into lamb fetuses, sampled their blood daily, and found that the animals' switch from fetal to adult hemoglobin was delayed.

Even a small increase in the production of fetal hemoglobin could be useful to sickle cell patients, Kan and Perrine realized, since cells containing fetal hemoglobin have a selective advantage: they survive up to 120 days in circulation, as normal red blood cells do, while the

Yuet Wai Kan wants to find out what triggers the switch from fetal to adult hemoglobin. He recently developed new, simpler tests for the thalasssemia mutations.

red blood cells of sickle cell patients generally live only 16 days. "In Saudi Arabia, only 4 to 8 percent of the hemoglobin that sickle cell patients make in cells that are entering the bloodstream is fetal, but that's enough," says Perrine. "Since the cells containing fetal hemoglobin survive longer, the patients end up with 25 to 30 percent fetal hemoglobin in their blood."

Butyric acid and a similar compound, sodium butyrate, did more than delay the switch from fetal to adult hemoglobin—they also reactivated the fetal gene after it had been turned off. As other researchers reported, the blood cells of chickens and baboons reacted to sodium butyrate in a similar way.

Could the biological clock for globin gene switching be stopped—or better yet, reversed? Perrine and her associates asked. They were eager to try it out.

The Food and Drug Administration gave

them a chance in 1992, when it authorized Perrine to do a short-term trial of butyrate in patients with sickle cell disease or betathalassemia, another genetic ailment caused by a different error in hemoglobin. The drug was given intravenously in the form of arginine butyrate, day and night, for two or three weeks.

Even in this short time—and even though it was only a test of the drug's safety, not its efficacy—the six patients' production of gamma (fetal) globin in newly made cells increased from less than 0.3 percent to 6 to 13 percent. They achieved "levels of fetal globin synthesis that would be predicted to completely alleviate their disease," said Douglas Faller of the Boston University School of Medicine, who collaborated with Perrine in developing the treatment.

Meanwhile, Perrine was trying to develop a more practical drug—one that could be taken by mouth rather than intravenously. As a food additive, butyrate is metabolized very quickly, but she produced a butyrate analogue in liquid form that lasts 10 hours in the body. She tried it out for efficacy and safety in sheep, monkeys, rats, and baboons. In each case it increased their fetal hemoglobin. Then she tried to figure out what would be an appropriate dose for humans—and "I just drank it," she says. So far there appear to have been no toxic effects from the drug at this dose in animals or humans, though the long-term effects are not known.

As news of her efforts to find an effective treatment for sickle cell disease spread, desperate parents began to contact Perrine, seeking help for their children. One of them, the mother of a very sick five-year-old boy, Maurice Harrell, had been warned that he ran a high risk of suffering a stroke (because sickled cells might block the flow of blood to his brain). The child had had painful crises almost every two weeks since he was five months old. He had been in and out of hospitals all his life—twice for severe bone infections and fractures, twice for blood infections, at least 10 times for pneumonia, innumerable times for bleeding ulcers from stress and other emergencies. In the past year his condition had worsened. He had nearly died of "acute chest syndrome" as his lungs became obstructed by sickled cells. He had required 12 blood transfusions, including a total blood exchange, and was fitted with a "MediPort"—a permanent opening to his blood vessels that was implanted in his chest, where it was barely visible as a small bump under his skin—so his veins wouldn't have to be punctured with a needle for every treatment.

"I got so desperate that I wanted him to have a bone marrow transplant, which might cure him," said Donna Harrell, his mother, "but his brother's bone marrow doesn't match his and we couldn't find the right match. So I started making phone calls all over the country. Finally I found Dr. Perrine."

Though Maurice couldn't be included among the six patients in Perrine's trial of butyrate, the FDA allowed her to treat him on a compassionate basis. In the fall, the little boy entered the San Francisco General Hospital for intravenous treatments with butyrate. He soon developed an allergic reaction to something, and fearing that it might be arginine (which was used to neutralize the butyrate in the intravenous version), Perrine decided to switch over to the new oral drug.

Very bright and worldly-wise for his years, Maurice willingly drank the bitter-tasting solution out of a large syringe. Normally Perrine would add a peppermint flavoring, but she feared to do so because of his allergies. When she worried about the taste, Maurice reassured her. "I can handle it!" he shouted proudly. She started him on an extremely small dose of the drug, to be safe, and increased it gradually. Maurice's mother spent every weekday and night in his hospital room to make him feel better; on the weekend, her husband Frank, a career noncommissioned officer in the Army, took over. All went well for a week. Suddenly Maurice cried, "Oh, Mommy, my back is hurting!"

"Then he started screaming and trying to run away from the pain," Donna Harrell

recalled the next day. "There's no warning when you have a sickle cell crisis. And the pain is incredibly intense. One woman told me her pain was so bad she wanted a doctor to cut off her limb to stop it. They need morphine, quickly. Sometimes they need a morphine drip into their veins. But this was a mild crisis. The morphine pills worked, and he managed to sleep part of the night."

As soon as Maurice recovered from the crisis, his butyrate treatment resumed. His fetal hemoglobin started to go up. But then he had a recurrence of another problem—chronic pancreatitis—that required him to eat only soft foods, and he was forced to stop the drug. "At least he's not allergic to it," said his mother, ever hopeful that it could be tried again at some later time. "Even if it doesn't work for Maurice, it may help others."

Most cases of sickle cell disease are not as severe as Maurice's. With aggressive use of antibiotics, blood transfusions, and other standard treatments, the life span of Americans who have sickle cell disease has gone up dramatically since 1960, when relatively few patients reached adult life. In Africa, where such treatments are not available, nearly all children with sickle cell disease still die before the age of two. But in the United States, 85 percent of patients now survive to age 20 and the median age at death is 42 for men and 48 for women. Nevertheless, "compared to the African-American population as a whole, this represents a 25-to-30-year decrease in life expectancy," points out Orah Platt, a professor of pediatrics at Children's Hospital of Boston, who has been studying the effects of sickle cell disease.

The fact that modern medicine still has no definitive treatments for sickle cell disease or thalassemia is particularly galling to researchers who are proud of a series of historic discoveries involving these widespread ailments. Hemoglobin disorders were the first to be understood at a molecular level. The very idea that disease may be caused by a flaw in a single molecule—a revolutionary concept when proposed by Linus Pauling and his colleagues in 1949—grew out of work on "sickle cell anemia," as it was called. Pauling, who later won a Nobel prize in chemistry, demonstrated that molecules of hemoglobin from sickle

In Africa, nearly all children with sickle cell disease still die before the age of two.

TESTING FOR SICKLE CELL GENES

"I knew I was a carrier," says Donna Harrell, "but my husband didn't know that he was; his mother thought it wasn't important to be tested for sickle cell. So we didn't find out until after Maurice was born."

As a newborn, while his blood still contained a great deal of fetal hemoglobin, Maurice was in fine health. But at five months of age he suffered his first sickle cell crisis. "He was screaming and crying, morning, noon, and night, for three weeks," his mother says. "His eardrums were swollen from the intense screaming. We were in Watertown, New York, where the local doctor had never seen such a case. He said it couldn't be sickle cell because the book says it doesn't start until they're 18 months old. I said, 'Well, my son didn't read the book.' "

An inexpensive blood test easily reveals whether a child or adult carries the sickle cell gene; it simply analyzes the hemoglobin in the red cells. Despite its simplicity, however, this test is not widely used. "Carrier testing has decreased in the past decade, and most people of child-bearing age who are carriers of the sickle cell gene don't even know it," says Clarice Reid, chief of the sickle cell disease branch at the National Heart, Lung, and Blood Institute, NIH.

In the middle 1980s, attention focused on newborn screening, and by now 42 states provide this service. These tests are designed to identify children who have the disease so they can receive proper treatment, including regular doses of antibiotics to prevent infections. As a byproduct, the tests also identify infants who are healthy carriers of the sickle cell gene, but this information is not always transmitted to the parents. And infants who are identified may not be aware of it later, when they grow up.

"Teen-agers and young adults should know whether they are carriers," says Reid. "They should seek out that kind of information—from private physicians or health units—because it's important in terms of family planning."

The sickle cell mutation is prevalent among African-Americans because it brings an evolutionary advantage: a single dose of this gene offers some protection against malaria, which is common in Africa. When malaria parasites invade the bloodstream, the red cells that contain defective hemoglobin become sickled and die, trapping the parasites inside them and reducing the infection. As a result, some 30 percent of the people in certain areas of Africa have the sickle cell trait. The trait is also widespread around the Mediterranean and in other areas where malaria used to be a major threat to life.

When two carriers marry, each of their offspring has a 25 percent risk of inheriting a double dose of the mutant gene and developing the disease. Prenatal tests have been available since 1978, when Yuet Wai Kan and Andrée Dozy of UCSF showed that a certain enzyme, which cuts normal DNA at a particular site, will not recognize—and therefore not cut—DNA that has the sickle cell mutation. The two researchers saw that this could provide a very precise test of the mutation. If DNA that is extracted from a fetal cell and exposed to this enzyme produces one long fragment (from one parent) and two shorter fragments (from the other parent), the fetus is a carrier. But if the DNA produces two long fragments, the fetus has inherited the disease. Since 1978 several newer tests have been developed on the basis of these observations. In the past few years, tests using the polymerase chain reaction (PCR) have proved particularly rapid, cheap, and sensitive.

The American College of Obstetricians and Gynecologists recommends that carrier tests be offered routinely to African-American couples, and that those who are at risk be told about the availability of prenatal tests for sickle cell disease. In many states, however, Medicaid will not pay for such tests. In general, only people with high levels of education request them.

When Donna Harrell became pregnant again a few years after Maurice's birth, she had some of the fetal cells tested for sickle cell nine weeks after conception. To her great relief, there was no sign of the sickle cell mutation. Now she also has a healthy younger son who inherited a normal gene from both parents and is not even a carrier of the disease.

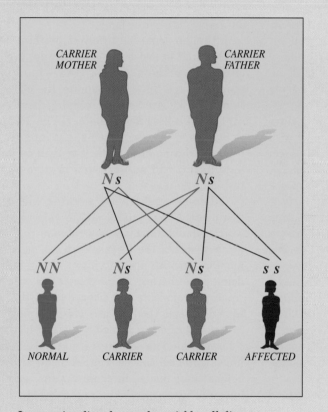

In recessive disorders such as sickle cell disease, both parents carry a single defective gene (s) but are protected by the corresponding normal gene (N). Two defective copies of the gene are required to produce the disorder. Each child has a 50 percent chance of being a carrier like both parents, a 25 percent chance of being free of the defective gene, and a 25 percent chance of inheriting the disease.

cell patients moved differently in an electric field than molecules of normal hemoglobin; so did some of the hemoglobin made by the patients' parents. Pauling deduced that sickle cell anemia resulted from an inherited defect in the hemoglobin molecule. He called it a "molecular disease."

A decade later, Vernon Ingram, a British biochemist, traced the defect to the substitution of a single amino acid (valine for glutamic acid) in the two beta-globin chains of the hemoglobin molecule. Also at Cambridge University, England, Max Perutz discovered that the faulty amino acid sat on the surface of the beta-globin chain, changing its shape so that each beta-globin became sticky when it was not carrying oxygen. The abnormal hemoglobins then linked together, forming fibers.

In 1978 Yuet Wai Kan showed that some inherited variations ("polymorphisms") in the patterns of DNA fragments from sickle cell patients could be used as markers of the disease. This enabled scientists to detect sickle cell disease and diagnose it prenatally by DNA tests. Current tests detect the precise DNA mutation—GTG instead of GAG—that caused one amino acid to be substituted for another and produced the disease.

Hemoglobin with the sickle cell mutation becomes insoluble after releasing its load of oxygen. It forms long fibers that create a rigid gel inside the red cell and twist the cell into a sickled or holly leaf shape. The process is reversible at first; as soon as the cells pick up oxygen again, they revert to a normal shape. But after repeated cycles of sickling and unsickling, the cells become permanently distorted. These irreversibly sickled cells, which make up about 10 percent of the red cells in a patient's blood, die young and are rapidly removed from circulation, leading to anemia.

Knowing this much about the disease, scientists tried many different treatments to prevent cells from sickling: hyperbaric oxygen, sodium nitrite, bicarbonate, alkali, urea, cyanate salts, and others, to no avail. It soon became clear that it would be extremely difficult to modify the enormous quantity of hemoglobin in a human body without damaging other proteins. Furthermore, sickle cell disease is so unpredictable that it would take a long time to determine whether a treatment works and whether organ damage and painful crises are really reduced.

The one trait that consistently predicted a mild form of sickle cell disease in many studies was a high level of fetal hemoglobin. Certain anticancer drugs such as hydroxyurea, which suppresses cell growth, were then found to raise the level of fetal hemoglobin in patients. When hydroxyurea was tried on patients with sickle cell disease or thalassemia, most patients' fetal hemoglobin increased after several months of treatment. However, at high doses the drug was not well tolerated by some patients. Doctors were particularly reluctant to give it to children. A large national trial of the drug on some 300 sickle cell patients was started.

To reduce the dose of hydroxyurea to a safer level, some scientists have tried to combine it with erythropoietin, a growth factor that tells stem cells they should make more red blood cells. The idea was that the two compounds might work together, multiplying their effect. This, in fact, proved true. Researchers are now investigating whether other combinations of drugs and growth factors might be even more effective against sickle cell disease.

"Each group thinks its approach is the best," points out Alan Schechter, of the National Institute of Diabetes and Digestive and Kidney Diseases, NIH, a pioneer of treatment with hydroxyurea. "It will take a few years to sort out— or maybe we'll use various drugs in new combinations, including butyrate."

Several researchers are trying to understand exactly how these drugs work. In order to develop the best treatment, says Yuet Wai Kan, "we need to see what triggers the production of fetal hemoglobin and what controls this switch."

"Most people believe the locus control region, or LCR, has something to do with the switch," Kan says. The LCR is a segment of DNA that lies close to the beta-globin gene cluster, a cluster of five genes on chromosome 11. The genes are activated at different times during development in response to varying combinations of proteins, called transcription factors, which act on the LCR and other DNA segments that are close to the gene cluster.

Kan's lab is comparing the transcription factors that bind to DNA in various kinds of red cells: cells that make fetal hemoglobin versus those that make adult hemoglobin, and bone-marrow cells extracted from some of Susan Perrine's patients before treatment with butyrate versus those that were extracted after the treatment.

At Harvard Medical School, Stuart Orkin, an HHMI investigator, is focusing on a transcription factor called GATA-1, which he discovered a few years ago while studying the development of blood cells. "This turns out to be a central, master regulator for genes expressed in red cells," Orkin says. "When GATA-1 is knocked out, red cells can't be made."

Although some 50 different proteins might be required to express a globin gene, probably only 2 or 3 of these, such as GATA-1, are really specific to red cells, he explains; the other proteins are likely to be housekeeping factors, present in all cells. Furthermore, GATA-1 appears to be "one of the few proteins that interact with the LCR during development to turn on embryonic, fetal, and adult genes in a sequential fashion," he says.

Other researchers are working with transgenic mice that model sickle cell disease— which should allow the researchers to analyze mechanisms of the disease and test new treatments. Some are looking into the possibility of doing gene therapy. After decades in which the treatment of sickle cell disease seemed to have hit a dead end, the field is evolving rapidly.

Perrine and her associates would like to see a large-scale trial of butyrate in patients. She hopes children can be treated at an early age, before their bones, internal organs, or brains become irreversibly damaged by sickle cell disease.

Some day, she speculates, butyrate or other agents might even be tried on mothers-to-be. As she points out, sickle cell disease can be diagnosed before birth, so it might be possible to give a pregnant woman some drugs to treat her fetus and prevent the switch from fetal to adult hemoglobin. Another approach might be to transfuse normal stem cells from the bone marrow of aborted fetuses into an affected fetus early in pregnancy, before the immune system is mature enough to reject foreign cells. "The potential for silencing sickle cell genes before they produce clinical disease is very exciting," Perrine says.

"This turns out to be a central, master regulator for genes expressed in red cells."

Human red blood cells

The Birth and Busy Life of a Red Blood Cell

Every minute that we live, we produce 120 million new red blood cells—more, if needed. Tiny, flexible disks, they make the full circuit of our arteries and veins 3,000 times a day, delivering oxygen and removing wastes. Somehow they manage to supply exactly the right amount of oxygen needed by tissues in different parts of the body, under extremely varied conditions.

Scientists are rapidly deciphering the sophisticated feedback systems that control the birth, development, and activities of red cells. Recent discoveries, such as the cloning of growth factors that speed red cell production, have brought new ways of treating anemias and other blood disorders.

Like other blood cells, red cells grow in bone marrow. But unlike the others, they expel their nuclei during development and become virtual bags of hemoglobin, almost entirely filled with the protein that binds oxygen. Then they slip through tiny openings in the walls of small blood vessels in the bone marrow and flow into general circulation.

In the bloodstream, red cells often twist their way through passages half the cells' normal width. They withstand severe buffeting and work nonstop for about four months. When they become too old and inflexible to do their job, they are gobbled up by scavenging white blood cells, the macrophages. About 200 billion red cells are replaced every day to maintain a steady state of some 25 trillion red blood cells in the human body.

First, a stem cell (right) in the bone marrow produces a cell that is fated to differentiate into a red blood cell.

This committed progenitor (BFU-E) cell then gives birth to a more specialized cell (CFU-E) that responds to signals from a growth hormone called erythropoietin. When the body needs more red cells, this hormone spurs the CFU-E cell to produce a proerythroblast—a cell that can be identified because, when stained with certain dyes, it has a large purplish nucleus surrounded by deep blue cytoplasm. The deep blue color reveals the presence of DNA and indicates that the cell is ready to make large quantities of protein—in this case, hemoglobin.

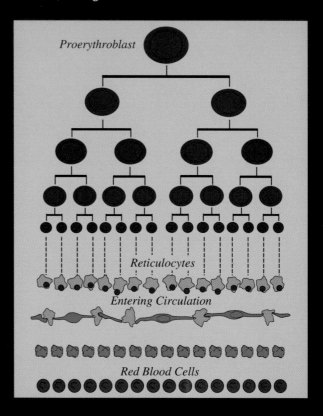

Proerythroblast

Reticulocytes

Entering Circulation

Red Blood Cells

The proerythroblast makes hemoglobin and undergoes four rounds of cell division over four days. This results in 16 cells with small, increasingly dense nuclei. The cells then eject their nuclei, which are swallowed up by macrophages in the bone marrow.

The immature red cells (reticulocytes) continue to make hemoglobin, dissolve other internal structures, and shrink in size. When they are small enough to pass through narrow openings in the vascular sinuses (channels that lead to a central vein in the bone marrow), they enter the bloodstream. Within a day they become erythrocytes—round, resilient, mature red blood cells.

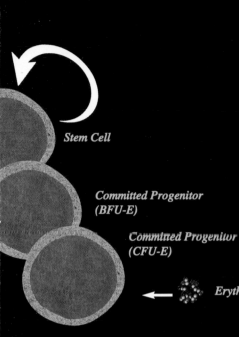

Stem Cell

Committed Progenitor
(BFU-E)

Committed Progenitor
(CFU-E)

 Erythropoietin

Proerythroblast

 Basophilic
Erythroblast

Polychromatic
Erythroblast

Pyknotic
Erythroblast

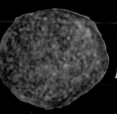

F

R R S
R

M

M

V

Bone

Developing red cells
form an island in the
bone marrow as they
cluster around a
macrophage, a large
white cell that provides
some nutrients for their
growth. Macrophages
obtain nutrients by eating
all kinds of debris—
including the nuclei that
are ejected during
red cell development.

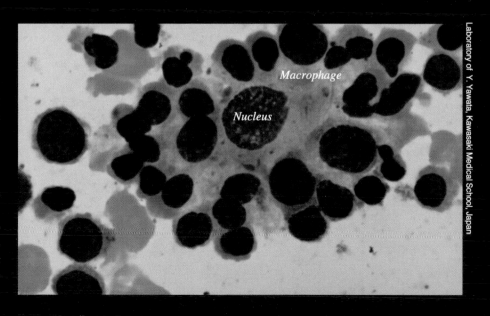

Macrophage

Nucleus

A Molecule to Breathe With

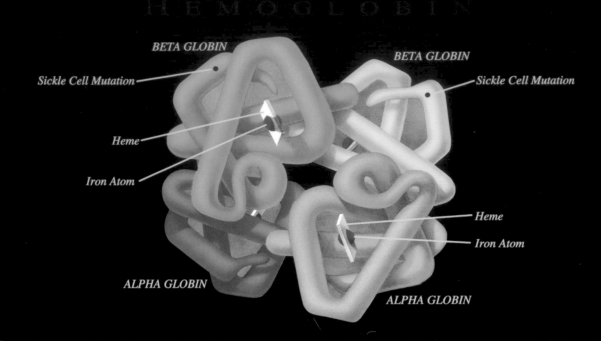

BETA GLOBIN

Sickle Cell Mutation

Heme

Iron Atom

BETA GLOBIN

Sickle Cell Mutation

Heme

Iron Atom

ALPHA GLOBIN

ALPHA GLOBIN

Riding snugly inside the convoluted structure of hemoglobin—one of the world's most complex and efficient proteins—the oxygen on which our lives depend can reach any part of our bodies within seconds.

About 280 million of these hemoglobin molecules are crowded into each red blood cell. Their three-dimensional structure was deciphered by Max Perutz of Cambridge University, England, in 1959 after 22 years of effort. Perutz, who later won a Nobel Prize for his work, used x-ray crystallography and performed millions of computations to determine the positions of 10,000 atoms in this enormous protein. When he first saw a crude model of one of the protein's four subunits, he was shocked and puzzled by its twisted shape. "A hideous, visceral-looking object," he called it. Only later, as details of hemoglobin's structure became visible at higher resolution, was its elegance revealed.

Hemoglobin is built around four atoms of iron (red in the picture above) that act as magnets for oxygen. Each iron atom sits on a small platform called a heme, which enables the iron to bind oxygen reversibly—firmly enough to hold during the bumpy trip from lungs to body tissue, but loosely enough to be released where needed. Most of the hemoblogin molecule acts as a shelter for the hemes; it protects them from the watery solution inside the cell and prevents the iron from rusting, as it would if the hemes touched water in the presence of oxygen.

When red blood cells in the bloodstream reach the lungs, where fresh air awaits, each hemoglobin molecule grabs four atoms of oxygen. It is a stunning display of cooperation among hemoglobin's four subunits—two alpha globin chains and two beta globin chains of amino acids, each produced by a separate gene (the alpha genes lie in a cluster on chromosome 16, while the beta genes are on chromosome 11). One of the four globin chains starts the process by binding a molecule of oxygen to its iron atom, which causes the chain to flex. This makes it easier for the three adjoining chains to bind their own oxygen molecules. Within 30 millionths of a second, the entire protein is transformed: more compact, loaded with oxygen, and bright red.

The bound oxygen travels safely during the red cells' difficult passage through the body's arteries and narrow capillaries. At exactly the right time and place—in parts of the body that need oxygen—the globin chains open up, releasing their precious cargo.

Freed of oxygen, the hemoglobin molecules turn dark purple. Their next job, on the return trip, is to help ferry carbon dioxide out to the lungs, where it is exhaled.

Mutations that alter even a single amino acid in a globin chain can be fatal, as in sickle cell disease. The sickle cell mutation produces a sort of sticky patch on the surface of the beta chains after the hemoglobin molecule has given up its load of oxygen. Because other molecules of sickle cell hemoglobin also develop a sticky patch, they adhere to each other and form long fibers that distort the red cell into a sickle shape.

More than 300 abnormal hemoglobins have been discovered so far. Some of these molecules are produced in normal amounts and have all four chains but do not function properly. In other cases—for example, in the thalassemias—an entire globin chain (either alpha or beta) may be missing or incomplete. Scientists are now defining the many genetic errors involved in these disorders in the hope of finding specific treatments.

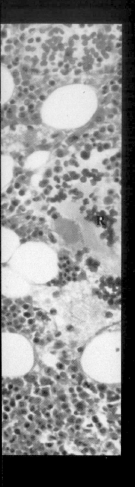

A rich mixture of developing blood cells forms in normal human bone marrow, which is dotted with fat cells (F). Red blood cells (R) that are almost mature clump together and prepare to enter into circulation through vascular sinuses (S). Large megakaryocytes (M) are the precursors of platelets. White cells predominate. A blood vessel (V) can be seen near the bone at the lower left edge of the marrow.

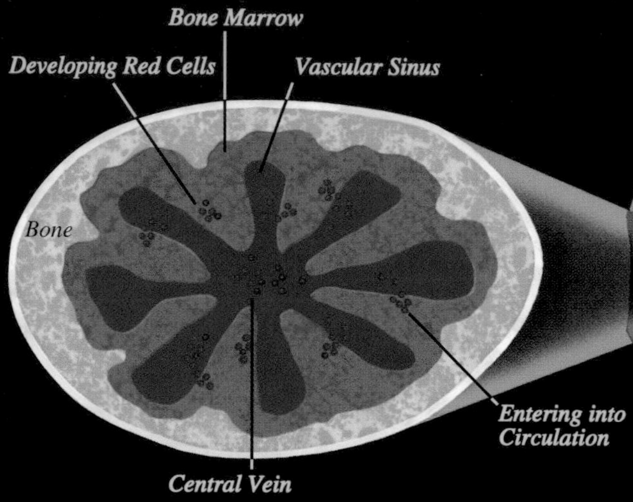

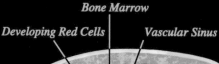

Bone Marrow

Developing Red Cells

Vascular Sinus

Bone

Entering into Circulation

Central Vein

Newly made red cells squeeze through openings in the walls of vascular sinuses to enter into circulation, as shown in this artist's conception of the cross section of a femur.

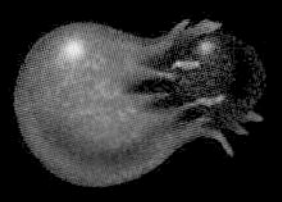

Ejecting the Nucleus

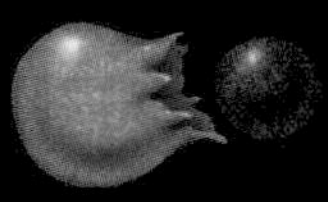

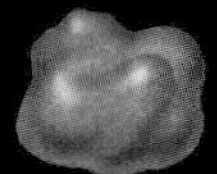

Reticulocyte

Red Blood Cell

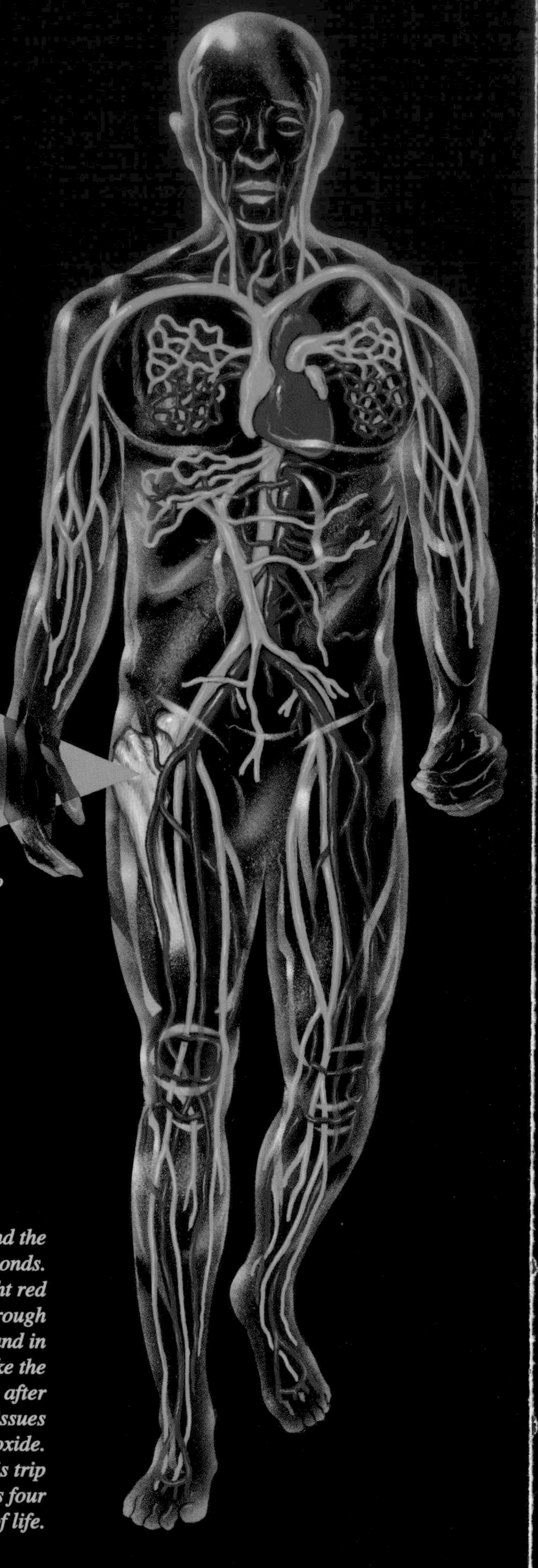

A blood cell can go around the entire body in about 30 seconds. Red cells are shown in bright red as they ferry oxygen through arteries and capillaries, and in dark purple as they make the return trip through the veins after delivering oxygen to tissues and picking up carbon dioxide. Each red cell makes this trip about 350,000 times in its four months of life.

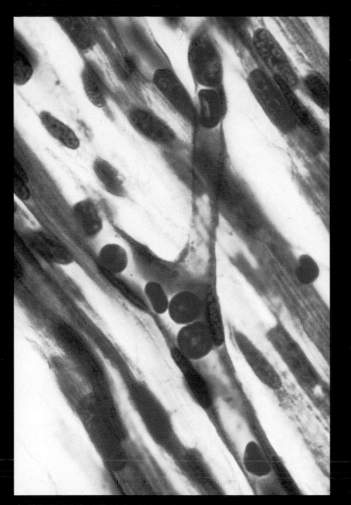

Red cells take a battering as they are forced through narrow capillaries, single file, by pressure from a pumping heart.

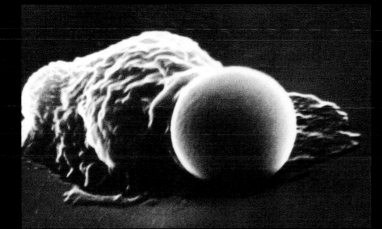

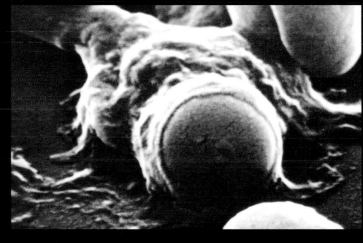

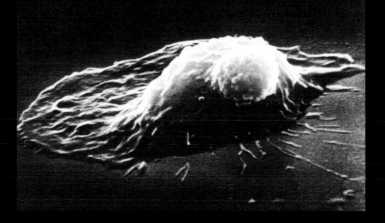

Advancing toward an aged red cell, a macrophage engulfs it, then digests it. The old cell's contents will be recycled to make new red blood cells.

TOO MUCH BLEEDING:

THE PERILS OF

HEMOPHILIA

It was four o'clock in the morning and Bob Massie, a 36-year-old hemophiliac, was wide awake, "mixing a shot" of factor VIII, the blood-clotting protein he needs to control bleeding episodes. Awakened by his five-year-old son, who had had a bad dream, Massie was unable to get back to sleep because of an ache in his right knee, an indication of a possible hemorrhage in the joint. Joint bleeds are the primary complication of hemophilia, and current clinical practice calls for treatment with factor VIII at the first sign of trouble.

Massie had received his first dose of factor VIII in a transfusion of whole blood in the summer of 1957, just before his first birthday. In the fall of 1992, 35 years and several thousand treatments later, he was still at it. Many of the details about the management of hemophilia had changed over the course of Massie's life, but the basic concept was still the same—and Massie was fed up with the whole business. "This is stupid," he thought as

he prepared the injection. "This is idiotic. Why do I have to do this? Why don't they just go ahead and fix hemophilia?"

That a person with a serious, chronic condition like hemophilia would wish for a cure in the middle of the night is not unusual. What is unusual is the fact that this wish for a cure may actually be granted. It won't happen tomorrow, and scientists working in the field are understandably reluctant to predict how many years it's going to take: you don't trifle with the hopes of people who have been through what hemophiliacs and their families have been through in the last 10 years. For them, two decades of progress in treating the disease seemed wiped out as nearly an entire generation of hemophiliacs was struck with AIDS. But the truth is, there's a definite feeling in the hemophilia community now, a kind of guarded optimism, that the younger generation of hemophiliacs alive today will see a cure in their lifetime. Told that

BY DAVID NOONAN

animal testing of some gene therapy techniques for the treatment of hemophilia is already under way, Massie, whose two-year-old nephew is also a hemophiliac, said, "Please, don't make my heart jump."

Hemophilia—Greek for "love of blood"—is one of the oldest known hereditary disorders. It was described in the Talmud in the second century A.D. by the Jewish scholar Rabbi Judah, who wrote: "If she circumcised her first child and he died, and a second one also died, she must not circumcise her third child." From that succinct, 1,700-year-old instruction, one can deduce the first essential fact about hemophilia: it's genetic. The caution against circumcision holds even if the woman has a child by a second husband, declared the physician and philosopher Moses Maimonides in the 12th century, pointing out the second essential fact about hemophilia: it is passed from mothers to sons.

Worldwide, hemophilia strikes one out of every 5,000 males. There are approximately 20,000 hemophiliacs in the United States, and according to the National Hemophilia Foundation, about 440 babies are born with the disorder each year. The most prevalent type of hemophilia, hemophilia A, is caused by a deficiency of functional clotting factor VIII in the blood; it affects 85 percent of all hemophiliacs. Hemophilia B, which accounts for 14 percent of all cases, is due to the lack of another clotting protein, factor IX. The remaining one percent are rare cases involving still other clotting proteins.

The severity of the disorder depends on the concentrations of the clotting proteins present in the blood. Seventy percent of all people with hemophilia A are classified as severe cases, showing concentrations of factor VIII that are just 0 to 1 percent of normal. Moderate cases, about 15 percent of those with hemophilia A, have 2 to 5 percent, and mild cases have from 5 to 30 percent of normal concentrations.

Interestingly, even a low level of factor VIII in the blood permits near-normal clotting function; some mild cases go undetected until adulthood, and people with just half the normal level of factor VIII don't have problems with bleeding at all. This means that even modest gains in the factor VIII levels of severe hemophiliacs would amount to major improvements. Just 5 percent more factor VIII, and a severe hemophiliac becomes a moderate, with far fewer day-to-day problems. This potentially large return from a small gain is one of the reasons that hemophilia is a prime target for gene therapy, the insertion of foreign genes into a patient's cells.

The blood-clotting process that goes awry in hemophilia is one of nature's more elaborate and elegant creations. It is a host defense system that evolved over millions of years. As the earliest living organisms evolved from single cells to more complex forms, they developed circulatory systems to supply oxygen and other vital nutrients to their multicelled bodies. Without a clotting response to protect against injury, those circulatory systems would not have been able to prevent fatal leaks; evolution as we know it would have been impossible; and only the simplest forms of life, such as algae and bacteria, would exist.

Any injury to the circulatory system, external or internal, triggers a sequence of molecular events that ultimately lead to the formation of a fibrin clot at the injured site. Twelve clotting factors become activated, each activating one or more factors in turn in a complex series of chemical reactions known as the clotting cascade. In the final stages of the cascade, the protein prothrombin is converted to thrombin, which then converts fibrinogen to fibrin.

The failure of a single link in this intricate chain may result in the failure of the entire blood-clotting system. Another detail that illustrates the fragile nature of the whole arrangement is the fact that factor VIII is remarkably scarce in the bloodstream, even at normal levels. Compared with the other proteins in the clotting

> The blood-clotting process is one of nature's more elaborate and elegant creations.

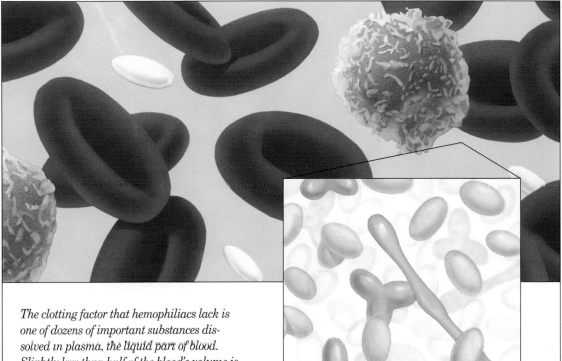

The clotting factor that hemophiliacs lack is one of dozens of important substances dissolved in plasma, the liquid part of blood. Slightly less than half of the blood's volume is taken up by blood cells. The rest is filled by plasma, a watery yellow fluid in which the cells are suspended. Plasma transports salts, nutrients, fats, hormones, antibodies, albumin, and clotting factors such as factor VIII and fibrinogen, only a few of which are shown here.

Antibody. The body can produce millions of antibody molecules that stick to specific microbes or foreign substances, labeling them for attack by other elements of the immune system.

Albumin. This simple protein contributes to blood viscosity and maintains osmotic pressure. It also ferries other small molecules by binding to them, which makes them more soluble in plasma.

Fibrinogen. Hemophiliacs have plenty of this clotting factor in their blood plasma but cannot convert it into strands of fibrin to strengthen blood clots for want of factor VIII.

cascade, factor VIII circulates at extremely low concentration, like a dash of salt in a vat of soup.

The cruel fact that a disorder as terrible as hemophilia has such a relatively simple cause is balanced by its much kinder corollary: treating hemophilia is also relatively straightforward. Replacement therapy, in which an individual's missing factor VIII is supplied from an outside source to control bleeding episodes, has been used in one form or another since 1840, when Samuel Armstrong Lane performed the first successful whole-blood transfusion in the treatment of hemophilia at St. George's School in London. Unfortunately, effective replacement therapy is a much more recent development. The only real progress in the treatment of hemophilia between 1840 and 1964, aside from the general medical progress in areas like blood-typing and antiseptic technique, was the improvement of transfusion methods and the substitution of fresh frozen plas-

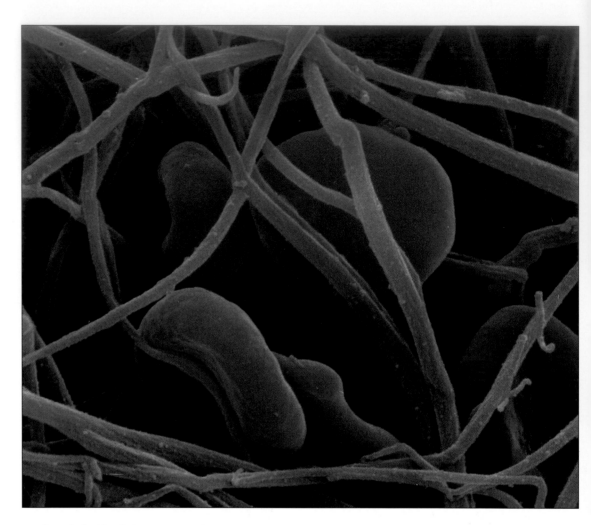

ma for whole blood during transfusions. The use of plasma didn't become widespread until the 1940s, and even that was only partially effective. Less serious bleeds were treatable, but the large volumes of plasma needed to control major bleeding episodes could put too much strain on the circulatory system. As a result, serious bleeds might lead to death by hemorrhage or, in some cases, by heart failure.

Just 30 years ago, daily life was a series of grim and difficult trials for hemophiliacs and their families. Nowhere is this better documented than in the book *Journey*, an account of the first 18 years of Bob Massie's life written by his father and mother, Robert and Suzanne Massie. It is a harrowing tale, particularly in the early sections, when the young parents are coming to grips with the gravity of their child's condition and the family's entire existence seems like one long bleeding crisis. During one difficult period, when her son

was six and overwhelmed by pain from a serious bleed in his left knee, Suzanne Massie moved him into her bed so she could comfort him. "After three such nights," she wrote, "all life became a blur, and there was no reality but The Pain. Each of these episodes was like a siege during which we tried to hold our fortress against an invisible enemy that battered and battered. There was nothing to do but just take the blows and wait for the enemy to withdraw as inexplicably as it had attacked."

"If I'd been born 50 years earlier I would have been dead, and if I'd been born 50 years later I would have been cured," Bob Massie says, chuckling wryly at the irony of his situation. As it is, Massie, who teaches ethics at the Harvard Divinity School, is a member of the most extraordinary generation of hemophiliacs of all time. Born in the 1950s, when more than half of those with hemophilia died before the age of 5 and only

11 percent made it to the age of 21, Massie and his contemporaries were initially condemned to the short, painful lives that had always been the lot of the hemophiliac. Their early years were filled with mad dashes to emergency rooms, extended hospital stays, and long nights of agony such as his mother describes above. Their bleeding joints were immobilized with plaster casts and braces, a misguided course of treatment that ended up doing them more harm than good. Their physical activity was drastically curtailed to avoid the little bumps of everyday life; in their scheme of things, "tag" was a dangerous game. Worst of all, perhaps, they were often stigmatized as "sick kids" or "crippled kids."

Then, in 1964, as these children were entering their teens, an accidental discovery revolutionized hemophilia treatment and transformed their lives. Judith Graham Pool, a hemophilia researcher at the Stanford University School of Medicine, decided to examine the thready-looking residue that collected at the bottom of plasma bags during transfusions. It turned out to be rich in factor VIII. Pool quickly developed a method for extracting the residue from fresh-frozen plasma. Called cryoprecipitate, it was easy to prepare and could be frozen and stored for as long as a year without losing its potency. Cryo, as it was called, eliminated the problem of having to infuse large amounts of plasma in order to provide enough clotting factor. And since cryo could be stored in sufficient amounts in home freezers, it freed hemophiliacs from their dependency on hospital emergency rooms and doctors' offices.

The so-called golden age of hemophilia treatment began just a few years later, when freeze-dried, concentrated forms of factor VIII became available. Bleeds were treated sooner and more effectively, reducing damage to the joints, and life expectancy soared to near-normal. The statistics tell the tale: the median age of hemophiliacs in 1970 was 11; in 1981, 20. As one professional working in the field put it, "Basically, no one died during the 1970s."

Then came AIDS. It came because each batch of the freeze-dried factor VIII concentrate was prepared from blood pooled from as many as 20,000 donors, and in the period from around 1977 until 1985 the blood supply was contaminated with the human immunodeficiency virus (HIV). Bob Massie's generation took the blow. Once again, the statistics tell the story. According to the Centers for Disease Control and Prevention, 70 to 90 percent of all severe hemophiliacs in the United States—more than 10,000 people—are HIV-positive. Through June 1992, 2,248 cases of AIDS were confirmed among persons with hemophilia and related bleeding disorders. More than 1,500 hemophiliacs have died of AIDS.

"To my mind, it's one of the greatest iatrogenic tragedies of all time," says HHMI investigator David Ginsburg, who studies blood-clotting proteins at the University of Michigan Medical School. "I think we owe these people a special debt. We have a special responsibility and duty to do something about it."

Massie, who was ordained an Episcopal priest in 1982 and got married the same year, learned that he tested positive for HIV in 1984. "I won't say it's not a big deal, but I've made my peace with it," he says. "Ultimately, we are all mortal." His two young sons are not infected with the virus.

Since 1985, new procedures—donor screening and viral inactivation, including heat treatment—have effectively eliminated HIV from the factor VIII concentrates. This has split the hemophilia community into two groups: those who were exposed to the virus and those born after 1985, who are safe.

It is a bitter irony that just as AIDS hit the hemophilia community, scientists were making a series of landmark discoveries that could ultimately lead to a cure for the ancient disorder.

In 1984 two teams of researchers working independently at two biotechnology companies—Genentech, Inc., and Genetics Institute—cloned (isolated and made multiple copies of) and

> The hemophilia community is split into two groups: those who were exposed to the AIDS virus and those born after 1985, who are safe.

sequenced (determined the sequence of DNA subunits in) the gene that codes for factor VIII. Since no one knew which cells made the factor, this took more than two years of intensive work. Their efforts uncovered a big, big gene.

With the factor VIII gene in hand, scientists began to explore the molecular genetics of hemophilia, opening a new era in research. HHMI investigator Jane Gitschier, who played an important role in finding the gene as a postdoctoral fellow in the Genentech laboratory, has established herself as a leader in the field.

As her eclectic collection of degrees implies— she has a B.S. in engineering science from Pennsylvania State University, an M.S. in

applied physics from Harvard University, and a Ph.D in biology from the Massachusetts Institute of Technology—Gitschier was still searching for her specialty when she joined the Genentech team in 1981. Though she had no particular interest in hematology, one aspect of the factor VIII project did intrigue her. "I knew nothing about blood," she says, "but I was really fascinated by the genetics of hemophilia—by the fact that new mutations were happening all the time."

That initial fascination deepened into a calling. Once the gene had been cloned, Gitschier, who left Genentech for the University of California at San Francisco in 1985, started

HOW HEMOPHILIA IS INHERITED

The flawed gene that causes hemophilia is located on the X chromosome, one of the two chromosomes that determine sex. Women have two X chromosomes. If they carry the defective gene on one, they are protected by the normal gene on the other and do not develop hemophilia. Men have one X and one Y chromosome. If they inherit the faulty gene on their single X chromosome, they have no protection and become hemophilic.

A woman who carries the defective gene has a 50-50 chance of passing it on to her children (see diagram). If she passes it to a daughter, the daughter, too, becomes a carrier. But if she passes it to a son, he will have hemophilia.

The daughters of male hemophiliacs are always carriers because they inherit their father's one defective X chromosome. The sons of male hemophiliacs are never carriers unless their mothers carry the flawed gene.

Though hemophilia is genetic, it cannot always be predicted or traced back genealogically, since approximately one-third of all cases are the result of new mutations. This remarkable phenomenon was first noted in 1935 by the British geneticist J.B.S. Haldane, who argued that

since only a small minority of hemophiliacs lived long enough to reproduce, natural selection would eventually eliminate the disorder unless it was balanced by spontaneous mutations. Haldane's theoretical calculations were confirmed more than 50 years later by direct study of the DNA of hemophiliacs and their families. These new mutations are known as "sporadic" cases. Queen Victoria, whose daughters spread hemophilia to the royal families of Germany, Russia, and Spain, apparently carried a sporadic mutation in the gene for factor IX (a cause of hemophilia B).

A much milder bleeding disorder, known as von Willebrand disease, strikes both men and women, since the gene that causes it is located on chromosome 12, which is inherited equally by both sexes. Von Willebrand disease is far more common than hemophilia and occurs in up to 1 percent of the population. (A different, severe form of the disease strikes 1 person per million.) It arises from errors in the gene for von Willebrand factor, a clotting protein that plays a key role early in the blood-clotting process. Von Willebrand factor also serves as a vehicle for factor VIII, sticking to it and ferrying it around in the bloodstream.
—D.N.

exploring the types of gene mutations that cause hemophilia.

The factor VIII gene is one of the largest yet isolated, containing 186,000 bases of DNA. It makes up about 0.1 percent of the X chromosome and is nearly six times bigger than the factor IX gene. This may account for the fact that there are roughly six times as many cases of hemophilia A as there are of hemophilia B; the big factor VIII gene seemingly presents six times as many opportunities for mutations.

The gene is made up of 26 coding segments (exons) separated by 25 noncoding sections (introns). The exons, which constitute less than 5 percent of the entire gene, contain the infor-

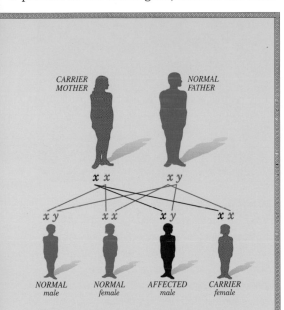

One normal copy (green x) of a gene on the X chromosome is generally sufficient for normal function. Women who have a defective gene (dark red x) on one of their two X chromosomes are protected by the normal copy of the same gene on the second chromosome. But men lack this protection, since they have one X and one Y chromosome. Each male child of a mother who is a carrier of hemophilia has a 50 percent risk of inheriting the faulty gene and the disorder. Each female child has a 50 percent chance of being a carrier like her mother.

mation for making the factor VIII protein; the introns are spliced out in the process.

Like the gene that codes for it, the factor VIII protein is remarkable for its size. It is organized into a complex series of functional units, or domains. By now, a good deal has been learned about the relationship between the factor VIII gene and the protein, including exactly which exons code for which domains. Gitschier attempted "to connect the level of severity with the type of mutation," she explains, "to see if there was any logic to it."

Two basic types of mutations of the factor VIII gene have been identified: deletions and point mutations. Deletions, in which part or all of the gene is simply missing, generally result in severe hemophilia because they prevent the gene from making a functional protein.

Point mutations, in which a single base substitution results in hemophilia, are more subtle and, to Gitschier, more interesting. Though the locations of these mutations vary, the switch is very often the same: a thymine (T) replaces a cytosine (C). Studies show that point mutations of this sort occur at certain "hot spots" up to 20 times more frequently than at random locations.

Some point mutations, known as nonsense mutations, interrupt the protein-coding process, resulting in a truncated factor VIII protein and severe hemophilia. Other point mutations, called missense mutations, cause a change in a single amino acid in the factor VIII protein—the protein is normal except for that one amino acid.

Gitschier has identified a variety of missense mutations, including one that causes severe hemophilia because the amino acid substitution occurs at precisely the site where the protein should be cleaved to become activated. "These patients have absolutely normal levels of the protein, but it can't be activated," she explains. "The protein is made, but it's just not doing anything."

Her laboratory also found two genes of unknown function buried in one of the large introns of the factor VIII gene—the noncoding regions that are normally spliced out when protein is made. "Mutations in these two genes could affect the splicing process, and thus account for some of the cases of severe hemophilia in which no other mutations have been found," says Gitschier.

HOW QUEEN VICTORIA SPREAD HEMOPHILIA AMONG EUROPE'S ROYAL FAMILIES

Queen Victoria of England, the most famous carrier of hemophilia, once wrote in her diary, "Our poor family seems persecuted by this awful disease, the worst I know." She had nine children. One son, Leopold, died of hemophilia after a minor blow to the head. Two daughters were carriers of the disease, spreading it to royal houses of Europe, and several of her grandchildren died of hemophilia in childhood.

This picture shows Queen Victoria and her family, including four children and six of her 34 grandchildren, on the occasion of a family wedding in 1894. Queen Victoria is seated in the foreground. Her granddaughter Princess Alexandra, a carrier of the disease, stands behind her (in a long fur boa) on the left side of the picture, next to her goateed future husband, soon to become the ill-fated Tsar Nicholas II of Russia. The Queen's youngest daughter, Beatrice, also a carrier of hemophilia, stands directly behind the pair. On the right side of the picture, another granddaughter, Princess Irene, wears the same kind of boa as her sister Alexandra; she, too, was a carrier. Queen Victoria's grandson Kaiser Wilhelm II of Germany, who is seated at the left foreground, and other members of her family who are shown here did not inherit the mutation.

One of her current projects is the creation of a hemophilic mouse. A line of such mice would be a major aid to hemophilia research across the board and might lead to ways of treating the disorder by gene therapy, which would mean inserting a normal gene into the patient's DNA. In the planning stages is a new project concerning the structure of the factor VIII protein, which should give clues to its precise function in the clotting cascade. Besides looking for amino acid substitutions that disrupt the protein's function, a laborious process, Gitschier wants to identify the changes that do not cause disruptions.

To do this, she plans to run computer comparisons of the factor VIII proteins of different

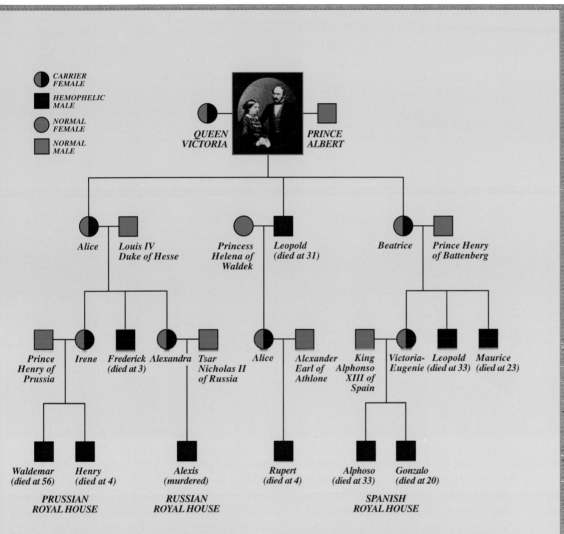

CARRIER
FEMALE

HEMOPHELIC
MALE

NORMAL
FEMALE

NORMAL
MALE

QUEEN
VICTORIA

PRINCE
ALBERT

Alice   Louis IV
Duke of Hesse

Princess   Leopold
Helena of   (died at 31)
Waldek

Beatrice   Prince Henry
of Battenberg

Prince   Irene   Frederick   Alexandra   Tsar   Alice   Alexander   King   Victoria-   Leopold   Maurice
Henry of   (died at 3)   Nicholas II   Earl of   Alphonso   Eugenie   (died at 33)   (died at 23)
Prussia   of Russia   Athlone   XIII of
                        Spain

Waldemar   Henry   Alexis   Rupert   Alphoso   Gonzalo
(died at 56)   (died at 4)   (murdered)   (died at 4)   (died at 33)   (died at 20)

PRUSSIAN   RUSSIAN   SPANISH
ROYAL HOUSE   ROYAL HOUSE   ROYAL HOUSE

This family tree shows only those descendants of Queen Victoria who inherited her faulty gene. Her daughter Alice, for instance, had seven children, three of whom are listed here. One child, Frederick, died at three after a fall. Princess Irene produced two hemophilic sons and one who escaped the mutation. Princess Alexandra, who brought the mutant gene into the Romanov line, bore a son, Tsarevich Alexis, who had hemophilia. When conventional doctors could not cure him, she turned to the fanatic monk Rasputin, whose influence on the Tsar's family hastened their overthrow and execution in 1918.

The present British royal family escaped the mutant gene, and in all the other royal houses the gene seems to have disappeared.

species. These comparisons should reveal variations in the amino acid sequence of the protein that nonetheless permit it to work properly. By doing enough such cross-species comparisons, as well as comparisons of factor VIII with two other plasma proteins that share some of the same amino acid sequences, Gitschier hopes to establish which parts of the protein are not involved in the mutations that lead to hemophilia. By logical extension, she can then determine which sections of the protein are critical to its activity.

Meanwhile, the cloning of the factor VIII gene has already made it easier to detect carriers of hemophilia and has enabled doctors to diagnose the disorder prenatally. It has also led

to the development of a new treatment: recombinant factor VIII, a genetically engineered form of the protein, produced in bacteria. Human clinical trials of this protein have been successfully completed and the FDA has approved it for use. The obvious advantage of the recombinant product is that it does away with the need for pooling blood from many donors and thus eliminates all risk of infection with AIDS or other viral diseases (some of which may not yet be recognized). On the downside, the cost per unit of recombinant factor VIII—at least initially—will be substantially higher than that of purified factor VIII made from human blood, which is also safe now that it is being treated with heat, detergents, and monoclonal antibodies to kill the virus.

The real excitement, of course, concerns the potential for a cure. Fred Rickles, the National Hemophilia Foundation's Vice President for Medical and Scientific Affairs, explains why hemophilia is such a good candidate for gene therapy. "The gene is known, and the levels of

factor VIII and IX do not have to be tightly regulated," he says. "There's no need to have the factors produced in a specific tissue, since they circulate widely in the bloodstream. If you get just a partial correction, you can go from a severe form to a mild form, and that's a vast improvement for the patient. Even a little success goes a long way."

Although the problems associated with gene therapy in general are enormous, the mere possibility has given new hope to people like Bob Massie. "Somewhere in the back of my mind, especially in the last four or five years, I've pretty much operated with the notion that my life span is constrained," he says. "But this morning I thought to myself, what if this HIV thing doesn't go anywhere? It's been 10 years and my T cells are normal. What if through some fluke they stay normal, and then someone finds a cure for hemophilia, and I'm still around when I'm 85?" Massie laughs happily at this idea. "I'm going to be one surprised guy." ●

Seven Years in Wheelchairs and Braces, Versus One Day on Crutches

How the treatment of hemophilia changed between 1962 and today can be seen from the experiences of two children who had similar joint bleeds then and now.

Bob Massie was five years old in 1962 when he developed a bleed in his left knee. It wasn't his first such problem, but it proved to be the worst by far in his young life. Thirty transfusions of fresh-frozen plasma were required over a period of three months to stabilize the bleed. By that time, the swollen joint was locked in a jackknife position. It was seven years before Massie was able to walk. His mother wrote in *Journey*: "That bleeding... dominated his childhood, caused him a thousand nights of pain, locked him for years in casts and braces and, eventually, destroyed his left knee."

Thirty years later, six-year-old Zack Dansker of South Pasadena, California, a severe hemophiliac, developed a bleed in his left hip. At the first sign of trouble, his mother, Stephanie, went to the refrigerator and got out freeze-dried factor VIII concentrate and sterile water. She quickly mixed up a shot and administered it to Zack herself. The bleeding stopped, and that was the end of the incident. To avoid further problems, Zack was supposed to use his crutches for a day or so after the bleed, but he had other ideas, and his mother had to keep reminding him. "Zack," she said as the busy boy roamed around the apartment, limping slightly, "if you want to walk tomorrow, please take good care of yourself today. Use your crutches. I know you're tired of them, but use them, please."

Bob Massie's survival kit: boxes containing a bottle of factor VIII concentrate and some sterile water. Each box costs $660. He uses two or three per shot, at least three times a week.

The cost of treatment for hemophilia today, however, is astronomical. Hemophiliacs generally pay between $60,000 and $100,000 a year for the factor VIII concentrate they need to survive. The cost may be somewhat lower for small children—Zack uses about $35,000 worth of factor VIII per year—but it can also be considerably higher for adults with more serious problems. Bob Massie, for instance, needs three to four treatments a week, and more if he's bleeding. Each of his treatments consists of at least 2,200 units of factor VIII at the cost of 60 cents per unit, or $1,320 a shot. That works out to more than $200,000 a year. The new, genetically engineered factor VIII—"not necessarily an improvement in my case," says Massie—would cost him more than $350,000 a year.

Before the blood supply was contaminated by HIV, the cost per unit was about 12 cents. The new technologies that were needed to establish and maintain a safe supply of factor VIII drove the price so high that it is jeopardizing the lives of patients. While some hemophiliacs, like Massie, are covered by private health insurance, many are not. Those who must depend on Medicaid and other state-supported programs are often forced to limit family income to maintain their eligibility, and it's not unheard of for husbands and wives to divorce in order to meet the requirements for this kind of aid.   –D. N.

magine a vampire bat, dining on the lifeblood of an unsuspecting cow. To break up any clots that might form at the wound, the bloodthirsty creature produces a substance in its saliva called bat-PA (plasminogen activator). This chemical caught the attention of pharmaceutical companies looking for new ways to attack the number one cause of death in this country: blood clots.

Clots that stop the flow of blood to the heart or brain kill about half a million Americans a year through heart attacks and strokes. Another 150,000 people die of clots lodged in their lungs. Most of these lethal clots form when fatty lumps, or plaques, that line the arteries crack and break open, producing a jagged surface. Immediately a clot caps off that rough patch and, like a clog in a drainpipe, blocks off the blood supply to the tissues downstream. Deprived of nutrients and oxygen, these tissues literally suffocate. Until recently, doctors could do little to help.

In the last decade, researchers have made major advances in treating these emergencies, mostly thanks to a handful of drugs that can be used in the crucial hours after a heart attack or stroke begins. Two of these drugs, streptokinase

A blood clot in a human coronary artery blocks the flow of blood from the heart.

and tPA (tissue plasminogen activator, a genetically engineered version of the natural substance), work by turning on the enzyme plasminogen, part of the body's own machinery for

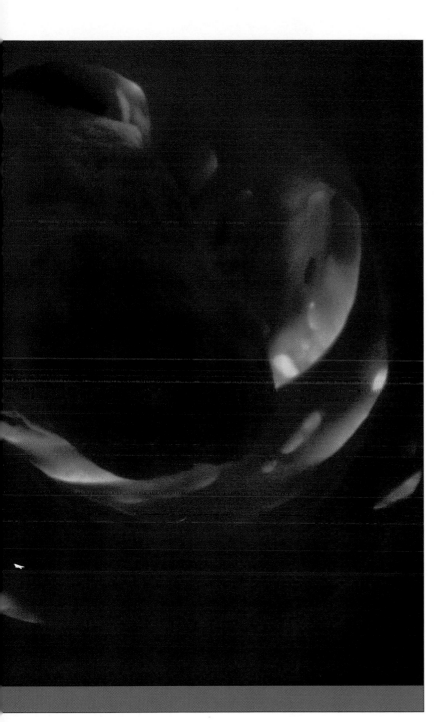

breaking down clots. These clot-busters have become a mainstay of clinical medicine.

Unfortunately, they are not universally effective. "You give tPA to some patients and—zip!—their blood vessels open and they're fine," says HHMI investigator David Ginsburg of the University of Michigan Medical School. "But with others, it doesn't work so well."

Researchers are busily seeking new medicines for the immediate treatment of acute heart attacks and strokes, as well as for the more important, long-term goal of clot prevention. Some of the newer drugs are coming from the animal kingdom. Bat-PA, for instance, is showing promise as a possible replacement for tPA because it

By Shawna Vogel

BLOOD CLOTS

OUR NUMBER-ONE KILLER

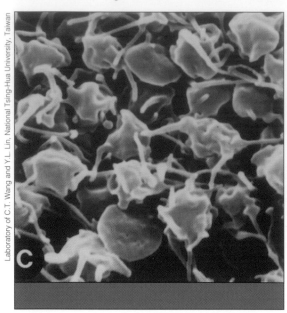

Laboratory of C.T. Wang and Y.L. Lin, National Tsing-Hua University, Taiwan

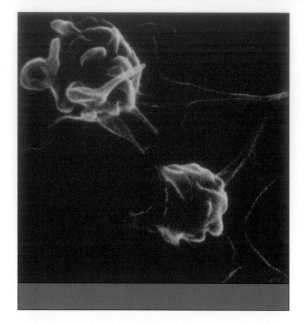

1 *Flat, smooth human platelets undergo a dramatic change in shape when activated by a chemical signal: they thicken and develop long, spiny filaments. This scanning electron micrograph shows platelets that have just become activated, mixed with others that are still quiescent.*

2 *Two activated platelets reach out to each other with newly made filaments that can stick together. This response to an injury in a blood vessel wall is their first step toward making a plug to stop the loss of blood.*

activates plasminogen right at the clot, rather than throughout the bloodstream. Other new approaches to dissolving clots are emerging from the extensive research now being conducted around the country on every aspect of blood coagulation.

"For a long time, people really didn't understand how blood clotted," says Evan Sadler, of the HHMI unit at the Washington University School of Medicine in St. Louis. Even after scientists learned that blood did not clot simply because of exposure to air, or because it had stopped moving, "there was the impression that maybe only a single enzyme was involved," Sadler says. "It was only in the 1950s, when clotting factors V, VII, and X were discovered and the inherited bleeding disorders hemophilia A and B were shown to be due to deficiencies of different proteins, that people began to realize things were more complicated."

In 1952, a patient named John Hageman turned up. He was a railroad brakeman who came to Cleveland seeking surgery for his peptic ulcer, and though he had no symptoms of a bleeding tendency, his doctors soon realized that his blood clotted abnormally slowly. They traced this fault to the lack of a substance now known as Hageman factor (or factor XII), which causes blood to clot when it comes into contact with a glass test tube, porcelain dish, or other foreign surface. Hageman factor opened the door to understanding the complicated chain of events in blood clotting. Oscar Ratnoff at Case Western Reserve University School of Medicine in Cleveland and Earl Davie, now at the University of Washington in Seattle, showed that foreign surfaces normally induce a change in this factor. Once changed, Hageman factor causes another protein to change. "And pretty soon," says Ratnoff, "we could see a pattern there."

The pattern was as elegant as a waterfall—a cascade. In 1964, Davie and Ratnoff and, independently, R. G. MacFarlane in England put their findings into print. The clotting cascade depends upon a series of proteins, or clotting factors, which move through the bloodstream in a dormant state. Once the first factor in the cascade is activated, it turns on some of the next factor. This turns on more of the next (each factor is present in the bloodstream in a larger amount than the one before, so only tiny amounts of the early factors are needed to start a chain reaction). Finally a factor called

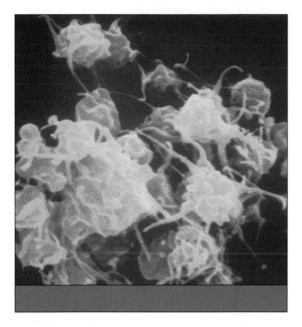

3 *More platelets join in to form a temporary plug at the site of injury. This plug, however, is very fragile—a stopgap until the clotting cascade produces stronger building materials for the clot.*

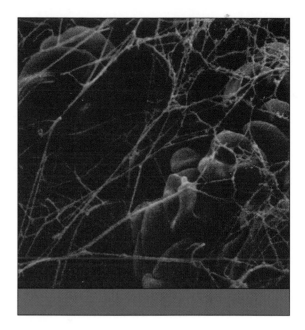

4 *Strong strands of fibrin appear and reinforce the platelet plug, forming a sturdy clot that traps red blood cells (as seen in the background) and stops the loss of blood.*

thrombin activates fibrinogen, the last factor in the cascade, converting it into strands of tough, insoluble fibrin. The fibrin strands then assemble into a mesh that forms the framework of the blood clot.

This clotting cascade provides a fail-safe mechanism—a series of possible cut-off points—ensuring that clots remain localized to the site of injury. It also creates a chain reaction that explains the remarkable ability of blood to flow freely and yet, almost instantly, generate an outburst of clotting wherever needed.

At the time the cascade was described, most of the clotting factors were known only by their effects. It took another decade for researchers to isolate and characterize these proteins and work out how substances in injured tissue—not just foreign surfaces—could initiate clotting along the same lines. Additional clotting factors in the blood (and links between them) are still being discovered.

Blood clots actually begin when some tiny, plate-shaped cells—appropriately called platelets—become activated. These platelets float by the millions through the bloodstream. Whenever the lining of a blood vessel is injured, the nearest platelets suddenly undergo a dra-

matic change in shape: their smooth exterior becomes studded with long, spiny filaments that they use to stick to the site. There they begin to release chemical signals. Attracted by these signals, more platelets then aggregate at the site of injury to form a temporary plug, which helps stanch the flow of blood at the wound. Meanwhile, the clotting cascade begins, leading to the formation of fibrin strands that combine with the platelet plug. As they trap red blood cells in their web, these fibrin strands turn the plug into a solid clot.

Such clots save lives by preventing blood from leaking out, but they can become deadly when fatty material, especially cholesterol, has accumulated within the walls of arteries and narrowed them. Recently Herbert Stary, a pathologist at Louisiana State University in New Orleans, examined the blood vessels of 1,286 infants, teens, and adults who had died of various causes. He found that, from the moment of birth, the walls of blood vessels thicken gradually as fatty deposits build up, creating bumps. At first these bumps do not obstruct the flow of blood. But after the age of 30, some of the bumps grow and become very fragile—particularly those with a high concen-

Bloodletting was believed to cure many diseases from the time of the ancient Egyptians until about 1850. Here a woman in 17th century Belgium applies leeches to draw her blood.

tration of cholesterol and certain other fats. Known as atherosclerotic plaques, these fatty bumps may crack repeatedly, accumulate some blood clots, and heal, growing bigger with each round until they cut off a large part, or all, of the blood vessel.

Blockages of this sort make it hard for tissues to get oxygen. They are the root of the heart ailment known as angina pectoris, in which limited blood flow to the heart, particularly during exercise, leads to a tightening or burning sensation in the chest. Typically this pain goes away with rest or if a tablet of nitroglycerine is taken to widen the vessels, allowing more blood to flow, and angina is not considered an acute heart attack. "There are plenty of people walking around with partial coronary blockages who are not having heart attacks," says Richard Nesto, director of clinical research in cardiology at Deaconess Hospital in Boston, Massachusetts. "But a small number of blockages can become almost volcanic. They blow their tops off." Large clots form on these jagged surfaces—large enough to completely stop the flow of blood. These are the clots that precipitate most heart attacks and strokes.

The clot-busters in use today (including streptokinase and tPA) exploit a mechanism that nature evolved to destroy and remove

Glass cups and brass suction pumps such as these were also used to draw blood, but first an incision had to be made with a sharp instrument.

unwanted fibrin after a clot has done its job: they activate plasminogen, which chews up the fibrin strands. Every clot contains substances

that will break it down in this way. But nature has also developed several inhibitors that keep such substances in check to avoid too much bleeding. One of the most critical elements in this balancing act is plasminogen-activator inhibitor 1, or PAI-1, a protein now being studied by David Ginsburg.

Originally Ginsburg thought that anyone who lacked this inhibitor would suffer very serious bleeding and might die. "But recently a patient was found at Indiana University School of Medicine who's deficient in PAI-1," he says. "It's a young Amish girl. She does bleed, but she's not all that sick." This gave him the idea that if PAI-1 were reduced just a little, it would probably cause few side effects and would allow the body's natural tPA to do its job more vigorously. "That might be a much safer way of turning the system down just a bit," says Ginsburg. He and other researchers are now hoping to develop a drug that would help to break down clots by damping the activity of PAI-1.

In any effective strategy against clots, researchers must also take into account the aftermath of clot busting. Breaking up a clot frequently leaves the body vulnerable to new clots, which form on exactly the same site as the first. It is essential to treat not only the clots along the vessel walls, but the blood itself.

In their search for drugs that can keep blood flowing freely past atherosclerotic plaques in the blood vessels, researchers have recently turned to the lowly leech. Using live leeches to draw blood from patients was a favorite remedy for almost any disease in the 19th century. After the leeches had drunk their fill of blood and dropped away from the patients' skin, the wounds went on bleeding for about 10 hours. Nobody understood why until 1884, when scientists discovered that these creatures' salivary glands contain a powerful anticoagulant.

Hirudin, as it is called, was much too difficult to extract to become a viable drug in those days. Today it can be produced in bulk by recombinant DNA techniques and is seen as a possible replacement for heparin, the standard anticoagulant drug. Heparin sometimes has undesirable side effects, such as severe bleeding, because different patients respond to the same dose in different ways. Hirudin may avoid such difficulties since it blocks thrombin directly rather than acting through an intermediate protein, as heparin does. Clinical trials of a hirudin-based drug, hirulog, are under way.

Other researchers are examining the body's own anticoagulants for clues to better drugs. At the University of Oklahoma Health Sciences Center and Oklahoma Medical Research Foundation in Oklahoma City, HHMI investigator Charles Esmon has been analyzing how a chain of reactions called the protein C anticoagulant pathway keeps unwanted clots from forming. In one of the odd economies of nature, the body uses thrombin—which causes clotting when it is activated in the clotting cascade—to prevent clotting in this pathway. This happens when the thrombin molecule undergoes a change that enables it to activate another enzyme, protein C. Protein C then chews up a couple of the proteins in the clotting cascade and shuts down clotting.

Esmon wanted to find out what triggers the change in thrombin. He started out with the idea that the triggering substance might be located on the cells that line the blood vessels. "Being that this is the Midwest, and being that there were a lot of slaughterhouses around," he says, he decided to see what would happen if he ran both thrombin and protein C through the rich set of blood vessels in pigs'

ears. "Needless to say, that idea was not greeted with a whole lot of enthusiasm from my colleagues," he notes. "But anyway we decided we'd line up pigs' ears. We had this rack of aluminum foil with these pigs' ears sort of hanging there and a little needle going into one side of the ears and the effluent dropping out of the other side."

When the results came back, he saw a tiny

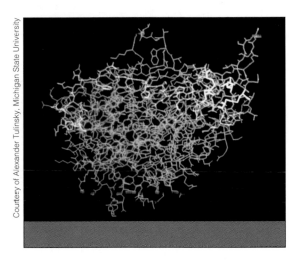

Hirudin (pink), an anticoagulant made by leeches, inactivates thrombin in this 3-D model of the two molecules bound together.

effect—a 2.5-second longer clotting time in blood that had passed through the blood vessels in the pigs' ears. The change could have been meaningless, but Esmon believed that the experiment had worked, that something in the cells of the vessel walls was indeed converting the thrombin into an anticoagulant, and he only needed to optimize the results. "I proceeded to ask everybody in the lab if they would pick this up and do it," says Esmon, who had his hands full with another project. "No way."

So the project sat for a year until the breakdown of a piece of machinery gave Esmon some free time. He used it to work with Whyte Owen, a friend from his days as a graduate student, who had set up a system for studying circulating blood using rabbits' hearts instead of pigs' ears. This time the results were astounding. They saw the anticlotting activity boosted 20,000 times. A few weeks later they identified the substance in the vessel walls

Bowls of this sort were designed to catch blood drawn from a vein.

To determine the best time for bleeding, barber-surgeons consulted astrological signs—for example, the markings on this lunar dial, made in Germany in 1604.

that transformed the thrombin into an anticoagulant and named it thrombomodulin.

In the years since, Esmon's group in Oklahoma, Evan Sadler's group in St. Louis, and others have been studying exactly how thrombin and thrombomodulin interact. If one could create a small fragment of thrombomodulin that binds to thrombin, "it would be a candidate for a drug that discourages clotting and activates protein C," Sadler says. Some companies are indeed trying to make such a drug. Esmon adds that the activated form of protein C might also be used as an anticoagulant. In baboons, it has been shown to prevent blood clots from forming without causing any excess bleeding.

Clot-busters and anticoagulant drugs are generally used after a person has suffered a heart attack or stroke. Scientists realize they cannot rely entirely on this approach. James Muller, codirector of the Deaconess Hospital's Institute for Prevention of Cardiovascular Disease, throws out some disturbing statistics: "I calculated the number of lives saved by dissolving clots per year in this country, and it's 3,000," he says. "And the number of deaths is 500,000. So all of this flurry of activity to study drugs for treatment is severely limited in what it's going to produce."

Obviously, prevention is the key. So far, though, the arsenal of preventive drugs has been extremely limited. "As far as the average Joe Doe out there walking around, aspirin is about the best we've got," says Richard Nesto. Indeed, the discovery of just how aspirin works was a giant step forward for preventive medicine.

What aspirin does is prevent platelets from sticking together by blocking a key enzyme in them. Even a low dose of aspirin has a profound effect on platelets, as Philip Majerus and his colleagues at the Washington University School of Medicine showed in 1975. "When you take aspirin for a hangover or a headache or whatever, the effect lasts only a few hours because most cells can make new enzyme," Majerus explains. "But platelets can't. So aspirin's effect on platelets is permanent; it lasts for the life of the platelet—basically two weeks."

To Majerus and other researchers, this meant they could use small doses of aspirin to wipe out platelet aggregation and prevent heart attacks without worrying about side effects in other cells. In 1989, a study of more than 22,000 physicians proved that taking a regular aspirin on alternate days reduced the risk of heart attacks by 44 percent. Since then a baby aspirin a day has become a habit for millions of people over the age of 50.

However, aspirin doesn't prevent all heart attacks. "Aspirin is a relatively limited antiplatelet agent," Muller says. "There are other ways platelets can be activated that are not blocked by aspirin."

This is why some researchers are now seek-

Fat cells such as these yellow giants, magnified 750 times, accumulate inside artery walls of people on high-fat diets and lead to atherosclerosis.

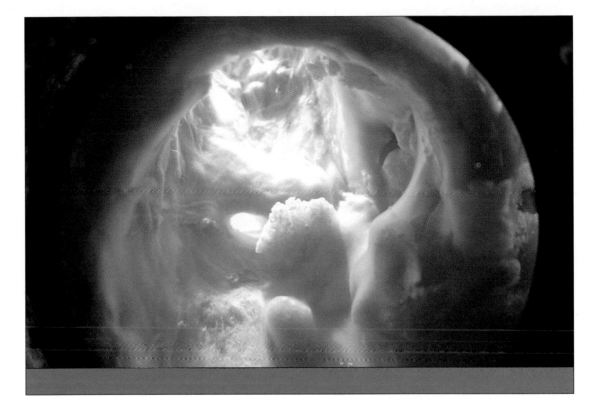

Yellow atherosclerotic plaques lining a human blood vessel leave little space for blood to flow.

ing drugs that would target other parts of the platelets. One of the new drugs—a monoclonal antibody against a particular receptor on the surface of platelets—is far more potent than aspirin in preventing lethal clots, but if it were given to a patient more than once it might trigger a dangerous immune response. So this drug is intended only for acute treatment. Seeing its success, several pharmaceutical companies are now using the same concept in an attempt to make drugs that might be used repeatedly—perhaps even on a long-term basis. Once again, Mother Nature has beat them to the punch: many of these drugs are based on snake venoms, which target the same receptor on platelets.

Other scientists are studying von Willebrand factor, a clotting protein that normally acts as a sort of glue over sites of injury on blood vessel walls. Von Willebrand factor anchors roving platelets and allows them to form an initial layer over the injured area. Other platelets then clump over this layer. David Ginsburg and Evan Sadler have focused on segments of von Willebrand factor that bind not only to the ves-

sel wall but also to a receptor on the surface of platelets, glycoprotein Ib. Zaverio Ruggeri of The Scripps Research Institute in La Jolla, California, has synthesized the fragment that binds to glycoprotein Ib and has used it as a drug to block this receptor on legions of platelets.

Taking prevention one step further, many researchers are analyzing how cholesterol and other lipids build up into unstable, fatty plaques that are vulnerable to breakage. "Theoretically, you might be able to prevent these dangerous blockages from forming in the first place," says Nesto.

In the end, people at risk of heart attacks or strokes may best be served by a mixture of drugs designed specifically for them. Some of these drugs would prevent the buildup of atherosclerosis. Others would break up blood clots as soon as they occur. Aspirin and other antiplatelet agents could be combined with drugs to beef up the body's natural anticoagulant pathways. In each case the goal would be to adjust the body's delicate balance between too much bleeding and too much clotting. ●

The Ultimate Therapy for Blood Disorders: NEW G

In a radical strategy against a hereditary disorder that clogged her arteries, a 29-year-old Canadian woman received gene therapy in Michigan in 1992: a normal gene was transplanted into her cells to serve a need her own faulty gene couldn't fill.

The Canadian patient suffered from familial hypercholesterolemia, a rare disease in which people accumulate catastrophically large amounts of cholesterol in their blood and develop heart disease at an early age. She had already undergone coronary bypass surgery, but the artery blockage had reappeared. Her brother had died of the same disease in his twenties. Both suffered from a missing or defective gene for LDL receptors, minute structures on the surface of liver cells that scoop up low-density lipoprotein (LDL), the "bad" cholesterol, from the blood and break it down.

Many doctors worry when a patient has a total cholesterol reading of more than 200 milligrams per deciliter of blood, but people with this disorder may have 500 to 1,000 milligrams of cholesterol. Some children with extreme forms of the disorder experience heart attacks before the age of six.

James Wilson, an HHMI investigator at the University of Michigan Medical School in Ann Arbor, led the effort to insert copies of a normal gene for the LDL receptor into the patient. It was a very difficult and ambitious procedure. First he removed 15 percent of the woman's liver. This supplied him with a "harvest" of about 6 billion liver cells. The cells were grown on 800 culture plates and infected with a genetically engineered, harmless virus that contained the desired gene. (The virus was designed to vanish after delivering its cargo of LDL receptor genes into the patient's cells.) About 20 percent of the growing liver cells actually took up the foreign gene.

Next, the gene-modified cells were fed back into the woman's body through a catheter in a vein that led directly to her liver, where the researchers hoped the cells would come to rest and divide.

Everything worked without a hitch. The patient and her husband actually went dancing about two weeks after the therapy. Several months later, Wilson was even more encouraged because a small liver biopsy showed that the foreign gene was functioning in some of the re-infused liver cells. Furthermore, the LDL in the patient's blood had dropped to a level about 15 to 30 percent below the previous high readings. It was certainly no cure, but a dramatic improvement.

Wilson said his patient's improvement was great enough to give hope that drug treatment would become useful in her case, reducing the levels even

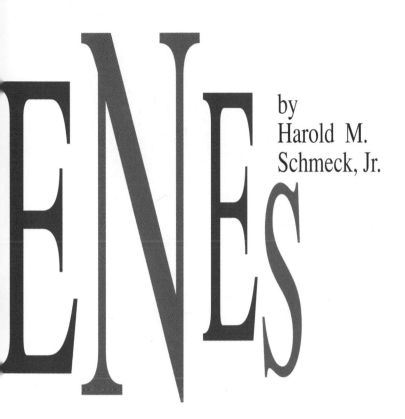

GENES

by
Harold M.
Schmeck, Jr.

more. And there was reason to expect that the patient's liver would restore itself, repopulating a substantial portion of its tissue with cells that have the needed gene. The human liver has great natural powers of regeneration.

The Canadian woman owes her treatment, and possibly her life, to two little girls in the United States who have an entirely different disease.

Human gene therapy began on September 14, 1990, when a frail, solemn-faced little girl of four received an infusion of her own white blood cells, which had been treated to make them carry a gene that she lacked. It was a simple procedure with a deceptive lack of drama, but it launched a new era.

The child's cells had been removed from her circulating blood and given copies of a gene for making an enzyme called adenosine deaminase, or ADA. Because her own ADA genes were either missing or defective, her T cells—white blood cells that play a critical role in the immune system—did not divide and proliferate at the normal rate and were in short supply. This left her virtually defenseless against infection. Even mild infections that a normal person would hardly notice were threats to her life. As one of her doctors, W. French Anderson of the National Institutes of Health (NIH), described her, "she was a very, very sick little girl."

A second child with the same disorder received similar gene therapy at NIH shortly afterwards. This disorder—severe combined immunodeficiency, or SCID—was made famous by an earlier patient: the "bubble boy" in Texas, who lived his entire short life inside a plastic tent that isolated him totally from the world, including his parents.

In the fall of 1992, Anderson addressed an audience of medical scientists convened by NIH in the very building where the four-year-old girl had been treated. He showed an anniversary picture of the gene therapy patients, their families, and the team of doctors, nurses, and other specialists involved. In the front row, surrounded by many adults with smiling faces, were the two girls, holding framed certificates honoring their places in medical history.

"The most remarkable thing," said Anderson, "is that the kids are there." A few years ago, no one would have dared expose them even to the modest risks of infection from sitting for a group picture.

Both children's immune defenses are now substantially reconstituted, according to Michael Blaese, an NIH scientist who, together with Anderson, led the team that pioneered the girls' treatment. The girls go to public schools, participate fully in school and social activities, and seem to have no more than the normal childhood quota of colds and other infections. There have been no detectable ill effects from the gene transfer or the presence of the foreign genes in their bodies.

The deficiency that affected the two girls is very rare. Doctors have joked that there are probably more scientists studying the disease than patients who suffer from it. But it was ideally suited for the first attempts at gene therapy. The ADA gene itself was well known and, for years, has been grown in laboratories. Though the enzyme it makes is vital to health, people can get along well with substantially less of it than the normal amount; furthermore, an extra ration seems to do no harm. So the scientists were freed from having to make the gene produce a precisely regulated amount of enzyme, at precisely the right time.

A drug called PEG-ADA had become available to replace some of the missing ADA in these patients, but its beneficial effects seemed to plateau after a year for both girls; it produced only modest improvement. The researchers decided to continue the drug treatment and add their own.

efore the gene therapy could begin, however, the NIH team had to develop ways to insert the ADA gene into the children's cells. They used specially rebuilt viruses, or "vectors," as their means of delivery. In nature the viruses they worked with—retroviruses— can be extraordinarily dangerous; such viruses have a great talent for invading cells and installing their own genes within an infected cell's genetic machinery. The infected cell is then forced to produce and assemble many new particles of the same virus so that a new crop of viruses can go forth to invade other cells. But in designing their vectors, the researchers stripped the viruses of virtually everything except the capacity for invading cells and depositing a cargo of genetic material— which they engineered to include a human ADA gene. These vectors cannot force the invaded cell to build any new virus particles. No viruses ever emerge.

Originally the NIH team hoped to deliver the ADA genes into the patients' bone marrow cells to ensure that the total spectrum of white blood cells made in the marrow would carry the ADA gene and make the enzyme. That would produce a permanent cure and a total reconstitution of

Preparing for gene therapy, James Wilson transfers cells taken from a patient's liver into petri dishes. Two days later he will "infect" these cells with a genetically engineered virus containing the gene for a normal LDL receptor the patient lacks. The gene-modified cells will then be returned to the patient's liver, where they will make the normal receptor.

the patient's immune defenses—as if the patient had received a bone marrow transplant from a healthy, matched donor. Thousands of such transplants are done every year to treat patients with some 40 deadly diseases, including cancers of the blood-forming system and the most severe forms of thalassemia and anemia, but bone marrow transplants require a very close match and are not always successful. Since transplanted bone marrow cells contain the donors' genes, in a sense "we've been doing gene transfer all along," said E. Donnall Thomas of the Fred Hutchinson Cancer Research Center in Seattle, Washington, who won a Nobel Prize in 1990 for his pioneering research on bone marrow transplantation. "We just wrap the genes in a cell membrane."

To their disappointment, however, the scientists who wanted to cure ADA deficiency through gene therapy were unable to make the patients' bone-marrow cells produce active ADA genes; if the transplanted genes functioned at all in these cells, they soon stopped. The scientists then decided to use the patients' circulating white blood cells as their target, rather than bone marrow cells. They would harvest T cells from the patients' blood, produce large numbers of these cells in the laboratory, insert ADA genes into them, and then infuse the cells back into the patients. It was assumed that this might have to be done periodically, perhaps every few months.

The strategy succeeded. But the circulating blood contains several different kinds of T cells and a host of different clones of each variety, with somewhat different defensive capabilities. Only by putting the foreign gene into the stem cells from which all these different clones arise could the doctors be sure of giving patients a complete immune defense system.

Blood and blood diseases have played a large role in the strategies and goals of gene therapy pioneers. French Anderson, who began thinking of gene therapy long before its name was even coined, first hoped to attack thalassemia, one of the most widespread of serious blood disorders. But he put that idea on long-term hold because he concluded that transplantation of corrective genes would not be effective against that disease without extremely precise control of the timing and amount of the gene product to be released. This degree of control is still beyond the capabilities of medical science.

The current plans in this field are focused on what is called somatic cell gene therapy, in which genes are inserted into specific body cells—somatic cells—to attack a disease that already exists. The disease may have done some damage before it is treated, and if it is a genetic disease the patient's potential offspring remain at risk.

In theory, there is another kind of gene therapy that might cure a disease before it starts and even eliminate the disorder from the person's genetic endowment so that future generations would be free of it. This utopian idea is called germline gene therapy. Certain genes would be inserted into the early embryo and would become part of its natural genetic endowment. Gene transfer of this sort is done routinely in the laboratory to produce "transgenic" ani-

mals. But the idea of using it in humans raises serious questions. What if a germline transplant accidentally produced a bad effect—susceptibility to early cancer, for example, or mental deterioration?

Even without such unexpected disasters, there are other difficulties. For example, germline therapy might lend itself to attempts not only to treat or prevent disease, but to "improve" the recipient. Parents might try for a child endowed with greater intelligence, athletic skill, or abilities of any imaginable kind. Today, no one has any idea what genes could be transplanted to achieve these wonders, but that doesn't guarantee that no one would try. The issue of enhancement, as opposed to disease treatment, also raises the troubling question of what really constitutes "improvement."

Presumably, germline gene therapy would require *in-vitro* fertilization and the insertion of the foreign gene into the fertilized egg in the test tube before reimplantation into the mother. At least, that's the way transgenic animals are produced. But in creating animals with foreign genes, large numbers of fertilized egg cells have to be treated for every animal that is actually born alive carrying the added gene in its germ cells. These numbers alone would most likely rule out the method for use in humans.

Practitioners of somatic cell gene therapy, who have trouble enough making their system work, view the problems presented by germline therapy as too vast to even consider seriously at present. French Anderson, among others, emphasizes that important technical improvements are still needed in the real world of somatic cell gene therapy. Notably, he sees the need for a simpler and better way of delivering the genes to the patients. His goal is what he calls an injectable vector. This would be an off-the-shelf gene preparation that would be injected directly into the patient, would migrate swiftly to the appropriate tissues, and would go to work there supplying whatever gene product was needed.

When and if an injectable, targeted vector becomes a reality, experts say, it might make gene therapy available not just to a few thousand patients with certain special diseases, but to millions with many kinds of illness—and possibly even for prevention. ●

The search for the "mother of all blood cells" continues, at least for human blood, and remains a subject of controversy. Meanwhile, the idea of using the body's own stem cells and growth factors to correct abnormalities is gaining power as a result of recent discoveries.

In the spring of 1999, scientists at a Maryland biotech company reported that they had isolated the "mother of all bone and connective tissues," mesenchymal stem cells, and that they had also discovered ways to make these pluripotent cells produce specific types of tissue. If those results are confirmed, researchers might nudge the cells to produce the stroma cells in bone marrow that provide support for blood-forming cells—an important factor in the success or failure of bone-marrow transplants.

Even earlier stem cells—the very first, primordial stem cells, which give rise to every type of cell in the body—have been derived from human embryonic or fetal tissue and grown in the lab (see p. 110). These totipotent cells might be produced in unlimited quantities and are universal, which means a patient's immune system probably would not reject them. But researchers don't yet know how to direct their development to make them become any particular type of cells, such as blood cells.

Hemophilia and HIV

Bob Massie, the hero of our story on hemophilia, remains free of AIDS more than 20 years after a blood transfusion infected him with HIV. Beating all the odds, he still has a normal T cell count. His children have no sign of infection. His case is so unusual, in fact, that researchers at Massachusetts General Hospital in Boston became very interested in him, as well as in some other "long-term non-progressors" whom they studied intensively (see p. 290).

A Brave Child's Struggle

Maurice Harrell, the little boy who struggled so bravely against sickle cell disease, was still holding his own when last seen by his California doctors in 1997. However, he still needed regular blood transfusions, had a liver ailment, and suffered frequent, painful crises. "He had many problems," says Keith Quirolo, a pediatrician who treated him in Oakland. "He's a very stoic kid; he never complained about anything. He's a sweetheart!"

Searching for Better Drugs

Although butyrate, the experimental drug that Maurice took for a while, did not help him in the oral form and in the fairly high daily dose he received at the time, the drug is now proving quite effective when given intravenously once or twice a month during a four-day hospital stay. In March 1999, George Atweh of the Mount Sinai School of Medicine and his colleagues, including Susan Perrine, reported that after this "pulse" regimen, nine of 11 adult patients with sickle cell disease started to produce fetal hemoglobin at high levels. These patients also had more fetal cells in their blood and fewer medical problems than before.

Another drug, hydroxyurea, was approved by the FDA in 1998 after a national clinical trial showed that it cut down the number of painful crises and hospital stays for adults. The drug is now used extensively in sickle cell patients around the world, but "it is of potential value to only half the patients, and we cannot predict which patients will respond and which won't," says Alan Schechter of the National Institute of Diabetes and Digestive and Kidney Diseases. Schechter says he is also concerned about the drug's effects on children if used over a long time. "This makes the search for other drugs important," he says, "and the new findings with butyrate seem very hopeful."

Another encouraging piece of news is that three different mouse models of sickle cell disease have been developed. These should help researchers analyze the effects of various anti-sickling drugs, as well as the molecular mechanism of the switch to fetal hemoglobin.

A Safer Blood Supply

Finally, in a major advance in clinical medicine, blood banks are phasing in new genetic tests that are beginning to eradicate all the viral infections, such as HIV and hepatitis C, that still occasionally show up in blood transfusions. These tests should yield a blood supply of unprecedented safety.

THE RACE AGAINST

A REPORT FROM THE HOWARD HUGHES MEDICAL INSTITUTE

THE
RACE AGAINST
LETHAL
MICROBES

Learning to Outwit
the Shifty Bacteria,
Viruses, and
Parasites That Cause
Infectious Diseases

Howard Hughes Medical Institute
4000 Jones Bridge Road
Chevy Chase, Maryland 20815-6789
(301) 215-8500

LETHAL MICROBES

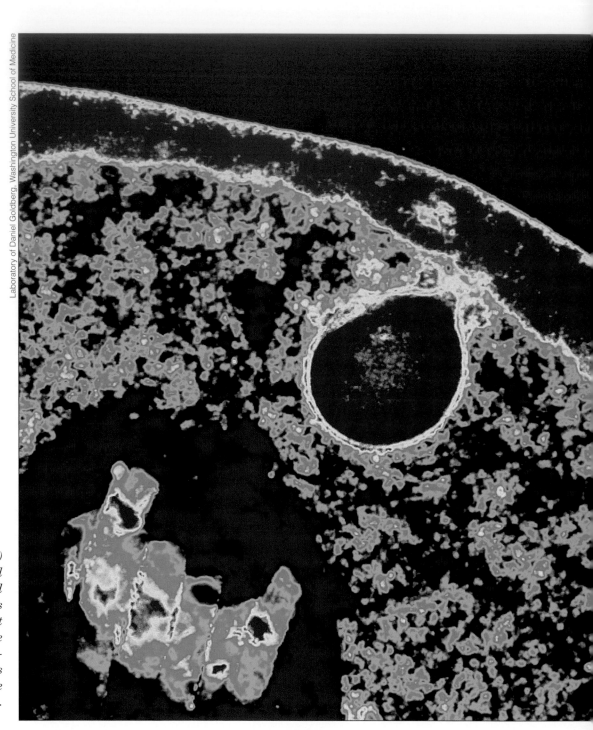

A malaria parasite (blue) that nearly fills the red blood cell it has invaded (red rim) sucks up the cell's hemoglobin into a transport vesicle (red pouch). The vesicle will dump its contents into the parasite's digestive vacuole (purple structure at lower left).

THE RACE AGAINST LETHAL MICROBES

Learning to Outwit the Shifty Bacteria, Viruses, and Parasites That Cause Infectious Diseases

INTRODUCTION
· ·

F lesh-eating bacteria, salmonella, HIV—even in highly
industrialized nations such as the United States we are
often at the mercy of microscopic organisms that can be
deadly. In the less-developed areas of the world, where
polluted water is often used for drinking and washing, microbes
clearly have the upper hand.

Infectious diseases are still the world's leading killers, wiping out
16.4 million people every year. We tend to forget that long-known
infections such as tuberculosis, measles, malaria, infant diarrhea,
and respiratory infections remain at the top of the list. While the
highly publicized Ebola virus epidemic took only 244 lives, measles
kills more than 1 million people a year and TB kills 2 million. Some
of these ancient plagues are now reemerging in the United States.

The spread of TB is closely linked to a newer threat, AIDS. While
nearly a third of the world's people have been infected at some time
with the TB bacterium, most of them have managed to subdue the
infection without ever becoming seriously ill. They suffer no harm
and cannot infect others as long as their immune systems remain
vigorous. However, should their immune systems weaken—as happens in AIDS—the dormant microbes erupt, and so does the disease.
AIDS thus brings huge increases in the number of people with active
TB, whose coughs and sneezes can infect those around them.

Before the AIDS epidemic, many people believed that infections
were no longer dangerous. The success of antibiotics after World
War II had led to the impression that all bacteria could easily be
defeated. Though many viral diseases remained unconquered,
some of the most frightening ones could be prevented by effective
vaccines. Much of the industrialized world basked in a feeling of
invulnerability, assuming that the only major health problems left
to worry about were the chronic ills of later life, such as heart
disease and cancer.

The rising death toll from AIDS punctured this comfortable
belief. At first, nearly everyone expected that a cure or a vaccine
would be developed rapidly. But instead, the disease exposed our
ignorance about the fundamental mechanisms of infection—how
microbes interact with the immune system, how they damage cells,
how they cause disease, and how they develop resistance to drugs
that originally crippled them.

Some bacteria can become lethal despite the strongest antibiotics
we can throw at them, simply by sharing a few drug-resistance
genes. This has already happened with some enterococci, bacteria
that can cause blood infections in hospitalized patients. Many
strains of enterococci have now acquired resistance to vancomycin,
the last antibiotic that was still able to fight them. So have strains
of the common bacterium *Staphylococcus aureus*, making these
organisms a very serious threat, especially in hospitals.

Furthermore, microbes don't respect national borders. A new outbreak in one part of the world can rapidly reach other regions. An infected person can travel across the globe during the incubation period, before any symptoms appear. A lethal microbe may be only a plane ride away.

The microbes that threaten us come in a stunning variety of shapes, sizes, and lifestyles. A rogues' gallery of some of these organisms appears on the large poster in the center of this report. The poster explains how such microbes enter our bodies and injure us, how we fight them, and how they fight back.

Scientists have made some great achievements in the battle against infectious organisms—even against those that cause AIDS. Until recently, infection with HIV left almost no hope of escape from an eventual slide into catastrophic disease and death. Now scientists hope that by using several new and more powerful drugs simultaneously, as is done in combination therapy for TB or cancer, they may be able to change HIV infection from a death warrant to a chronic condition. Such therapy cannot stop the AIDS epidemic—a vaccine will be needed for that—but it may offer some patients a fighting chance at survival.

The only complete victory over any infectious disease so far, however, has been the eradication of smallpox. The final pages of this report show the last man in the world to have been infected with this once-devastating virus. They also describe scientists' first steps toward identifying the complete genetic blueprints of certain bacteria and parasites.

The DNA sequences revealed in this way contain a wealth of information that will be mined in the next few years. Molecular biologists hope this work will bring new victories against our microbial enemies—and perhaps lead to a longer-lasting golden age of safety.

Maya Pines, *Editor*

La Quinta High School in Southern California does not offer any courses in infectious diseases or epidemiology, but dozens of students and staff members there earned unwanted credits when they were exposed to a drug-resistant form of tuberculosis.

TU

Orange County officials now believe the disease was spread by a single, misdiagnosed student.

By February 1994, 11 students had developed active TB.

THE RETURN OF

By
Stephen
Hall

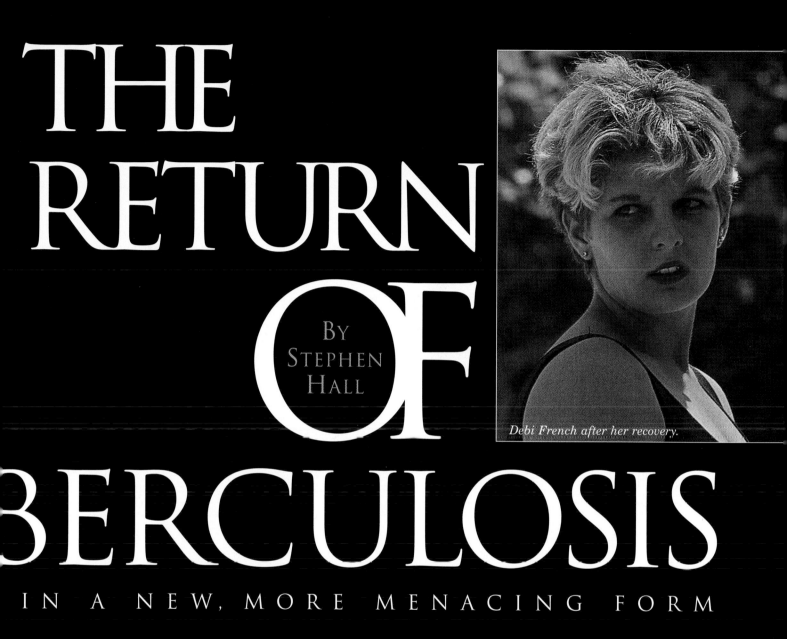

Debi French after her recovery.

3ERCULOSIS

IN A NEW, MORE MENACING FORM

One of them, Debi French, developed a particularly resistant form of TB and had to be flown to a special treatment center in Denver, Colorado, where doctors removed part of her right lung to save her life. In this California landscape of palm trees and ranch houses, far removed from the stereotypical TB setting of penniless immigrants in urban tenements, another 178 high school students and staff soon tested positive for infection. While some of them might have been exposed to the disease before, it was a thoroughly suburban epidemic.

Similar stories have cropped up with disturbing frequency in recent years. In Hawaii, state health officials launched an investigation when an airline passenger died of TB several days after arriving in Honolulu. Epidemiologists determined that in the previous month, the sick woman had traveled in the confined quarters of commercial airliners from Honolulu to Baltimore via Chicago, returning a month later by the same route. With the help of the airlines, health officials tested 802 of the 925 passengers and crew members on those airplanes who lived in the United States. The good news is that no one has developed TB yet, but the sobering reality is that at least six passengers seated near the victim on the final Chicago-to-Honolulu leg tested positive for infection.

And in Long Island, health officials tested more than 3,000 children for TB after learning that five bus drivers serving several elementary and high school districts in Nassau and Suffolk Counties had been diagnosed with TB. At least 41 students were found to be infected, probably from this exposure.

Some of the TB strains in these episodes were drug resistant, some not. The worrisome point, especially to anyone with impaired immune function, is that it's impossible to tell which is which until the microbes have had a good chance to multiply and spread. And spread they do: on buses and planes, in the suburbs as well as the inner city, in high school classrooms as well as crack houses, in healthy teenagers as well as AIDS patients. A disease once thought to be vanquished continues to pop up in the unlikeliest of places, reminding us that in the war against infectious disease, there are no permanent treaties, just pauses between battles.

"Tuberculosis is the leading cause of death in the world from a single infectious disease," says immunologist Barry Bloom, an HHMI investigator at Albert Einstein College of Medicine in the Bronx, New York. Bloom has seen the Third World come to New York's northernmost borough, watched with alarm as the number of cases began to rise in the mid-1980s after decades of decline, and recognized as part of this alarming comeback the spread of multi-drug-resistant tuberculosis, which accounts for about 20 percent of cases in New York City and threatens to change the nature of the disease. "We had our vaccine, we had our drugs, and it was supposed to disappear," Bloom says. "It didn't."

Each year, there are 8 million new cases of TB worldwide, and 2 million people perish of the disease. The World Health Organization recently issued a report predicting that global deaths could reach 4 million annually by the year 2005—or even more if drug-resistant strains continue to spread.

Where two decades ago TB seemed on the verge of extinction in the United States, it has discovered a new niche in a world of AIDS and suppressed immunity, homelessness, and the inappropriate use of antibiotics. Failure to kill off all the microbes that cause a person's disease allows the surviving microbes to mutate, and those that

are resistant to the drug begin to multiply. Two decades ago, the problems of TB in the Third World were literally and figuratively a world away. By 1991, the rate of tuberculosis in some neighborhoods of New York City exceeded those of cities in central Africa. These rates have dropped recently, but the danger persists.

"The ominous change in the face of this disease is multidrug resistance," Bloom continues. Drug resistance develops in patients who do not take their medication regularly for the long period—usually 6 to 12 months—required to knock out the organism that causes TB.

For many years, some states used federal funds to institute a highly effective program known as Directly Observed Therapy.

"**W**e had our vaccine, we had our drugs, and it was supposed to disappear," Bloom says. "It didn't."

But when federal TB funding became a budget casualty in 1972, drug-resistant strains of TB began to flourish. Within a decade, they had become entrenched in precisely the patient populations where they can most easily spread: the homeless, prisoners, drug users, and people suffering from AIDS, who are extremely vulnerable to the drug-resistant forms. By 1992, with emergency funding from the government, the states reinstituted Directly Observed Therapy programs, and partly for this reason the number of TB cases in New York City declined 22 percent in two years. As long as AIDS is with us, however, drug-resistant forms of TB remain a serious threat.

"You don't know whether a patient's microbes are resistant to the major anti-TB drugs until weeks after you start treatment, and that's worrisome at a number of levels," Bloom says. "First of all, we don't have other drugs that are terrific. The fatality rate of untreated tuberculosis, almost anywhere in the world, is 50 percent. It's a very powerful disease. If people with drug-susceptible forms of TB are treated appropriately, the cure rate is about 96 percent. But for multidrug-resistant TB, the case fatality rate is as high as 40 percent in people whose immune system is not impaired. In HIV-positive people, it's greater than 80 percent."

And, as Bloom never tires of pointing out, TB is everybody's problem. "The major risk factor for acquiring tuberculosis," he notes, "is breathing."

Barry Bloom outlines his strategy for creating a new vaccine against TB. Graphs on the computer monitor at right show that mice given the experimental vaccine are protected against TB and survive.

AN ALL-AMERICAN GIRL MEETS TB

Debi French, a glowingly healthy teenager, developed a life-threatening form of TB just by breathing the air in her school. "I never even met the student who infected me," says Debi, referring to a young woman who attended La Quinta High School in Orange County, California, at the same time she did. For two years this young woman walked around with an active case of TB.

At first Debi knew only that she felt tired all the time and could not get rid of a bad cough. She was also losing weight. When an x-ray taken in 1993 revealed that she had contracted TB, the news was greeted with some relief by her worried family. "I thought, at least this is something we can cure with medicine," recalls her mother, Patti French.

While it remains true that most cases of TB can be cured rapidly with antibiotics, the bacteria that lodged in Debi's chest had become

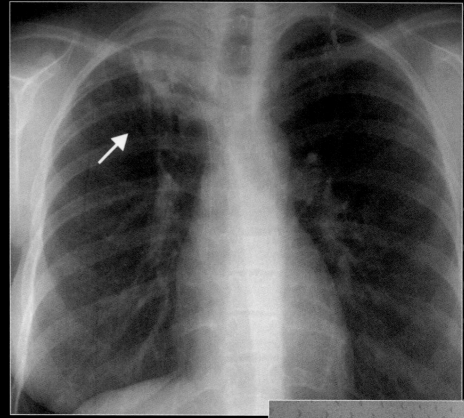

resistant to the drugs that are used as the first line of treatment. For a while she seemed to respond, but soon she relapsed and was hospitalized for treatment with additional medications. Even this proved insufficient, however.

"They told us there was nothing more they could do," says Patti French, remembering how frightened she was at the time. "They said her only hope was to go to the National Jewish Center for Immunology and Respiratory Medicine in Denver, Colorado, for specialized treatment."

Debi and her mother then flew to Denver by private jet, to avoid spreading her germs. By the time she arrived there, the disease had chewed a large hole in Debi's right lung, according to Michael Iseman, chief of the team of doctors who treated her at National Jewish. The team reinforced her medical treatment, adding some older TB drugs that were seldom used because of

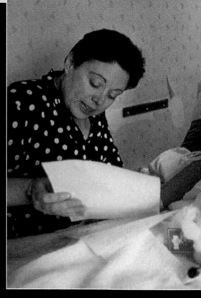

potentially serious side effects. But so many bacteria had lodged in the upper lobe of Debi's right lung that there was a real risk she might not overcome the infection—in which case her germs would become resistant even to the few drugs that were still useful.

"This would be virtually a death sentence," Iseman explains. Even if Debi survived, she would remain so dangerously contagious that she could never mingle with other people—never go to school or work, never marry, not even live with her family. To avoid this fate, the team removed nearly half of her right lung, giving the antibiotics a better chance to knock out the remaining bacteria.

Two months after her surgery, Debi came home to flowers, balloons, and a hero's welcome. She slowly resumed her normal life—though with a year's delay, since the rest of her senior class at La Quinta High School had already graduated.

She reenrolled as a senior and took up, once again, her many extracurricular activities: she kept statistics for the baseball team, competed with the flag team, and sang in a church choir. But she still needed to take 14 pills every day—and continues to do so, to make sure any surviving TB bacteria in her left lung remain in a dormant state. In addition, she must return to Denver for frequent checkups.

The whole family remains sharply aware of the fact that Debi had a narrow escape, and they clearly see the dangers presented by new, drug-resistant forms of TB. "If it can happen to us," says her father, Fred French, a former policeman, "it can happen to anyone." *– M.P.*

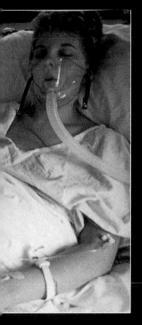

An x-ray of Debi's lungs before her surgery (top left) shows a large white patch (arrow) filled with fluid, debris, and TB bacteria. Surgeons removed this highly diseased area, making it possible for antibiotics to clear out the remaining infection. While recovering after the operation, Debi (right) was unaware that her mother had come to see her.

The root of all this sickness and death is a strange little bug known generically as *Mycobacterium*. Originally a grazing resident of ordinary soil, known for its sluggish growth and a cell wall that one biologist has likened to "a steel-belted radial," at some point in history the strain now known as *Mycobacterium tuberculosis* made the leap to human lungs. In this accommodating environment, the microbe can lie dormant for decades, multiplying slowly. But it can become active any time that its host's defenses are down. Then it typically forms the lung nodules that 17th-century anatomists first called "tubercula" (from the Latin for "swellings").

The disease has been with us for a long time, for several well-preserved Egyptian mummies dating back to 2000 to 4000 B.C. bear unmistakable traces of the type of physiological erosion that gave TB its colloquial name, consumption. *M. tuberculosis* has shortened many a brilliant career, among them Emily Brontë's at age 30 and Frédéric Chopin's at 39. Inspecting the bright red blood he had coughed up in the fall of 1820, John Keats, a physician as well as a great Romantic poet, knowingly remarked, "That blood is my death warrant." He died at age 25.

For much of its (and our) history, tuberculosis has been a familial and largely urban disease. In the most common mode of transmission, an infected adult passes on the disease to a child, usually by coughing, in the overcrowded and underventilated living conditions of the urban poor. Only 1 infected person in 10 actually develops the disease, and even then it can take many decades. The other 9 battle the microbe to an immunological stalemate. Still, as recently as the 1940s, tuberculosis claimed as many as 5 million victims a year worldwide, 80,000 in the United States alone.

That all changed with the discovery of several potent drugs, beginning in the 1940s with streptomycin. No sooner had researchers finished patting themselves on the back for this fine accomplishment than the TB microbe showed its ability to mutate, developing resistance first to strep-

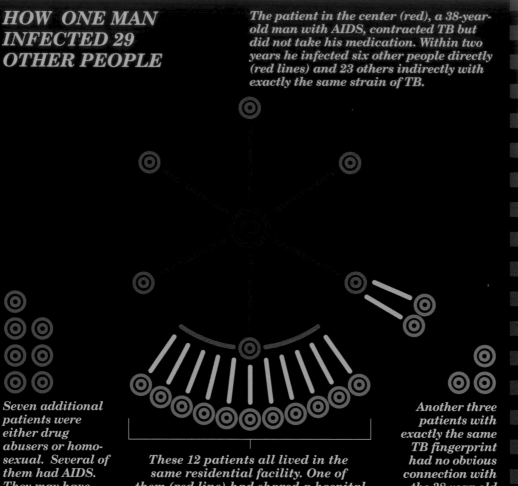

HOW ONE MAN INFECTED 29 OTHER PEOPLE

The patient in the center (red), a 38-year-old man with AIDS, contracted TB but did not take his medication. Within two years he infected six other people directly (red lines) and 23 others indirectly with exactly the same strain of TB.

Seven additional patients were either drug abusers or homosexual. Several of them had AIDS. They may have mingled with the 38-year-old man or his contacts in homeless shelters or medical clinics.

These 12 patients all lived in the same residential facility. One of them (red line) had shared a hospital ward with the 38-year-old man and infected the others. Two other people (upper right) were linked to one patient who had been in the same medical clinic as the 38-year-old man.

Another three patients with exactly the same TB fingerprint had no obvious connection with the 38-year-old man or his way of life. No one knows where they may have been exposed to the infection.

tomycin, then to para-aminosalicylic acid. Only in the 1950s did the medical community realize that it required the combination of those two antibiotics, plus a third drug, isoniazid, to decisively rout TB.

At that point, researchers began to talk about eradicating the disease. Rates of TB dropped (although they had begun to drop even before the drugs had been discovered), funding for prevention was cut, monitoring of patients' compliance with drug therapy eased, and soon drug-resistant strains began to re-emerge. Indeed, there are now seven major drugs for TB, and some TB strains are resistant to all seven. After federal funding for TB programs stopped, "it took 10 years for TB to come back," Bloom points out. When it came back, it discovered a new, more hospitable world.

The changes in the epidemiology of tuberculosis transmission have been brought brilliantly into focus since the early 1990s, when HHMI investigator Gary Schoolnik and his colleagues at Stanford

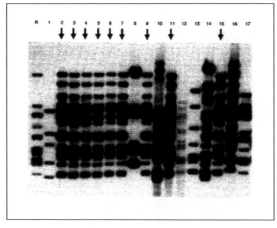

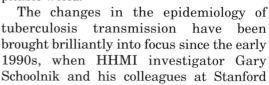

These telltale "fingerprints" show that DNA fragments from TB bacteria infecting nine patients formed exactly the same pattern (arrows)—evidence that these patients were infected with the same strain of TB. Eight other patients were infected with different strains of TB. The column on the left, labeled R, serves as a control.

University School of Medicine undertook an ambitious program to map the spread of TB with modern molecular tools. Their laboratory was nothing less than San Francisco, their mandate "to find every patient with TB in the city and county of San Francisco."

Using recombinant DNA technology, Schoolnik's group screened the DNA of *M. tuberculosis* and created a molecular fingerprint of the microbe from each person who contracted the disease. The researchers were able to do this because one portion of the microbe's DNA—an inconsequential bit of padding between genes—happened to feature a great deal of variation from one strain to another. Biologists can excise this region of DNA, cut it up into smaller fragments with enzymes, and then sort these pieces by size in a procedure known as a gel electrophoresis.

Each gel displays a distinct pattern of fragments. Indeed, the "fingerprint" more accurately resembles a plump version of the universal bar code or a ladder with oddly spaced blobs for rungs. Different strains of TB possess distinctly different patterns of rungs.

"For most of human history," says Schoolnik, "if we were to take a TB bacillus (a rod-shaped bacterium) isolated from an older adult who had clinical TB and do a molecular fingerprint of it, we'd be looking at the fingerprint of the organism that man or woman acquired maybe 50, 60, or 70 years ago. And in the United States, where so many people come from other parts of the world, the molecular fingerprints would reflect the places they were in as children. So in the pre-AIDS era, if we had done a molecular fingerprint analysis of all the strains isolated from every patient in San Francisco, which is what we're doing now, probably 90 percent of those strains would be different, indicating that they were from reactivated latent disease acquired early in childhood from people who came from all over the world."

The Stanford researchers found an altogether more troubling pattern, however. Working with the San Francisco Department of Public Health's Division of Tuberculosis Control, they tracked down 85

Robert Koch discovered the cause of tuberculosis in 1882 when he identified Mycobacterium tuberculosis in tissue taken from people who had died of the disease. He then grew the bacterium in his lab and showed that animals injected with it developed TB.

percent of all TB cases in San Francisco, collected more than 1,000 molecular fingerprints, and programmed a computer to compare each and every one of them. By the time they had analyzed all the data—part of which was published in the *New England Journal of Medicine*—they were able to document a disturbing departure from transmission patterns of the past: between 25 and 30 percent of all new strains of TB in San Francisco occurred in clusters.

"Now the significance of this," Schoolnik explains, "is that all of the patients in the cluster, with the exception of one, must have been recently infected. One of the patients in the cluster may have had latent disease that reactivated, but all the rest had to have been recently infected, because otherwise they would all have different strains. So if you have a cluster of 30 patients, all with the same strain, maybe 1 of those 30 got it from a childhood experience. The other 29 got it from each other."

That conclusion, in turn, raised a critical question of medical geography: if they got it from each other, where did they get it? Urban anthropologists and TB case officers looked at the group and asked, What do these people have in common, and where do they intersect? The answers showed that no locale is too mundane or too bizarre to play a role in the transmission of TB. Airplanes and school buses are the least of it.

Of all
infections
common to
AIDS patients,
TB is the only
one that can
spread fairly
easily.

Take one very perplexing group. It included an unmarried man from Oakland and three male prostitutes taking female hormones in preparation for transsexual surgery. None of the four men had ever met the others, but they all had the same strain of TB. Through interviews, the Stanford group uncovered the surprising point of intersection. "The one place where they all went and could interact with each other, although they didn't know each other, was in the private viewing rooms of a pornography studio," Schoolnik says. "These viewing rooms are poorly ventilated, and if anyone were to cough, the concentration of bacilli would rise continuously." A perfect environment, in other words, for TB to spread.

A second cluster of 14 patients consisted of homeless people with social problems. The common point of intersection? All attended a crowded social clinic that provided counseling to the homeless.

A third cluster announced itself toward the end of 1990, when 12 people in a residential AIDS care facility developed TB over a short period of time; another 18 people were found to carry the same TB fingerprint (see p. 246).

This last, ongoing outbreak is particularly significant, according to Schoolnik, because it demonstrates how TB "telescopes" its normal course—how it progresses from infection to active disease in a matter of months or years rather than decades—in people with compromised immune systems.

"The other thing of public health significance," Schoolnik says, "is that for every person found to have disease, there are probably another 10 who have been exposed and will now have a positive skin test for TB. They have not progressed to disease, but they remain at risk for the rest of their lives."

This accelerated course of disease may foreshadow the future of TB, especially in the developing world. If WHO predictions of 30 to 40 million HIV-infected people by the end of the century are correct, there will be limitless possibilities for clusters and amplification among all those immunocompromised hosts.

Undertaking a similar study at a Bronx hospital, Bloom's group found a depressingly similar pattern of transmission. Their high-risk population differed from that of San Francisco in some details (in New York they tended to be Hispanic, low income, HIV positive, and young), but the bottom line was the same: about 40 percent of the TB cases occurred in clusters. "What do we know about these people?" Bloom asks. "Two-thirds of them are HIV positive, 50 percent have microbes that resist one or more drugs, and 24 percent are multidrug-resistant. That's what's being transmitted every day in the Bronx."

The overall message of these two studies

is that in the age of AIDS, tuberculosis has become a quicker, more resilient, more opportunistic killer. "First of all, using this kind of analysis, we have proven that the natural history of the disease has been telescoped," Schoolnik says. "The second thing we've learned is that when people whose immune systems are suppressed become infected with TB and are treated for it, they do not develop resistance to reinfection with the TB bacillus. Normally, TB treatment and cure confer partial immunity against the disease—but not in people with AIDS.

And the last thing we've learned is that casual contact can lead to tuberculosis."

Of all infections common to AIDS patients, he points out, TB is the only one that can spread fairly easily to immunologically healthy people.

In the context of multidrug-resistant TB, where diagnosis is slow and the risk of transmission is high, researchers feel pressed for solutions. There is an urgent need for a way to identify patients with drug-resistant strains immediately, so they can be treated effectively before they infect

Bill Jacobs' "turn-on-the-light" assay shows which drugs will be effective against a particular strain of mycobacteria. He holds two test tubes containing samples of the same bacteria. One tube glows because the bacteria in it remain alive, resisting one type of drug. The other tube is dark because the bacteria were killed by a different drug.

others; similarly, there is a need to develop new drugs for those found to be resistant to all the regular drugs. Both these tasks are being pursued at Einstein.

Visitors may notice a lot of armadilloes on the fourth floor of the Forchheimer Building at Albert Einstein College of Medicine in the Bronx. Not real ones; more what one might call sentimental curios—stuffed animals and statuettes and rubber stamps. But HHMI investigators William Jacobs and Barry Bloom have a genuine affection for these animals, and it stems from the scientists' exasperation with the finicky bacteria they need to study.

Robert Koch's discovery of *Mycobacterium tuberculosis* in 1882 arguably ranks as the greatest microbial discovery of the 19th century, but generations of TB researchers since then have also viewed it as something of a curse. Mycobacteria as a group—which includes those that cause leprosy (*M. leprae*) and a form of TB in cows (*M. bovis*)—are a balky, problematic, and just plain badly behaved family of bugs for laboratory research. Where it takes the common stomach bacterium *Escherichia coli* only 20 minutes to double and 8 hours to go from a single organism to a visible lump of billions of bacteria in a lab dish, it takes *M. tuberculosis* 24 hours simply to double and three weeks to form colonies.

M. leprae, discovered in 1873 by G. Armauer Hansen as the cause of leprosy, is even less well behaved. Not until the 1960s did researchers get it to multiply in the lab, and then only in the footpads of mice, where it took six months to expand from a thousand to a million organisms. Ten years later, two biologists with the U.S. Public Health Service in Carville, Louisiana, reasoned that since leprosy tends to occur in the coolest parts of the body, primarily the face and hands, perhaps they should try to grow it in animals with low body temperatures. A quick look at the encyclopedia yielded the fact that armadilloes have one of the lowest body temperatures of all mammals.

The researchers then discovered that *M. leprae* would grow in the nine-banded armadillo. Nothing like gangbusters, of course—it took 14 days for the cells to double and a year for them to multiply sufficiently to cause disease. But that was enough to draw a lapsed mathematician and born-again geneticist named William Jacobs into the field of mycobacteria research, and ultimately to the Bronx.

For years, Bloom had entertained the dream of creating a one-size-fits-all vaccine for use in the Third World, and he hoped to do it by genetically engineering the well-known Bacille Calmette-Guérin (BCG) strain of mycobacteria. Since 1948, this weakened strain of *M. bovis* has been used to immunize 2.5 billion people safely and cheaply against some of the effects of TB. Bloom had the idea that a clever molecular biologist could drop bits of genes from other pathogens into this vaccine, thus stimulating the immune system to produce antibodies against other common Third World scourges—measles, leishmaniasis, and schistosomiasis, to name a few. To do so, he needed someone with experience with mycobacteria. In 1985 he induced Jacobs to come to Einstein, where Jacobs is now an HHMI investigator. (The vaccine dream, incidentally, still lives: the researchers have tested a dozen antigens in BCG, and a Maryland-based biotech company called MedImmune, Inc. has begun safety testing in humans of a BCG-based vaccine against Lyme disease.)

Jacobs and Bloom make an engaging team. Bloom is an immunologist, a public health moralizer of evangelical lyricism, a polyglot who writes beautiful, refreshingly opinionated papers; Jacobs is burly and disheveled, blue-collar by way of Pittsburgh, with a puckish sense of humor and an enthusiasm every bit as infectious as the organisms he studies. Bloom makes stirring speeches about the need for enlightened public health and puts his money where his mouth is, having devoted three decades to the conquest of infectious diseases that afflict developing countries; Jacobs spends his time as molecular biology's version of a grease monkey, working with his hands under the hood, tinkering with the genes of pathogens as if they were

Jacobs dug up some dirt at home and isolated some of these viruses. He named the first one "Bronx Bomber-1."

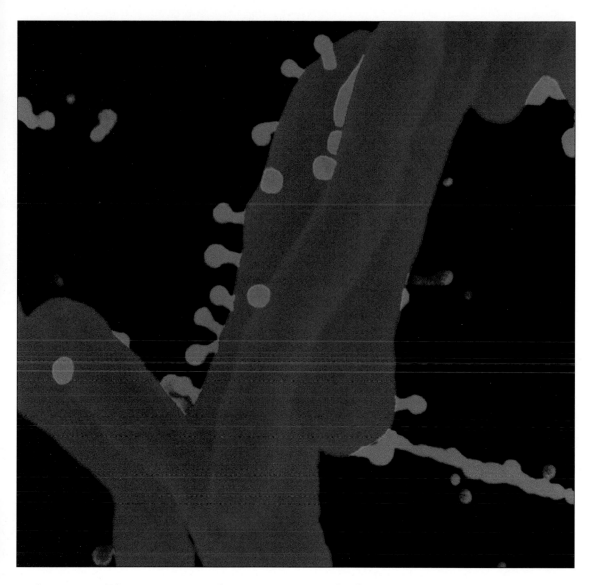

Tiny viruses (aqua) that have been engineered to contain a firefly gene infect a live mycobacterium (blue) and shoot this gene into it. The mycobacterium will then glow like a firefly as long as it remains alive.

carburetors. Bloom goes to the opera; Jacobs proudly displays a Steelers cap on a bookshelf in his cramped and disorganized cubicle.

Jacobs complements Bloom's scientific interests perfectly. He studied the leprosy bacterium during graduate school at the University of Alabama, so he was one of the few molecular biologists who had the experience—and patience—to work with these difficult microbes. His job was to determine the genetic makeup of *M. tuberculosis*, especially what genes made it so virulent. He planned to do that by adding or subtracting genes—inserting certain genes into nonvirulent bacteria, for example, and then look-ing for bacteria that had become virulent after the transaction. But in order to slip genes into mycobacteria, he needed a way to puncture their steel-belted radial coat. This led to a project that began in, of all places, Jacobs' backyard in the Bronx.

Mycobacteria thrive in dirt, and so do the tiny viruses, called "phage," that infect them. Indeed, phage represent millions of years of accumulated expertise, honed by evolution, at sneaking into mycobacteria. So Jacobs dug up some dirt at home and isolated some of these viruses. He named the first isolate Bxb1, for Bronx Bomber-1.

In the lab, Jacobs found that he could introduce DNA from a similar virus into

The Dangers of Attachment

The bacteria that cause TB have the nasty habit of invading human cells and hiding there, safe from the body's attempts to get rid of them. Most other infectious bacteria do not enter our cells; they just sit on the cell surface and spew out toxins.

These toxins can produce fever, diarrhea, and shock, as well as cardiovascular damage. They are responsible for such diseases as cholera, diphtheria, tetanus, toxic shock syndrome, scarlet fever, and gangrene. It takes only one molecule of diphtheria toxin to kill a cell in the human heart or nervous system, for instance—and each bacterium can produce 5,000 such molecules per hour.

The very first thing all infectious bacteria must do, however, is attach themselves to selected cells. These cells—in our airway, gut, or an open wound—must have specific receptors that fit particular molecules on the bacteria's surface.

An intestinal cell (orange) erects a pedestal for an EPEC bacterium (purple), enabling the bacterium to stick to the cell securely. The pale knobs are microvilli, normal features of epithelial cells' surface.

Bacteria are very choosy in this way. Thus *Helicobacter pylori*, the spiral-shaped bacteria that are now known to cause stomach ulcers, will ignore cells from the nervous system or urogenital tract and stick only to epithelial cells from the stomach. Having found a proper target, the bacteria need to attach themselves firmly to it. (Without firm anchors on the slippery mucus lining, the bacteria might be washed away by various body fluids.)

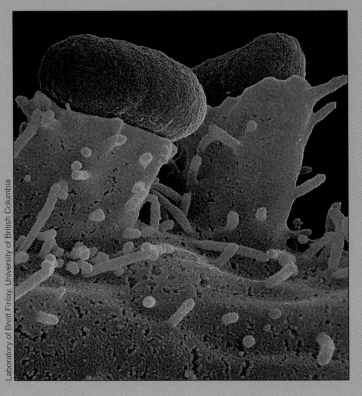

Laboratory of Brett Finlay, University of British Columbia

E. coli as a plasmid—that is, as a loose, free-floating ringlet of DNA, a kind of appendix to the main genetic text of the bacterium, which would get copied along with each division of the *E. coli* cells. Indeed, Jacobs was using *E. coli* as nothing more than a high-speed, biological copying machine. He could then manipulate the DNA of the virus, add or subtract genes, and use the

reengineered virus to insert novel genes into mycobacteria. The more copies he had, the better his odds of hitting the genetic jackpot—getting a desired gene in and getting it to work. Jacobs called his little genetic creation a "shuttle phasmid"—part phage, part plasmid, it shuttled DNA back and forth between everyday *E. coli*, the single-celled workhorse of genetic research,

Then they can begin to multiply. As they grow, they produce and spread their toxins. Only a few kinds of bacteria use the same strategy as the TB microbe and actually invade human cells. These intracellular bacteria tend to produce serious infections, such as bubonic plague, Legionnaire's disease, and typhoid fever. They have proved far more difficult to study than the noninvasive kind. But they, too, start the process of infection by attaching themselves to specific receptors on human cells.

In recent years, Brett Finlay, an HHMI International Scholar at the University of British Columbia in Vancouver, and his associates have focused on this critical step, analyzing the various adhesive molecules, or adhesins, that bacteria use as they latch onto human cells. The researchers are also studying the cells' response to attachment by bacteria. "Most diseases are caused by the initial adherence of bacteria to cells," Finlay says. "We want to define the factors involved in this attachment."

Finlay believes attachment results from a kind of "cross talk" between the microbe and the cell. His group found that enteropathogenic *Escherichia coli* (EPEC) bacteria—the leading cause of bacterial diarrhea in the world—use at least two different types of adhesins to attach themselves to intestinal cells. Each type of adhesin sticks to a different kind of receptor on the cell. When the bacteria attach themselves with the first adhesin, "they seem to send a signal, or ring a doorbell, in that host cell," Finlay says. This signal releases calcium inside the cell, and the increased calcium affects filaments of actin, a protein involved in cell structure. Actin filaments then accumulate beneath the bacteria. Next, a pedestal begins to grow out of the cell, enabling the bacteria to attach themselves securely to the cell with their second adhesin. Only then are the bacteria tightly bound and ready to multiply.

A different form of cross talk allows intracellular bacteria to dive into target cells after attachment. Ralph Isberg, an HHMI investigator at Tufts University School of Medicine in Boston, Massachusetts, has shown that a protein called invasin, located on the surface of *Yersinia* bacteria, actually forces the cell to engulf these bacteria. He points out that an understanding of these events might lead to ways of interrupting them.

Bacterial attachment is dangerous. If researchers succeed in identifying the key factors involved in it, they may learn how to derail the process. "There are currently no drugs that block bacterial adherence," Finlay says. "We may be able to design a new kind of drug that would do so—and thereby prevent or stop infections." — *M.P.*

and the tough-coated mycobacteria.

Those shuttle phasmids have now provided an answer to an urgent medical question relating to multidrug-resistant TB. Normally, when patients come to the hospital with fever, chills, and a bad cough, a quick and simple test will reveal whether they have been infected by mycobacteria, providing an instant diagnosis of TB that enables doctors to immediately initiate treatment with antibiotics (the current treatment of choice is a combination of isoniazid, rifampicin, and pyrazinamide). But what if this particular strain of TB is resistant to some or to all three drugs? There is no quick way to tell.

"If you live in a Third World country such as the Bronx," says Jacobs, "you culture a patient's strain and then it takes five weeks to grow it up and another five weeks to test its susceptibility to drugs. But within a month, the patient comes back to the hospital. He's doing worse. You suspect drug-resistant TB, so you start him on two other drugs. Another month later, he's dead. Two weeks after that, you find out he was infected with a particular kind of multidrug-resistant TB—and if you had known which drugs he was resistant to and which ones he was still sensitive to, you could have put the patient on a very different set of drugs. The CDC (Centers for Disease Control and Prevention in Atlanta, Georgia) has documented this sort of thing."

Beyond the personal tragedy, there lurks a potential public health disaster, too, because this infectious patient, believing the medication worked, could be spreading multidrug-resistant TB to his doctors, to his family, and to acquaintances, all the while thinking he was getting better. "It's a huge problem," says Jacobs.

But his Bronx Bombers have provided a nifty solution. Using gene-splicing technology, Jacobs inserted the gene for luciferase, the protein that puts the glow in fireflies, into the viruses that infect *M. tuberculosis*. The gene turns on—and the chemical glows—when the phage multiply, and the phage multiply only when their hosts, the mycobacteria, are themselves alive and growing.

These glow-in-the-dark viruses enabled Jacobs and Bloom to devise a quick and ingenious diagnostic test. Starting out with test tubes containing mycobacteria from a patient, they add a different TB drug to each test tube and mix in the engineered phage, which promptly infect their host. Drug-sensitive TB bacteria will die off and the light will never go on, while drug-resistant bugs will continue to grow and the light will switch on. "I call it the 'turn-on-the-light' assay," Jacobs says with a smile. The rub, still, is the sluggish growth of mycobacteria; they need to be nurtured a week in culture before the phage can be added. "From there, it's two hours," Jacobs says. If he finds an industrial partner to develop the test, he says, it could be mar-

keted within two years.

What if the test reveals a strain of TB resistant to every existing drug? The other main thrust of Jacobs' efforts is to find new drugs for TB, especially the drug-resistant strains. His team recently uncovered a promising candidate while investigating the activity of an old friend: isoniazid.

Isoniazid, first synthesized in 1912, sat on the shelf for 40 years until researchers tested it against TB. It then proved so successful that it has been a part of standard treatment since 1952. Yet no one quite knew how it worked. In the past two years, Jacobs, enzymologist John Blanchard, and crystallographer James Sacchettini at Einstein have teased apart its mechanism of action and discovered that isoniazid blocks the activity of one of *M. tuberculosis's* genes. This gene, called *inhA*, plays a role in the construction of chains of fatty acids. As the researchers picked this mechanism apart even further, they learned that the gene encodes an enzyme that functions a bit like a mason, helping the organism build long, monotonous chains of fatty acids that get incorporated into that nearly impenetrable mycobacterial cell wall.

The three-dimensional structure of this inhA enzyme has now been determined by Sacchettini and his colleagues, who plotted the location of all its atoms. Then the researchers sat down, rearranged a few atoms, and designed a new chemical that they thought would block this key masonry enzyme even better than isoniazid. After a bit of tinkering, they came up with kynolic acid (KOA for short). In the test tube, it knocks out *Mycobacterium tuberculosis*.

"We have a drug!" Jacobs roars in delight. "Coming off the damn blackboard." Indeed, even as he was speaking, a colleague barged into the office with culture plates showing that KOA-17 had stopped a lab version of mycobacteria cold in a test-tube experiment. Many, many more such experiments will be necessary to establish efficacy and safety, but the lab is sufficiently encouraged that it has begun an even more difficult task: trying to convince a drug company to develop and test the substance for what is considered a piddling market—the 28,000

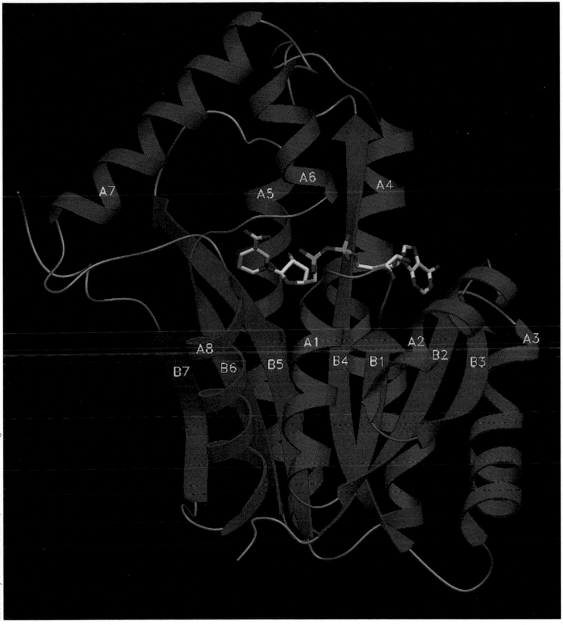

Scientists found the basis of the TB bacterium's resistance to isoniazid, a powerful anti-TB drug, after deciphering the 3-D structure of the drug's primary target, the bacterial enzyme inhA (left). Resistance develops when inhA changes either of two amino acids, located on A1 and B4 in the center of this 3-D model.

people in the United States who contract TB each year.

Given the enormous toll it has exacted—an estimated 1 billion lives over the last two centuries—surprisingly little is yet known about how tuberculosis actually causes the damage that makes it so lethal. Biologists are still in the dark about TB—its virulence, its latency, why other pathogenic bacteria are vanquished in two weeks by antibiotics while some mycobacteria persist for six months against drugs like isoniazid.

Fortunately, today's molecular biologists have uniquely powerful new tools to attack these problems.

Jacobs is particularly hopeful about the prospects of finding more powerful drugs through molecular research. "If we could figure out how to kill these mycobacteria in two weeks," he says, "we'd give everybody in the world a drug and we could wipe out TB. We'd say, 'Hey! Twelve noon on May 23, 2020, everybody takes the drug. It's the end of TB.' That's what we're shooting for." ●

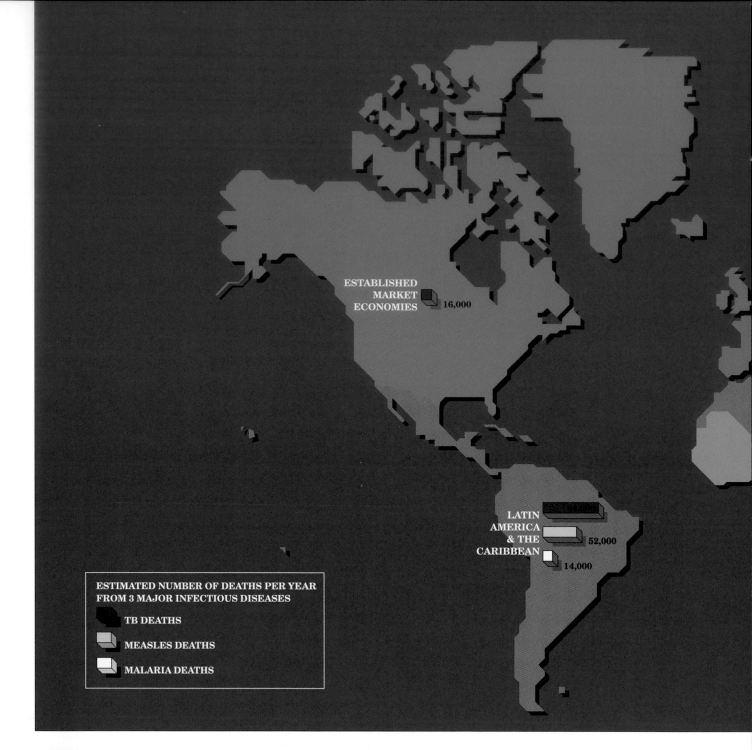

ESTIMATED NUMBER OF DEATHS PER YEAR FROM 3 MAJOR INFECTIOUS DISEASES

■ TB DEATHS

■ MEASLES DEATHS

■ MALARIA DEATHS

ESTABLISHED MARKET ECONOMIES ■ 16,000

LATIN AMERICA & THE CARIBBEAN 52,000 / 14,000

The champion killer among infectious microbes today is the tuberculosis bacterium, which takes 2 million lives each year. As shown on this map, it kills people at a high rate all over the world—except in such highly developed areas as western Europe, North America, Australia, and Japan.

People who live in the industrialized nations, where measles has nearly disappeared, may be surprised to learn that the measles virus is the second-deadliest infectious organism in the world. However, the ferocity of measles—the most contagious of all human infections—is no secret to

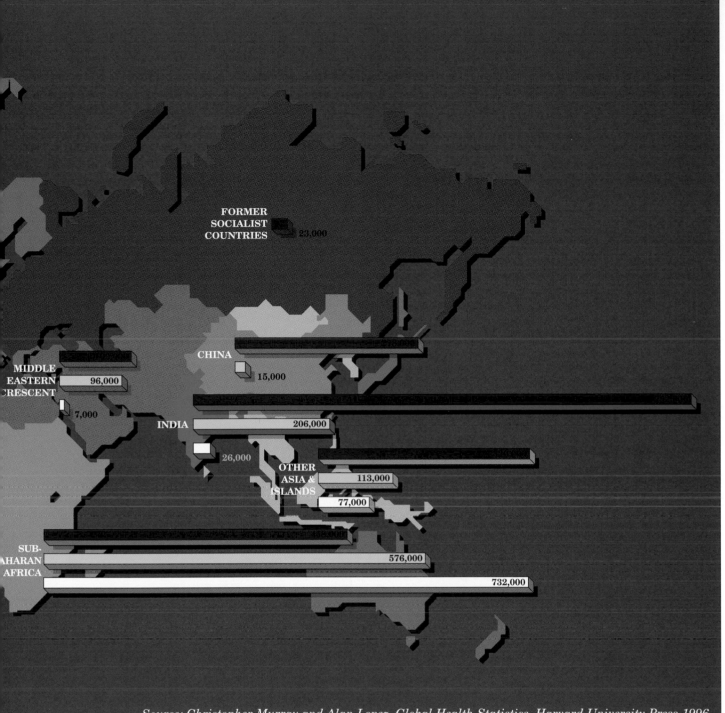

FORMER
SOCIALIST
COUNTRIES 23,000

MIDDLE
EASTERN
CRESCENT 96,000

7,000

CHINA 15,000

INDIA 206,000

26,000

OTHER
ASIA &
ISLANDS 113,000

77,000

SUB-
SAHARAN
AFRICA 576,000

732,000

Source: Christopher Murray and Alan Lopez, Global Health Statistics, Harvard University Press 1996.

THE MAJOR KILLERS

millions of people in the developing nations. Though the measles vaccine is very effective, it does not provide immunity if it is administered at birth, since a newborn is still protected by its mother's antibodies; the vaccine must be given between the time these antibodies wane, around 9 months of age, and the child's exposure to the measles virus. Vaccinating infants during that narrow window of time has proved very difficult. As a result, about 206,000 children die of measles in India, and as many as 576,000 die of it every year in Africa. Worldwide, the disease kills more than 1 million people every year.

Another major killer is the malarial parasite *Plasmodium falciparum*, which invades the bloodstream and may become fatal when it reaches a victim's brain. Nearly 1 million people die of malaria every year, primarily children and pregnant women. Such deaths are especially prevalent in sub-Saharan Africa.

These are the worst infectious diseases produced by a single organism—so far. But AIDS is catching up rapidly. Infection with HIV, the virus that causes AIDS, is spreading around the world. By 2005, deaths from AIDS are expected to reach more than 2 million a year and to rival the death toll from TB.

Children under the age of 5 are at particularly high risk of dying from infectious diseases, especially in the Third World. Their most dangerous enemies are respiratory infections and diarrheal diseases, both caused by a variety of bacteria and viruses. Taken together, acute respiratory infections kill more than 4 million children under 5 every year. The various diarrheal diseases take the lives of another 3 million young children every year. The many microbes that cause these diseases are always ready to strike. In the developing nations, unsafe water, poor sanitation, malnutrition, and lack of treatment allow these microbes to gain the upper hand.

Two Dangerous Epidemics Start to Overlap

Two life-threatening epidemics—TB and HIV—are speeding towards each other throughout the world, bringing disaster as they meet.

Nearly 2 billion people carry the bacterium that causes tuberculosis. Though their risk of developing active TB persists throughout their lives, they generally keep the infection in check and have no symptoms of the disease as long as their immune system remains strong. Should their immune system weaken because of cancer, malnutrition, old age, or infection with HIV, however, they become very likely to develop active TB.

In the past decade, the AIDS-causing virus has infected more than 15 million people worldwide. As it weakens their immune system, it makes such people highly vulnerable to other infections, such as TB, which can flare up in their bodies and rapidly become fatal. The two diseases fuel one another. A TB infection can also knock out a person's ability to cope with HIV and speed the development of AIDS.

The two epidemics have begun to overlap. More than 5.4 million people are now infected with both TB and HIV. In countries where people who have these infections live close together, a fuse has been lit.

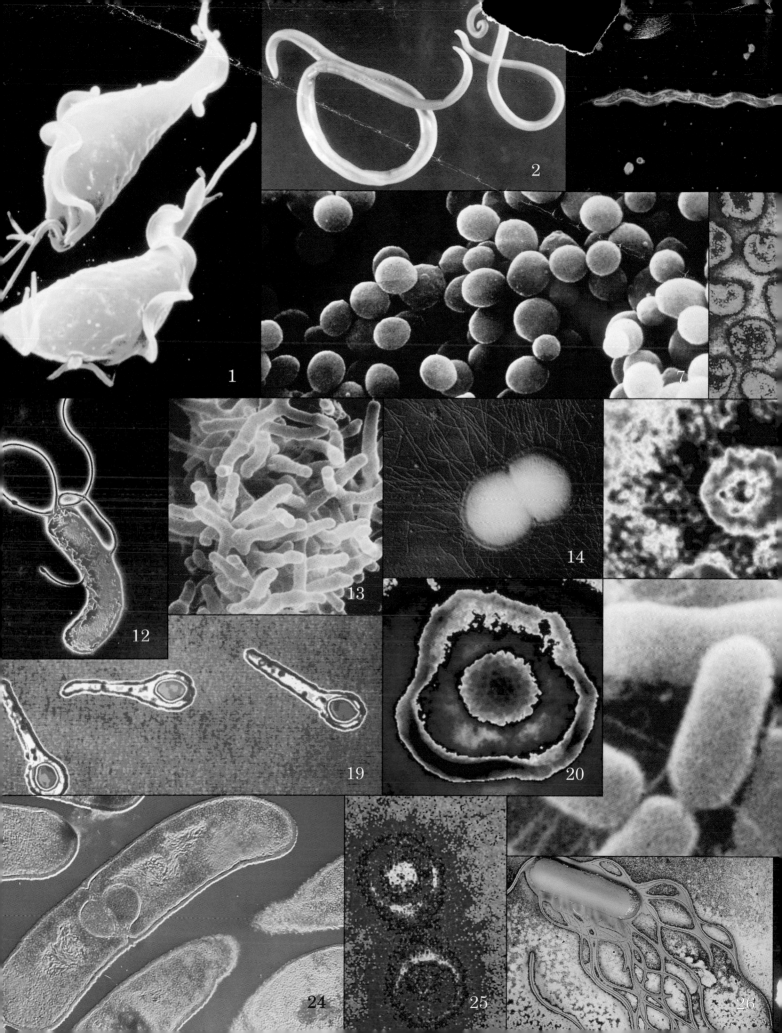

THE NEVER-ENDING BATTLE

The never-ending battle between disease-causing microbes and the body's defenses generally results in a draw—a delicate balancing act, symbolized by the gymnast in the center of the page. But sometimes our defenses break down and the microbes gain entry into the body. This poster shows how such microbes harm us, how we fight them, and how they fight back.

1 How They Enter The Human Body

Through the eye
(trachoma)

Respiratory route
(TB, flu, measles)

Oral route
(salmonella, hepatitis A)

Injections
(AIDS, hepatitis B)

Genitourinary route
(AIDS, herpes)

Cuts or bites
(tetanus, malaria)

2 HOW THEY HURT US

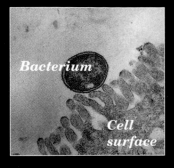

Bacterium

Cell surface

The microorganism must first attach itself to the surface of a cell by binding to specific receptors on the cell surface.

Once attached, some bacteria just stay outside the cell and multiply there, often producing toxins that kill the cell (as in cholera, strep, or staph infections).

Other bacteria—and all viruses—actually invade the cell. Once inside, the microbes multiply, robbing the cell of its nutrients. This usually destroys the cell. The newly made microbes go out to invade other cells.

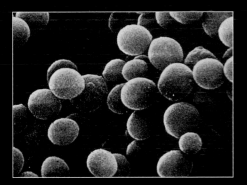

Cluster of Staphylococcus bacteria

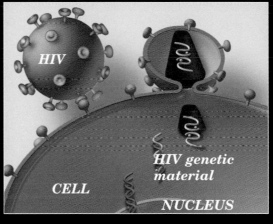

HIV

HIV genetic material

CELL

NUCLEUS

HIV, the AIDS-causing virus, enters a human cell.

Gymnast: Barika X. Williams

MICROBES THAT THREATEN US

An extraordinary variety of viruses, bacteria, and parasites stand ready to attack us and feed off our bodies' cells, incidentally producing disease. Photographers in many parts of the world have been fascinated by these tiny invaders. Using microscopes of varying power, they show us a teeming microbial world that we could barely imagine without their help.

The approximate sizes of the microbes shown on the next two pages can be figured out by using the following rule of thumb:

Viruses are the smallest of all infectious agents, averaging about 100 nanometers (100 billionths of a meter) in length. They have so few genes and proteins of their own that in order to reproduce they need to commandeer the machinery of the cells they invade.

Bacteria vary widely in size and shape, but tend to be at least 10 times larger than viruses, or at least 1 micrometer (1 millionth of a meter) long. They are single-cell organisms that reproduce independently.

Single-cell parasites tend to be at least 10 times larger than bacteria, or about .01 millimeter long.

Multicellular parasites are so large they can usually be seen with the naked eye. Tapeworms, for instance, can reach a length of 6 meters (20 feet).

Here is a list of the microbes shown on the following pages:

1. **Trichomonas** (protozoan parasite)
2. **Ascaris** (multicellular intestinal roundworm), female and male
3. **Treponema pallidum** (spirochete bacterium that causes syphilis)
4. **Adenovirus** (virus that causes colds)
5. **Influenza virus** (virus that causes flu)
6. **Streptococcus pneumoniae** (bacterium that causes lobar pneumonia)
7. **Staphylococcus** (bacterium that causes "staph" infections)
8. **Rubella virus** (virus that causes German measles)
9. **Mycobacterium tuberculosis** (bacterium that causes TB)

(continued on page 266)

Virus

Bacterium

Single-cell parasite

Multicellular parasite

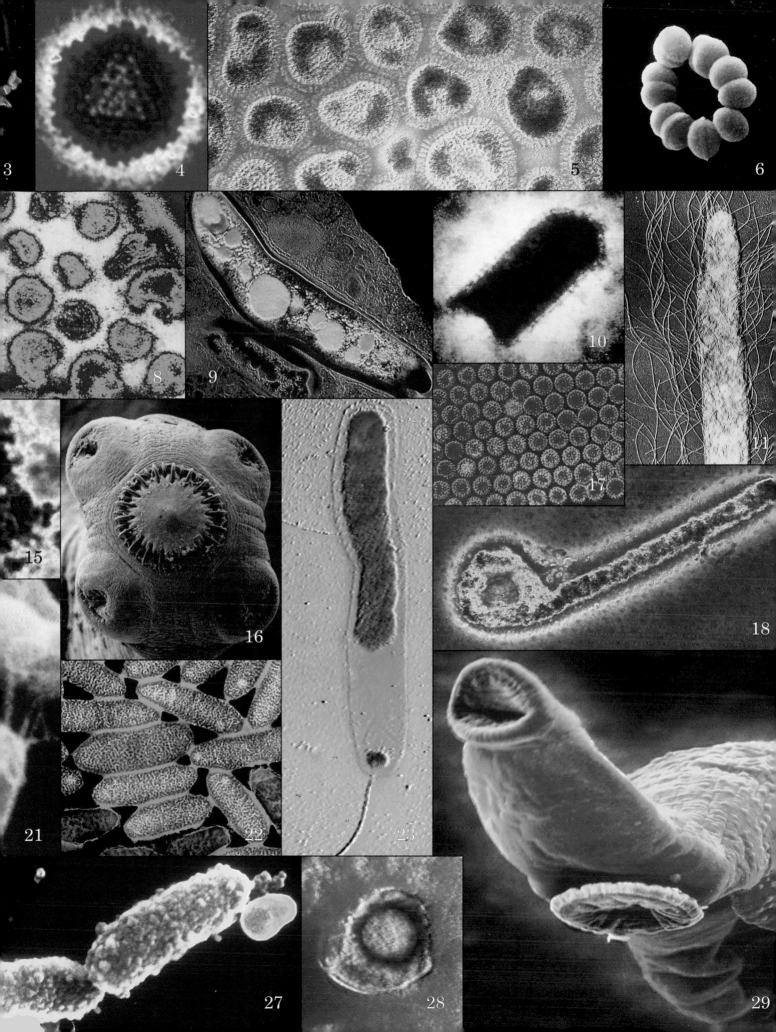

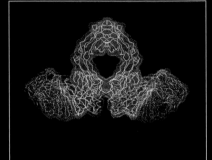

3 How We Fight Them

NATURAL DEFENSES

Skin and mucous membranes form the body's first line of defense. If these barriers are breached, other nonspecific defenses appear rapidly, including macrophages, white blood cells that swallow up some bacteria upon contact.

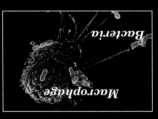

Macrophage

Bacteria

Several days later, if the infection persists, the immune system produces weapons that are tailored against particular antigens (molecules that the body recognizes as foreign).

B cells (white blood cells) release specific antibodies that bind with particular antigens, tagging them for destruction.

T cells (another kind of white blood cells) attack any cells that bear antigens they can recognize.

Structure of an antibody molecule

T cell

DRUGS

Penicillin and some other antibiotics kill bacteria outright (see below); they prevent the bacteria from making new cell walls or interfere with the production of bacterial DNA. Other types of antibiotics do not kill, but cripple bacteria enough for the body's immune system to finish the job.

Staphylococcus bacteria before and after

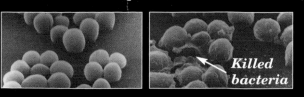

Killed bacteria

VACCINES

Vaccines prevent disease by stimulating the immune system to produce specific antibodies against particular bacteria or viruses, so that the body is always prepared to attack these invaders. Vaccines may consist of live but harmless microorganisms, killed microorganisms, inactivated toxins, or components of bacteria or viruses.

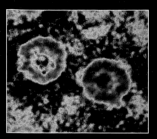

Polio virus used in vaccine.

4 HOW THE MICROBES FIGHT BACK

AGAINST THE IMMUNE SYSTEM

Organisms such as HIV, the virus that causes AIDS, or African trypanosomes, the parasites that cause sleeping sickness, constantly vary some of the antigens on their surface to stay one step ahead of slowly developing antibodies.

To avoid being recognized and destroyed by a person's antibodies, green-glowing trypanosomes alter their surface antigens. The orange trypanosome in the center is still in the process of switching from a green to red antigen type.

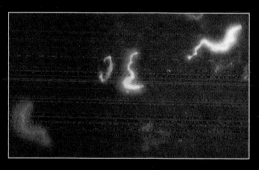

AGAINST DRUGS

Many microbes rapidly become resistant to drugs through mutations. This enables them to destroy or inactivate antibiotics or to prevent the drugs from reaching key targets inside them. Once they become resistant to a drug, they can share this ability with other microbes.

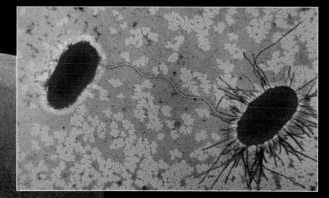

Bacteria sharing resistance plasmids, little packets of genes that enable them to resist several different antibiotics.

(continued from page 259)

10. **Rabies virus** (virus that causes rabies)
11. **Proteus mirabilis** (bacterium that causes infections in urinary tract and wounds)
12. **Helicobacter pylori** (bacterium that causes peptic ulcers)
13. **Actinomyces** (bacterium that causes dental disease)
14. **Neisseria gonorrhoeae** (bacterium that causes gonorrhea)
15. **Polio virus** (virus that causes polio)
16. **Scolex, or head of tapeworm** (a multicellular parasitic intestinal worm), showing 4 suckers and many hooks for attachment to host
17. **Rotavirus** (virus that causes acute intestinal infections in young children)
18. **Ebola virus** (virus that caused recent epidemic in Africa)
19. **Clostridium tetani** (bacterium that causes tetanus infections)
20. **Herpes simplex virus** (virus that causes herpes infections)
21. **Escherichia coli** (bacterium that infects intestinal tract, usually without harm but sometimes causing disease)
22. **Legionella** (bacterium that causes Legionnaire's disease)
23. **Vibrio cholerae** (bacterium that causes cholera)
24. **Listeria** (bacterium that causes meningitis)
25. **Hepatitis B virus** (virus that causes hepatitis B)
26. **Salmonella** (bacterium that causes intestinal infections, including typhoid fever)
27. **Bordetella pertussis** (bacterium that causes whooping cough)
28. **Cytomegalovirus** (virus that can harm a fetus)
29. **Schistosoma** (a multicellular parasite that causes schistosomiasis)

In Hot Pursuit of a Deadly Parasite

by Deborah Franklin

In 1965, parasites seemed neither exotic nor intriguing to John Donelson—just an ever-present nuisance or danger to be endured or avoided wherever possible. Now a leader in the field of molecular parasitology, Donelson was then a skinny, idealistic kid from the Midwest who had been plunked down for a Peace Corps teaching stint in rural West Africa. He had at the time, he says, absolutely no interest in biology or medicine. But the shiftiest of all parasites would soon change that.

"I had been a biophysics major at Iowa State, but had never even taken a college biology course," says Donelson, an HHMI investigator at the University of Iowa. "My training was all physics, chemistry, and math, and that's what I taught to students my own age at a junior college, way up-country in Ghana."

Though he was spared a personal encounter with any of the half dozen or so parasites endemic to the region, many of his students were not so lucky. The water they bathed in and drank came from freshwater streams contaminated with *Schistosoma*—a family of blood flukes that bore their way through human skin to invade and damage internal organs. Malaria, caused by a bite from a mosquito infected with another parasite, *Plasmodium*, was also a recurrent plague, bringing

THAT ALWAYS KEEPS
ONE STEP AHEAD

with it fevers, chills, and anemia.

One parasite in particular sapped life from everyone in the village, whether they themselves were among the 60,000 Africans infected each year or not: the trypanosomes, a family of hardy protozoa transmitted to people and domestic cattle through the nip of a tsetse fly. For generations there had been very little meat and few dairy products available in the protein-poor region because trypanosomes decimated herds with the fatal neurological disease called "ngana," a Zulu word meaning "loss of spirits." In people, the disease is known as sleeping sickness. Ten million square kilometers of otherwise rich grazing land—essentially the middle third of the African continent, including the Ghanaian village where Donelson taught—cannot support the breeding of cattle because of the trypanosome-infected fly. Despite the promise Donelson's bright and lively students showed inside the classroom, he says, illness always lurked in the background, gnawing at the standard of living and the quality of life.

One of his few ties to home in those two years was the single copies of *Scientific American* that regularly arrived at the school via sea freight—six or eight months late. "It was really from reading those articles about the genetic code, transfer RNA structures, and the like that I realized there was a revolution going on in molecular biology," Donelson says. "And I knew I wanted to be a part of it."

Parasitology was ripe for that revolution. While antibiotics have helped quell many bacterial illnesses around the globe, or at least turned the battle in our favor, the weapons against parasites have always been fewer. Powerful poisons aimed at eliminating the insects that transmit disease once seemed the best bet. But by the late 1960s it was clear that the widespread spraying of insecticides, while helpful, would never be enough. The known and suspected effects on people and the environment were mounting, even as the insects were becoming resistant. Better, thought Donelson and a growing number of scientists, to go after the parasite directly,

IT IS CALLED "SLEEPING SICK- NESS" IN PEOPLE, AND IN ANIMALS "NGANA," A ZULU WORD MEANING "LOSS OF SPIRITS."

track the ways it evades immune defenses, and find and exploit any weakness.

They couldn't have picked a wilier foe than the African trypanosome. Unlike microbes that burrow inside host cells to do their damage, all five clinically important, closely related trypanosome variants in Africa—*Trypanosoma gambiense* and *T. rhodesiense*, which afflict people, and *T. congolense*, *T. vivax*, and *T. brucei*, the scourges of livestock—brashly sail in armed flotillas through the bloodstream, resisting the immune system's frontal assault. Quick-change artistry is the key to the success of these single-celled parasites. When the going gets tough and antibodies have identified and wiped out 99 percent of the invaders, one of the few remaining trypanosomes suddenly strips off all identifying markers, sprouts a new coat, and begins reproducing madly, looking for all the world like a brand-new enemy. By the time the immune system has retooled to fight the refurbished foe and launch a second attack, one of the new generation of parasites has already switched to yet another coat and escaped again.

The precise timing and physical mechanism of that switching, known as antigenic variation, remain quite murky to this day. But studies from a number of labs in the mid-1980s showed that once in a great number of cell divisions—perhaps one in a hundred, perhaps one in a million—a parasite will express a protein coat that is quite antigenically different from its twin's. Interestingly, while the pressures of pursuit by the host immune system do indeed encourage the growth of creatures who make such a change, they by no means trigger the switch. Even trypanosomes grown in the lab, far removed from host immune factors, repeatedly undergo the same sudden changes in their coat proteins, which have come to be called variant surface glycoproteins, or VSGs.

As early as 1910, colonial doctors tending patients had an inkling of that vicious cycle, though they couldn't know what was happening at the molecular level. When British investigators Ronald Ross and David Thomson took daily samples of blood

John Donelson studies the switching mechanism that enables African trypanosomes (shown on the slide projected behind him) to keep changing their coats.

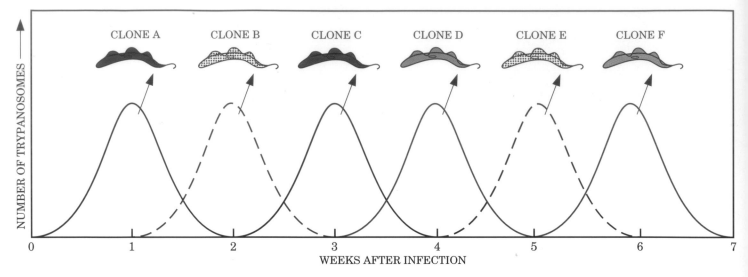

CLONE A   CLONE B   CLONE C   CLONE D   CLONE E   CLONE F

NUMBER OF TRYPANOSOMES →

0   1   2   3   4   5   6   7
WEEKS AFTER INFECTION

A week after infection with African trypanosomes, a person begins to produce specific antibodies that will soon destroy most of the parasites carrying a particular type of variable surface glycoprotein, or VSG, on their surface (clone A). But a few parasites survive by expressing a different coat protein (clone B). These camouflaged trypanosomes start to multiply,
reaching a peak during the second week. A new wave of antibodies tailored specifically to attack them soon eliminates most of them, but meanwhile some trypanosomes that carry yet a different VSG (clone C) start to proliferate. The cycle is repeated many times as the trypanosomes keep changing their coats to stay one step ahead of the immune system.

from one patient and actually counted the number of parasites in his blood, they discovered a curious pattern. Every seven or eight days, they reported, their patient experienced a sharp rise and subsequent fall in the number of blood-borne trypanosomes—slender creatures about twice the length of a red blood cell, each with a delicately undulating, fin-like ridge that tapers to a whip of a tail, called a flagellum. The man's fever, which kept fluctuating, tended to rise and fall in parallel with the number of parasites. Eventually the waves of parasites advanced to invade the central nervous system, leading to lethargy, slurred speech, coma, and death.

"A few parasites escape destruction..." Ross and Thomson wrote, quoting the theory of an Italian contemporary. "These are the parasites which cause the relapses." It took nearly six decades and a new generation of molecular tools to confirm their turn-of-the-century prescience.

The secret is indeed in the coat. About the same time Donelson was setting out for the Peace Corps, Keith Vickerman of the University of Glasgow suggested that there was something about the proteins in the

African trypanosome's thick coat that enabled it periodically to assume a new antigenic identity. In 1975, George Cross, then at the Medical Research Council's Molteno Institute in Cambridge, England (and now at the Rockefeller University in New York), confirmed Vickerman's hunch when he biochemically purified *T. brucei* coat proteins for the first time and was able to nail down their amino acid constituents.

Most living cells boast tens or hundreds of types of outer coat proteins—a gangly forest of intricate loops and branches that poke and protrude at various, highly specified angles to help the cell snag particular molecules floating past, thereby enabling the cell to interact with the other cells and substances in its environment. But the trypanosome's proteins have an unusual feature. On any single organism, Cross found, the VSGs that stand branch to branch to form the outer coat—all 10 million of them—are identical. Yet when he compared the coats of several populations of trypanosomes descended from a single organism, he saw they were distinct.

Five years later, Richard Williams at the International Laboratory for Research on

Animal Diseases in Nairobi, Kenya, and, independently, Piet Borst of the Netherlands Cancer Institute in Amsterdam, along with Cross, cloned the first genes of what would soon be recognized as a repository of perhaps as many as a thousand VSG genes scattered throughout the coils of DNA in each trypanosome. Amazingly, that means that fully 5 to 10 percent of the organism's genetic endowment is devoted to antigenic variation. And of those thousand genes, one and only one seems to be expressed at any single moment during most of the creature's life.

If the action on the trypanosome's surface seems extraordinary, the genetic machinations that power those changes are even more so. In the last decade, several research teams, including Donelson's, have determined that each trypanosome has not 46 distinct chromosomes, as humans do, but 150 or so, with VSG genes scattered throughout. About a hundred of these chromosomes are tiny. In addition to these minichromosomes, each trypanosome has about 20 chromosomes that are 5 to 50 times larger, and it now seems clear that, at least in terms of antigenic variation, that's where the action takes place. On each end of those chromosomes is an expression site—a place where copies of VSG genes from throughout the genome are installed and activated to make these coat proteins.

Why is only 1 of these 1,000 genes on 40 sites active at any one time? The trypanosome employs several alternative strategies, and the metaphor that seems to describe the process best, Donelson says, is the operation of a video cassette recorder.

"Imagine," he says, "that I have a library of shelves containing 1,000 video cassettes—VSG genes—each with a different movie. To watch a western, I make a duplicate of 1 cassette and put it into 1 of 40 VCRs—expression sites—on the table, while leaving the original on the shelf. Then, if I want to watch a new movie, I remove the copy of the western, throw it away, and replace it with a copy of a comedy."

It gets more complicated. Trypanosomes in the blood double in number every 5 to 10

SOMETIMES THE FIRST PART OF THE TRYPANOSOME'S "MOVIE" WILL SEEM TO BE A WESTERN, AND THE SECOND HALF A COMEDY...

hours, and at least every six minutes a cell with a completely different coat pops up. At that rate, the full repertoire of even a thousand VSG genes would soon be exhausted. Yet the relentless cycle of infection can last many months in a sick person or animal, usually ending only with the host's death. Clearly the parasite has developed strategies for increasing the size and diversity of its "film library" even further.

Researchers have noticed that sometimes the first part of the trypanosome's "movie" will seem to be a western and the second half a comedy—as though two tapes got spliced together in the duplication process, Donelson says. The genes for these composite VSGs, which exhibit segments that seem to arise from different donor genes, are called mosaics and probably arose when the organism tried to put two copied "tapes" in a "VCR" at the same time. The resulting protein may be an antigenic hybrid and quite unlike the two originals.

In the last two years, Donelson and his group have focused on what he calls a mutation generator—a mechanism by which the trypanosome systematically introduces point mutations into the duplicated copies of its VSG genes. Careful analyses of both strands of the DNA in each gene have shown that such mutations aren't scattershot: some kinds of mutations are much more likely to occur in one strand than the other. That indicates, Donelson says, that the mutations occur during the duplication process itself. Though the exact mechanism is still mysterious—and could occur during any one or in several steps in the process—one possible model might be found in HIV, the AIDS virus. HIV also changes its coat proteins markedly over the course of an infection. The instigator of change in that case is a sloppy enzyme—reverse transcriptase—that doesn't faithfully transcribe viral genes.

Donelson suspects that a similarly sloppy enzyme may be at work in trypanosomes. What if, for example, such an enzyme garbled the RNA message it creat-

The "Kissing Bug"

You don't have to go to Africa to run afoul of a trypanosome. *Trypanosoma cruzi*, a sneaky cousin of *T. brucei*, stalks people in rural parts of Latin America and has recently found its way into the U.S. blood supply, too.

When it first turned up in the gut of a bug in the backlands of Brazil in 1908, *T. cruzi* was a parasite in search of a disease. Carlos Chagas, an ambitious 29-year-old doctor and skilled entomologist, had traveled to a steamy wilderness town north of Rio de Janeiro to battle an epidemic of malaria among workers building the Central Brazilian Railway. Chagas had long before learned to pay attention to insects; he knew that the female *Anopheles* mosquito, lurking near puddles of water and around homes at dusk, was the key culprit in the spread of malaria.

Those same thatch-roofed, mud-walled homes, he soon discovered, were rife with another type of insect—blood-sucking reduviid bugs, or "kissing bugs," that lived in the roofs and wall cracks by day and emerged at night to quietly feast on sleeping humans. On a hunch that the roachlike insect might transmit disease, Chagas checked the intestinal contents of one under a microscope. He found a swarm of tiny, arrow-shaped parasites. The organisms, it turned out, belonged to a previously unknown species of *Trypanosoma*, which Chagas named *T. cruzi* after his mentor, the legendary Brazilian biologist Oswaldo Cruz.

Chagas eventually isolated the same organism from the blood of a 3-year-old girl suffering from the first stages of what would come to be called Chagas' disease. Though the initial infection can be fatal, especially if left untreated in young children, it more often gives rise to high fever and swollen face and glands. These symptoms disappear after several weeks. At that point, the number of *T. cruzi* in the bloodstream decreases, but the parasites persist inside cells of a variety of organs. Such an infection may be silent for years as the parasites multiply inside cells, periodically wiggling out to invade and destroy a neighboring cell or to cruise in the bloodstream. Eventually, in 10 to 30 percent of all cases, the infection severely damages the heart and gastrointestinal tract. This may produce irregular heartbeat, congestive heart failure, and stroke, or debilitating gastrointestinal disease. Death from Chagas', sometimes decades after infection, can be swift—or slow and ugly.

Drugs are of little use after the first several weeks of infection. No vaccine exists, and some scientists doubt one can be developed. Though *T. cruzi* doesn't keep changing its coat, as African trypanosomes do, it evades the immune system by diving into human cells. So most countries have concen-

---

ed from the VSG gene and then, occasionally, not only produced a garbled protein, but also reincorporated the mish-mash of a message back into the parasite's genome? Presto, says Donelson, a new genre of "films" is added to the organism's "library," producing still more variation in its coat. The genetic possibilities may be endless.

Now that so many "VCRs" and "tapes" have been located in the trypanosome genome, a central question remaining is,

Who has the remote control? In other words, where is the master switch that determines which one of the dozens of expression sites and thousands of genes will suddenly become active while all the rest remain quiescent? George Cross and other researchers have recently inserted marker genes into the trypanosome's expression sites in tandem with VSG genes, which allows them to analyze the switching events and may bring about a

The blood-sucking "kissing bug," or reduviid bug, emerges at night to bite people in thatch-roofed, mud-walled houses in Latin America. It often carries—and transmits—a trypanosome called T. cruzi, *which causes Chagas' disease.*

trated their money and energy on eradicating the insects that carry *T. cruzi*. Despite their efforts, however, Chagas' disease continues to be a problem. Estimates of the number of people infected with the parasite today range from 10 to 18 million in Latin America. Once almost exclusively a disease of the rural poor, Chagas' has in the last decades followed urban migration. And once inside cities, it easily jumps all class and economic barriers to infect the general population via blood donations from healthy-looking people who don't know they are infected.

The few countries—Brazil among them—that do widespread screening in blood banks for *T. cruzi* rely on tests that detect antibodies to the parasite. Unfortunately, antibodies to other parasites may cross-react with antibodies to *T. cruzi*, which means that such tests for Chagas' are clouded by false positives.

The problem of keeping *T. cruzi* out of the blood supply is no longer a solely Latin American concern, says Louis Kirchhoff, a Chagas' specialist and colleague of John Donelson's at the University of Iowa. Though the disease is only very rarely transmitted to people by insects in the United States, a few surveys conducted among recent immigrants from Latin America lead Kirchhoff to believe that there are now 50,000 to 100,000 people infected with *T. cruzi* north of the Mexican border. That makes the parasite more prevalent in the pool of potential blood donors than human T cell leukemia virus, which U.S. blood banks already take great care to screen out.

A few cases of transfusion-associated Chagas' disease have already turned up among people with suppressed immune systems, Kirchhoff notes. Such people are at particular risk of developing *T. cruzi* infections that are rapidly lethal. One patient whose immune system had been weakened by leukemia treatments died of the infection. Kirchhoff believes

much more detailed understanding of the mechanisms involved.

In 1993, using x-ray beams to probe the nooks and crannies of several VSGs, crystallographer Don Wiley, an HHMI investigator at Harvard University, and his colleagues were able to actually catch a glimpse of the changeable proteins from every angle and figure out their three-dimensional structure. This led to a surprising discovery.

The scientists knew that the sequences of amino acids making up the various VSGs were quite different. Nevertheless, they found that about 60 percent of the VSG's overall structure was the same in all cases. Most of the differences in shape were concentrated near the exposed outer edges of the VSG—parts that are "visible" to the host's immune system as the parasite swims through the bloodstream. Perhaps, Wiley says, maintaining the same struc-

such cases are merely sentinels of many others in whom the infections are still silent.

What would be helpful, both in the United States and abroad, he says, is a cheap, easy, and accurate test that can directly detect the very few parasites circulating in the blood of infected people. Kirchhoff, in collaboration with Donelson and others, has developed a prototype of what they hope will be just such a test, using the amplifying power of the polymerase chain reaction, or PCR, to reveal sequences of the parasites' DNA. And another HHMI investigator, Larry Simpson of the University of California, Los Angeles, has independently developed an analogous test that amplifies a different part of the parasites' DNA.

Simpson, working with collaborators in Los Angeles and Brazil, has focused on the DNA found inside *T. cruzi's* "kinetoplast," a large, rod-shaped organelle that sits at the base of the creature's whip-like flagellum. The genetic material of the kinetoplast, divided into tiny loops called minicircles, is both quite specific to *T. cruzi* and variable from strain to strain of the organism. This means that eventually it may be possible, Simpson says, to figure out whether different

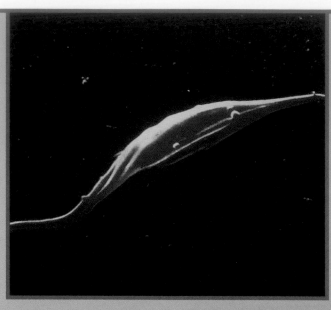

This trypanosome, T. cruzi, *recently found its way into American blood banks.*

strains of *T. cruzi* lead to different sets of symptoms and might be susceptible to different drugs.

As a distinctly controversial sidelight, Simpson thinks his test for *T. cruzi* could help settle a long-standing historical question: did Charles Darwin die from Chagas' disease, picked up on his 1830s voyage on the H.M.S. *Beagle*? In his later diaries, Darwin regularly complained of extreme fatigue and vomiting attacks after meals, which compelled him to give up dinner parties altogether. Darwin scholars have for years debated the source of the mysterious symptoms, suggesting diagnoses that range from psychosis to a pigeon allergy. Simpson, who is not the first to suspect that at least part of Darwin's problem may have been Chagas', is only partly kidding when he makes what he calls "a modest proposal."

"I propose we remove a sample of Charles Darwin for PCR analysis of *T. cruzi*," Simpson says. "After all, he was interred in a very accessible location—in Westminster Abbey, right next to Isaac Newton—and we only need a little piece." – *D.F.*

ture from protein to protein facilitates the trypanosome's ability to pack VSGs tightly. Indeed, the primary function of most of the protein seems to be to serve as an impermeable wall to prevent hit squads of antibodies from ripping into the organism's outer membrane.

Given this similarity in shape from protein to protein, the myriad differences in the VSGs' amino acid sequences seem—at least at first blush—like overkill. Cold and

flu viruses, which also change the configuration of their coats to elude immune defenses, are able to stay ahead of antibodies with a mere dozen or so point mutations. Why does the trypanosome employ so many more? Wiley thinks he may have an answer for that, too.

While the rampant variation may be just the result of an overly enthusiastic mutation generator, Wiley suggests it may also be crucial in helping the trypanosome escape cellular immunity, the second arm of the mammalian immune system. Helper T cells, key players in that branch of the defense, are activated only when they "see" bits of chewed-up antigenic protein, properly presented as short peptide sequences by other specialized immune cells. By sprinkling genetically programmed alterations throughout the VSGs of each generation, the trypanosome may be ensuring that every bit of chewed-up protein presented to a T cell from one trypanosome looks very different from the analogous markers of its ancestors.

Donelson, too, has recently become quite interested in cellular immunity. "All of us in the field have been so mesmerized by this VSG switching phenomenon," he says, "that I think we may have been misled into thinking that's the only way the trypanosome evades the immune system." In fact, starting with T cells, a host of other defenders in the blood must be foiled.

If the African trypanosome has an Achilles heel, it may well reside in the flagellar pocket. This small invagination is located just ahead of the point where the flagellum attaches to the trypanosome's central trunk and is the only spot on the organism's surface that isn't completely covered by VSGs. It is through this pocket that the trypanosome conducts all housekeeping functions; nutrients and wastes pass in and out and so do a number of potentially interesting chemical messengers, according to Donelson. Identifying these chemicals could lead to new ways of interfering with the trypanosomes' growth in the blood—perhaps just enough so that other immune factors could

eliminate the infection.

The human body is already able to disarm at least three common strains of African trypanosome. One of them—*T. brucei*—is morphologically indistinguishable from the strains that cause fatal illness in people. *T. brucei* infects humans but disintegrates in the blood before it becomes established. Researchers have long been eager to figure out exactly what it is in human blood that does the trick, and Stephen Hajduk and his colleagues at the University of Alabama now believe they may have done exactly that.

Cell biologist Mary Rifkin, now at Mount Sinai Medical School in New York,

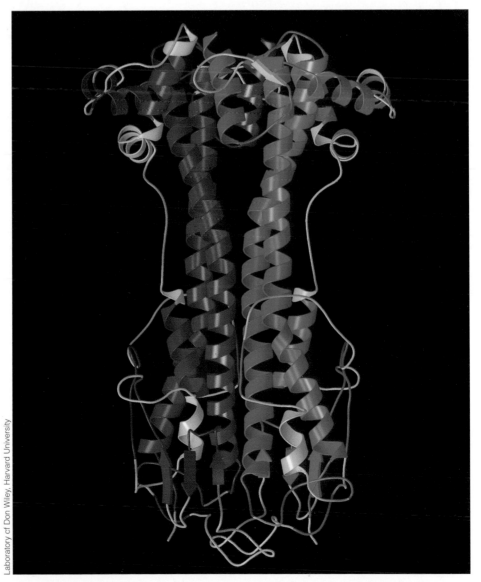

Same shape, different components: Only the areas shown as thin green ribbons in this model of a VSG from a trypanosome's coat vary from one VSG to the next, even though the amino acids that make up the VSGs have strikingly different sequences.

Laboratory of Don Wiley, Harvard University

determined some years ago that whatever the killing factor was, it seemed to reside in the fraction of human blood serum that contains high-density lipoproteins, or HDLs, molecules much better known for their role in ferrying cholesterol from the blood. Hajduk and his team then localized the killing factor to a subclass of HDLs they called trypanosome lytic factor, or TLF. Next they purified a constituent of TLF that they think is the actual killing factor—a protein similar in structure to haptoglobin, a substance in the blood that binds to hemoglobin.

The Alabama researchers now believe that this newly identified protein also binds to hemoglobin and that the mixture is taken up by receptors on the surface of *T. brucei's* flagellar pocket. From there it travels to the parasite's lysosomes, small sacs filled with digestive enzymes. Once inside a lysosome, the mixture of haptoglobin-related protein and hemoglobin seems stimulated by the acidic enzymes to generate destructive free radicals; these, in turn, break holes in the lysosome. Within minutes, in a ghoulish turnabout, the acids spill out, and the parasite is digested from the inside out.

Final proof that the haptoglobin-related protein is the key difference between infective and noninfective trypanosome strains awaits cloning of the gene for the protein, Hajduk says. Then, if the researchers can put the gene into mice that are vulnerable to *T. brucei* and thereby protect the mice from the parasite, they will have opened the possibility of eventually using genetic engineering to confer trypanosome resistance on cattle.

For Donelson and many other researchers whose interest in parasitology started with friendships forged in countries plagued by the diseases they study, this potential for real-world applications of their work has fired new resolve.

"Everything we've learned about this organism in the last decade suggests that it is just too clever for a conventional vaccine against it to be successful," Donelson says. To thwart the parasite in animals and in people, researchers will have to use strategies that are just as subtle, and it is only in the last few years that the new tools of genetic engineering have offered them a fighting chance. •

SHIVERING AND SHAKING TO DEATH

Back from a trip to South America, a Canadian filmmaker suddenly began to feel very cold. Friends he was visiting brought him a jacket, then a blanket, but nothing stopped his shivering.

"Minutes later, they watched, horrified, as I lay shaking and twitching uncontrollably on their couch," the filmmaker recalled. When the chills stopped, he became hot and woozy, with a fever of 104° that left him drenched in sweat. The same cycle of violent chills and burning fever repeated itself 36 hours later. This sent him to a hospital, where doctors diagnosed a classic case of malaria, transmitted through the bite of an infected mosquito.

Actually, he was lucky. Had he been infected with the deadliest of the four strains of malarial parasites that afflict humans—*Plasmodium falciparum*, which is rapidly becoming resistant to drugs—he could have died. The strain that he acquired, *P. vivax*, was not as dangerous and could still be treated with high doses of chloroquine. Nevertheless, he ran the danger of having a relapse within a year.

Throughout the world, about 300 million people now suffer from malaria and close to 1 million die of it every year, especially in tropical areas. Most of these people die from lack of treatment, but some die because the drugs no longer work. The disease has had a soaring revival in Asia, Africa, and Latin America in the past 20 years, even though public health experts had done everything in their power to wipe it out and thought they had won the battle.

Right after World War II, a two-pronged attack was launched on both the *Anopheles* mosquitoes that transmit malaria and the parasite itself. DDT, which had just become widely available, seemed the ultimate weapon: it could be sprayed on the inside walls of houses, where it killed mosquitoes on contact for an entire season, as well as on swamps and other watery breeding places. The World Health Organization coordinated an international campaign to eradicate the parasite in this way, and at first it proved immensely successful. In Sri Lanka, for instance, where there had been 3 million cases of malaria in 1946, only 29 cases were reported in 1964. But that same year, the spraying stopped because people assumed they had eliminated transmission of the disease. Four years later, there were again half a million cases of malaria in Sri Lanka, and by 1970, the number had swelled to 1.5 million. Meanwhile, as farmers around the world used DDT excessively to keep insects away from their crops, the *Anopheles* mosquitoes became resistant to it.

Heavy spraying of DDT along the Tennessee River in 1945 did take care of malaria in the United States, where the disease had once been a very serious problem. Each year, about half the soldiers on either side in the Civil War came down with malaria. Even though we still have some mosquitoes that could transmit the disease, the United States has remained nearly free of malaria ever since the DDT spraying. Europe, too, cleared itself of malaria, mainly because of its cooler climate and a tamer strain of mosquito. But when one looks at the world as a whole, the number of cases of malaria has doubled in the past 10 years. The disease is, once again, a major killer of children under the age of 5, who die when the parasites reach their brains and block off the flow of blood.

To make matters worse, drug therapy—the second front in the war against malaria—is also failing. "The parasite can mutate to escape the action of drugs," explains Alan Cowman, an HHMI International Scholar at the Walter and Eliza Hall Institute of Medical Research in Melbourne, Australia, who has been studying the mechanism of this resistance in the hope of finding ways to thwart it. He points out that in Asia and Latin America, 95 percent of the malarial parasites have become resistant to chloroquine, the cheapest and most potent antimalarial drug. This resistance is then passed on to other malarial parasites, "and it spreads like an invading army. You can see it on a map," he says. Nor are there many alternatives to chloroquine. "We've had to go back to quinine—30 years back-

wards—to a drug that has bad side effects and is not all that great," Cowman says. If the parasites become resistant to that and a few other drugs, there may be nothing left.

Now that molecular biologists are tackling parasitic diseases, however, they have begun to redefine the terms of engagement, both in the war against the mosquitoes that spread malaria and in the struggle against parasites in the human body.

Instead of trying to kill off all mosquitoes, for example, some researchers are focusing on the most dangerous strains and planning to alter them by introducing new genes, so that the mosquitoes

A bite from a female Anopheles mosquito such as this one may transmit malaria.

become unable to transmit the malarial parasites. "Most strains of mosquito cannot transmit malaria to humans," explains Louis Miller, chief of the Laboratory of Malaria Research at the National Institute of Allergy and Infectious Diseases (NIAID) in Bethesda, Maryland. "And there is no advantage to the mosquito in transmitting it. So in theory it should be possible to put certain genes into malaria-bearing strains to make them stop. This would not eliminate the mosquitoes, but it might eliminate the parasite."

Molecular biologists are also devising new kinds of drug therapy. One idea is to starve the parasite to death by preventing it from using the hemoglobin that it sucks up in red blood cells. This may actually become possible as a result of recent research by

Daniel Goldberg, an HHMI investigator at the Washington University School of Medicine in St. Louis, Missouri. "Malaria parasites have a voracious appetite for hemoglobin," says Goldberg. "In just a few hours, they can chew up a quarter pound of hemoglobin in a heavily infected patient." Goldberg was studying how *P. falciparum* manages to break down all this hemoglobin so rapidly when he discovered that one type of digestive protein, an aspartic protease, does the main job by snipping through a crucial section of the hemoglobin molecule. This hinge-like section normally keeps the molecule stable after it has picked up a cargo of oxygen, but when the protease cuts through it, the hemoglobin molecule breaks open and is easily digested by other enzymes. Goldberg's lab is now designing inhibitors that might block this protease and thus prevent the hemoglobin from being broken down. "If the parasite can't chew up hemoglobin and can't grow, it will die rapidly," Goldberg says.

The best weapon of all would be an effective anti-malarial vaccine. Several labs are working on it, but "this is one of the biggest scientific challenges we have today," says NIAID's Louis Miller. The problem is that malarial infections do not produce the kind of immunity that we associate with measles or polio, where a single bout of disease—or a single shot of vaccine—is enough to give lifelong protection against reinfection. "You can be infected with malaria 100 times a year," Miller explains. "Though you no longer get very sick, you still get a fresh infection." This makes it very difficult to develop an effective vaccine against the disease, he says.

Another curious aspect of the malarial parasites is their baroque life cycle. Malarial infections begin when a female mosquito that carries parasites in her salivary glands bites a human being and feeds on blood that will nourish her eggs. Immature, threadlike parasites enter the human body with the mosquito's saliva. Taking refuge from the body's immune system, they dive into liver cells, where they multiply. Then the parasites pour into the bloodstream and invade red blood cells. This is when people become ill—when the para-

sites gobble up hemoglobin, multiply rapidly inside the cells, produce toxins, and destroy the infected cells as they burst out of them. Sexually mature parasites then circulate in the bloodstream, waiting for another blood-seeking mosquito to swallow them up; once this happens, the parasites reproduce in the mosquito's stomach, their offspring migrate to the insect's salivary glands, and the cycle starts all over again.

Throughout it all, malarial parasites show an amazing ability to outsmart the immune system. For years nobody could understand why the immune system fails to destroy cells that are infected with *P. falciparum*. Scientists knew that a certain telltale protein called PfEMP1, which is made by the parasite, is involved in anchoring infected cells to the walls of blood vessels. But since this is clearly a foreign protein, they were puzzled by its ability to dodge immune responses.

Malaria parasites (yellow) burst out of infected red blood cells.

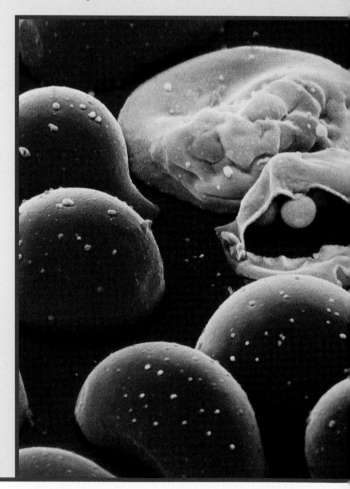

The answer came serendipitously last year. While looking for genes that make malaria resistant to chloroquine, Thomas Wellems of NIAID came across 5 related genes whose function was at first a mystery. He soon found as many as 150 similar genes scattered over the 14 chromosomes of a single parasite. Since each parasite may have a unique set of these genes, which he named *var* for "variation," "there may be a million genes of this sort in malarial strains all over the world," Wellems says. Then Russell Howard of the Affymax Research Institute in Santa Clara, California, who was looking for the PfEMP1 protein, cloned two genes that actually produce such proteins on the surface of infected red cells. He found that the genes belong to the *var* family. Finally, Chris Newbold of the University of Oxford, England, and Louis Miller of NIAID showed that *P. falciparum* can turn on different *var* genes to produce different versions of the protein. In this way, the parasite ensures that the infected blood cells will remain unharmed by antibodies designed to combat an earlier version of the protein. Like trypanosomes, *P. falciparum* evades the immune system by staying one step ahead of its enemies. Now that the *var* genes have been identified, however, scientists hope that drugs may be developed to prevent or counteract the protein's effects.

One reason that progress in fighting the malarial parasite has been so slow is that until this year, it was impossible to manipulate the *P. falciparum* parasite genetically—to introduce foreign genes into its DNA or to modify existing genes so as to learn what they control. But Thomas Wellems has just developed a method of stably "transfecting" genes into the parasites during the stage when the parasite is inside red blood cells. "We can target specific areas," he says. "We are on the threshold of gene knockouts and specific modification of genes." According to Alan Cowman, who is working on the same problem in Melbourne, Australia, this is an important step toward understanding drug resistance and designing new drugs. "It will make a major difference to the whole field," Cowman says. "The next three to four years will be a boom time."

In another sign of progress, a dozen laboratories around the world are collaborating in an effort to make a detailed map of all the genes in the parasite at every stage of its life cycle. This effort is coordinated by David Kemp, an HHMI International Scholar at the Menzies School of Health Research in Darwin, Australia, and funded by the Wellcome Trust of England.

Some researchers are going one step further: they are deciphering the DNA sequences of the parasite's entire set of genes. "If only we could get all these DNA sequences in our hands, it would be fantastic!" says Louis Miller. "It would save years of painstaking effort." Miller believes that a rich database of DNA sequences would give scientists a bonanza of information about the parasite's biology and would enable them, at last, to develop effective ways of preventing—or curing—this debilitating and often fatal disease.   – *M.P.*

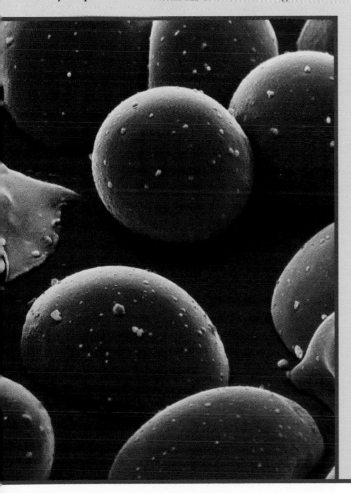

CAN AIDS BE TAMED?

Hundreds of newly made, lethal viruses (small blue balls) emerge from an HIV-infected white blood cell, ready to invade other cells.

Scientists believe they now know why all treatments for AIDS failed in the past—and how the disease might be tamed in the future. They base their views on a series of experiments conducted in 1995 with a crop of new, more powerful drugs against the human immunodeficiency virus, HIV, which causes AIDS.

Any treatment with a single drug is "doomed to fail," declares David Ho, the dynamic director of the Aaron Diamond AIDS Research Center at the Rockefeller University, New York, who led one of the ground-breaking studies. By the time such treatment is started, "a billion to a trillion variants" of HIV may be circulating in the patient's blood, and almost inevitably one of these variants will prove resistant to the drug. That particular strain will be selected for survival and will then multiply rapidly, defeating the treatment.

"This virus is not like anything we've seen before," Ho comments. Some other viruses also mutate rapidly, but they reproduce in the body for only a few days or a week, while HIV infection may last for 10 years or more before symptoms appear. "What makes HIV unique is its high mutation rate multiplied by the very large number of rounds of replication that go on in the infected person," Ho says. "If a person is making a billion virus particles a day, and there are thousands of replication cycles,

by Maya Pines

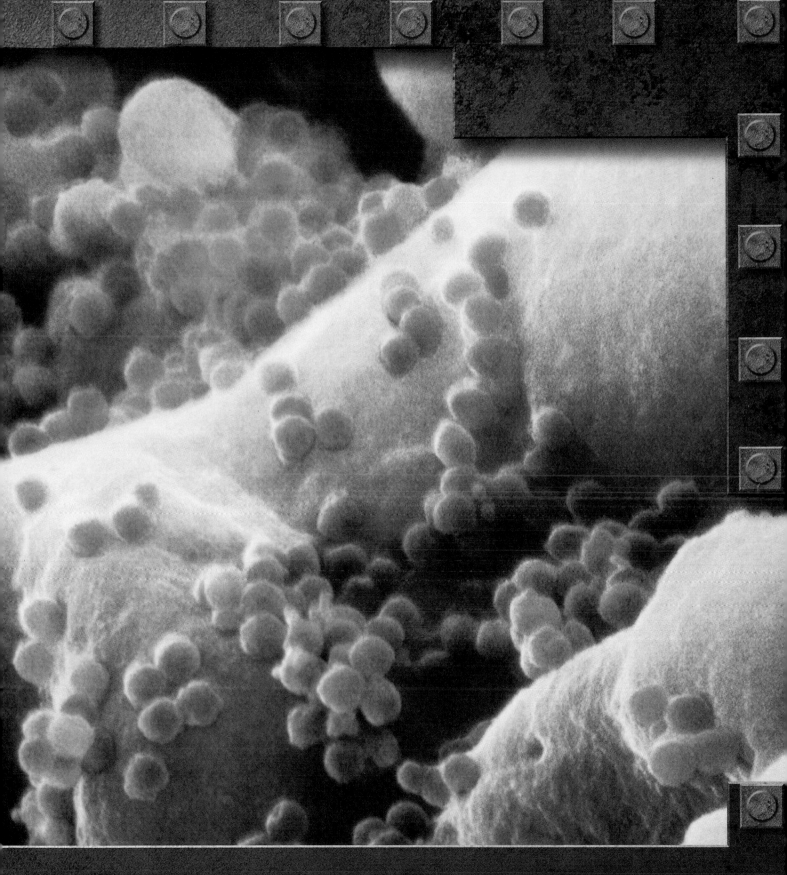

and we know the mutation rate for HIV is about 1 in 10,000 bases of DNA, an escape from drugs should not be a surprise at all."

Therefore Ho and others have suggested hitting the virus with several different drugs at once. Two drugs are not enough

"The 1 in 100 million chance of a virus developing two different mutations that make it resistant to two different drugs is not statistically unlikely," Ho says. "With a billion to a trillion variants, it happens all the time." But with three or more drugs

HOW HIV COMES TO LIFE AND MULTIPLIES

HIV, the virus that causes AIDS, springs to life after it enters a human cell. Until then, it is just an inert package—a few genes wrapped in a cocoon of proteins, lacking most of the machinery required by living organisms. Its life and a chance to multiply hang on a set of sticky proteins that protrude from its outer envelope. If some of HIV's sticky proteins adhere to an appropriate cell, the virus will fuse its envelope with the cell surface and spill its contents into the cell.

Mission accomplished! The virus has found a home. Like a skilled robber, it proceeds to strip this home of all the items it finds valuable. It takes key parts of the cell's machinery for making proteins and uses them immediately. The first order of business is to melt down the thick core, or capsid, that contains HIV's genetic material and free the genes to do their work inside the cell.

The next step is critical—and explains why HIV mutates so rapidly. It consists of a quick and dirty conversion of HIV's genes, which are single strands of RNA, into double strands of DNA that can be spliced into the chromosomes of the invaded cells. This conversion takes place without any of the proofreading that provides quality control when DNA is copied into more DNA, as happens during cell division. HIV is a retrovirus, a member of a strange group of viruses whose genetic material is inserted permanently into the host cell's chromosomes after a rapid conversion from RNA to DNA. The viral enzyme reverse transcriptase, which first copies the RNA strand into a single strand of DNA and then copies this DNA strand into a complementary strand, producing DNA's characteristic double helix, has a notoriously high error rate. It makes about one error per cycle, so each retroviral DNA genome generated by this enzyme can be expected to contain at least one mutation.

The conversion is completed about four hours after the cell was invaded. The newly made viral DNA then enters the cell's inner sanctum, its nucleus, and is permanently integrated into host chromosomes. From then on, it is processed by the cell's machinery, under the control of viral regulatory genes. When the cell is activated, the viral DNA is translated into messenger RNA that moves out of the nucleus into the cell's cytoplasm. There,

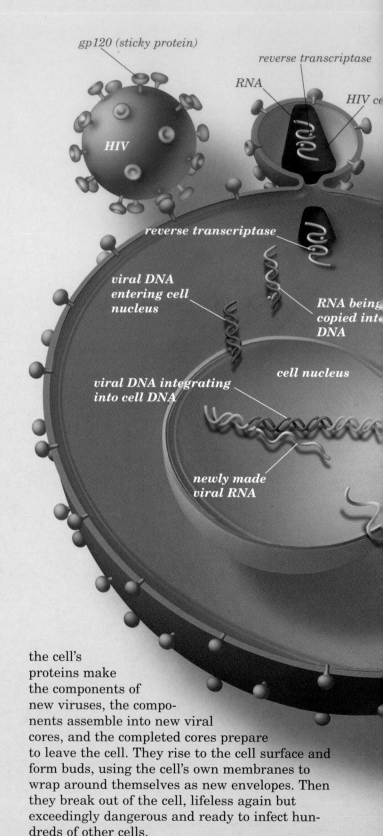

gp120 (sticky protein)

reverse transcriptase

RNA

HIV c

HIV

reverse transcriptase

viral DNA entering cell nucleus

RNA being copied int DNA

cell nucleus

viral DNA integrating into cell DNA

newly made viral RNA

the cell's proteins make the components of new viruses, the components assemble into new viral cores, and the completed cores prepare to leave the cell. They rise to the cell surface and form buds, using the cell's own membranes to wrap around themselves as new envelopes. Then they break out of the cell, lifeless again but exceedingly dangerous and ready to infect hundreds of other cells.

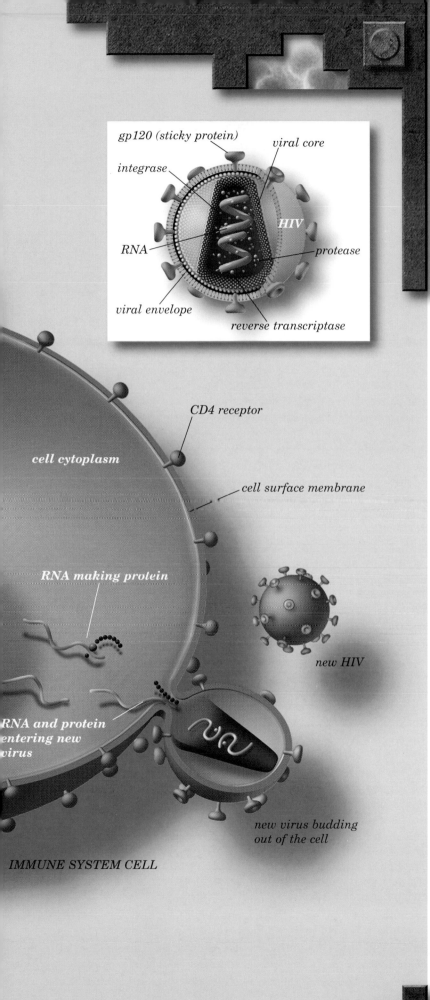

gp120 (sticky protein)

viral core

integrase

HIV

RNA

protease

viral envelope

reverse transcriptase

CD4 receptor

cell cytoplasm

cell surface membrane

RNA making protein

new HIV

RNA and protein entering new virus

new virus budding out of the cell

IMMUNE SYSTEM CELL

there is hope. "If we ask the virus to mutate simultaneously at three different positions, what are the odds? Well, the odds are not high," Ho says. "It would be about 1 in a trillion—a very, very low number, or a very improbable number."

This means that the virus might be kept in check by therapy with multiple drugs. People who took such drugs would not be cured of their infection, Ho points out, but with relatively small amounts of virus in their blood they might never develop AIDS. They might live almost normally for many decades, as do people who have diabetes or hypertension.

The most widely used drug so far, AZT, is not particularly potent, but it could be given with more effective drugs. Many new drugs are now being developed to target different phases of the viral life cycle. Several clinical trials using various combinations of drugs have started. In addition, scientists are exploring some totally new approaches to treatment based on their growing understanding of how the virus operates. They are also trying to develop vaccines. Compared to the gloom and doom among AIDS researchers only a few years ago, there is some optimism in the air.

Nevertheless, the death toll keeps rising. AIDS is now the leading cause of death among people between the ages of 25 and 44 in the United States, surpassing accidents, cancer, and heart disease. Since 1980, more than a million people, including infants, have been infected with HIV in the United States alone, and the number is growing by about 40,000 a year. Half of these people have already died of AIDS. In large parts of Africa and Asia, the AIDS epidemic is raging out of control, decimating the young and wiping out entire villages.

The mountains of knowledge acquired by scientists in the past 15 years ("We know more about HIV than about any other virus," says Ho) and the huge sums of money spent on AIDS research will do people who have been infected with HIV little good if effective therapies do not come soon enough. As Robert Livingston, chairman of the appropriations committee of the House of Representatives, noted, "We spend $295 per patient on cancer research, $93 on heart disease, $54 on Alzheimer's, $26 on Parkinson's—and $36,000 per AIDS patient on research." Yet until 1995, these enormous efforts had failed to provide either powerful drugs or a workable strategy on how to combat the disease.

When a few experimental drugs that seemed more potent than AZT came out of the research labs, Ho's team and a team led by George Shaw of the University of Alabama at Birmingham launched the studies that would radically change scientists' views of how the human body responds to infection with HIV and how AIDS develops.

The first sign of infection with HIV is often a flu-like illness that does not seem too unusual or frightening. People get over it and think they are well. Unless their blood is tested for antibodies to HIV, which appear a few weeks or months after the initial infection, they may have no idea that they carry the seeds of death and can transmit them to others.

The 3 to 10 years after infection are generally a period of deceptive calm before The Horror, as it has been called, begins. Then the dreaded symptoms of AIDS become evident: the purple blotches of skin cancer, bouts of life-threatening pneumonia, intractable diarrhea, swollen lymph nodes, fevers, weight loss, meningitis, and a plague of "opportunistic" infections—viral, bacterial, or fungal.

Such illnesses usually develop after a sharp decline in the number of white blood cells called CD4-positive T lymphocytes, or CD4 cells, circulating in the infected person's blood. These cells are the shock troops of the immune system. They normally bear the brunt of battle against viruses and other infectious agents. But they are also the main target of HIV. The virus invades CD4 cells and uses the cells' machinery to make thousands of new copies of the virus—thereby killing the CD4 cells. The new viruses then go out and infect more cells. Some infected CD4 cells are killed by other immune system cells, which recognize the HIV-bearing cells as abnormal. A drop in the number of CD4 cells is very bad news.

Until the Ho and Shaw studies, scientists believed that the virus remained quiescent in the blood for many years after infection, until something activated it and the disease began. But in fact, as the two studies showed, HIV and CD4 cells wage an epic battle inside every infected person's body every day.

Both Ho and Shaw gave AIDS patients large doses of experimental drugs that stop HIV from reproducing inside infected cells. The researchers also used very sensitive new techniques to measure the number of viruses circulating in the patients' blood before and after the treatment. They took blood samples up to 12 times a day and made extensive calculations.

In both hospitals, the researchers were stunned to discover the extent of warfare that went on in the patients' blood before treatment began. Every day, about a billion freshly minted, lethal viruses rose in the bloodstream to attack the body's immune defenses. Every day, a billion uninfected CD4 cells surged to replace the dead white cells and to kill the viruses. Both sides strained to their utmost in this relentless struggle. Over short periods of time, the viruses and CD4 cells seemed to reach a steady state, but in the long run the number of viruses went up while the CD4 cell count went down.

"It's like a sink that has an increasingly low water level," explains Ho. "We used to think it was a supply problem—that there wasn't enough CD4 flowing in from the tap. But in fact, the tap is pretty wide open and there is plenty of water coming out of it. It's just that the drain is draining slightly faster. And that drain is the destruction induced by HIV. So what you need to do is to attack the virus and plug up the drain."

The immune system works so hard to replenish its CD4 cells that it actually goes into overdrive, producing 25 to 78 times more CD4 cells in advanced stages of the disease than before, Ho says. But "the source of replenishment cannot be infinite," he notes. In the end, the mechanism for regenerating CD4 cells appears to become exhausted, and the number of CD4 cells dwindles to zero.

The experimental drug that Ho used in

Every day, about a billion freshly minted, lethal viruses rose in the bloodstream to attack the body's immune defenses.

his study, ABT-538 (now called ritonavir), is a protease inhibitor; it blocks the activity of HIV protease, an enzyme that cuts up a large precursor of essential viral proteins. Without HIV protease to set them free, these viral proteins cannot function and the virus cannot reproduce. The drug was so powerful that the number of viruses in the patients' blood dropped rapidly within two days; within two weeks, the viruses were down to only 1 percent of their previous count. At the same time, the patients' CD4 counts rose. But then the virus became resistant to the drug and the two trends were soon reversed.

In Alabama, Shaw had a similar experience with both ABT-538 and another experimental drug, nevirapine, which attacks HIV's reverse transcriptase (see p. 286), but far more potently than AZT does. His patients' virus counts fell rapidly, but the virus became drug resistant within two to four weeks.

To avoid such drug resistance, researchers have begun several trials of combination therapy, most of them with patients who have low CD4 counts and full-blown AIDS. But at the Aaron Diamond Research Center, a dozen otherwise healthy people are undergoing what Ho calls "early and aggressive treatment" with high doses of three different drugs only three months after their infection with HIV. Ho believes this is the most effective time for treatment because there are fewer variants of HIV in a patient's blood right after infection and therefore fewer chances that one of these variants can resist the drugs.

In addition to ABT-538, this trial uses two drugs that interfere with the virus's reverse transcriptase, AZT and the newer 3TC. When used alone, each of the latter drugs provokes the emergence of drug-resistant HIV strains containing mutations in their reverse transcriptase. But a recent study of 3TC by Vinayaka Prasad at the Albert Einstein College of Medicine in the Bronx, New York, suggests that 3TC has a most interesting property: it seems to provoke the appearance of one particular HIV strain whose reverse transcriptase—usually so error prone—copies its genes faithfully, thereby preventing any other drug-resistant mutants from being generated. The chosen mutant is resistant to 3TC but not to other drugs. Having only one strain of HIV to fight would make it much easier for the immune system (and other drugs) to win the battle against the virus.

Even assuming that the three-drug trial works as well as Ho and his associate Martin Markowitz hope it will, the scientists do not yet know "whether we will be able to stop the drugs at the end of the year," Markowitz says, or whether this expensive treatment, with its various side effects, would have to go on forever to be effective.

Meanwhile, the search for other drugs, new approaches to treatment, and an AIDS vaccine goes on in laboratories around the world.

Though hundreds of excellent antibiotics have been developed against bacterial diseases, very few antiviral drugs of any kind exist. Bacteria are generally self-sufficient organisms whose numerous genes code for proteins that are quite unlike those of human cells. For this reason, an antibiotic aimed at bacterial proteins is likely to be safe for people. But viruses travel light, with so few genes and proteins of their own that they cannot even reproduce unless they steal some of the machinery of the cells they invade. This makes them much harder to attack, since most drugs that would stop viruses would damage human cells as well.

It is particularly difficult to deal with retroviruses such as HIV, which splice their genetic material right into the chromosomes of the infected cell. Most of the drugs that have been developed so far attack HIV's reverse transcriptase, which copies the virus's RNA into DNA that can be inserted into the cell's DNA. Obviously if a drug can prevent this from happening, the virus will not be able to reproduce. But HIV's nine genes make other important proteins as well. Much effort is now going into deciphering precisely what these proteins do, how important they are to HIV's survival, and how best to throw a monkey wrench into them. The three-dimensional structure of HIV's reverse transcriptase has been known for some time, but recently scientists revealed the detailed structures of the viral integrase and protease as well. This is helping researchers to design drugs that may block them.

Molecular biologists are examining each stage of HIV's life cycle to find a possible opening and to develop a better understanding of this resilient virus. At Duke University Medical Center in Durham, North Carolina, for example, HHMI investigator Bryan Cullen focuses on Rev, a key regulatory protein made by HIV; he recently identified the cellular protein that Rev binds to. Other researchers are studying the

Tat and Nef regulatory proteins.

Scientists are also trying to understand the many ways in which the immune system fights HIV. In addition to CD4 cells, the white blood cells called CD8 lymphocytes may be extremely important in this battle, for they produce certain chemical factors that seem to suppress HIV's ability to copy itself. Several teams are now exploring the use of such factors as treatments for AIDS.

The most effective way to stop the AIDS epidemic, of course, would be to develop a vaccine. A major effort is under way, but it has been hampered by the lack of a good animal model of the disease. As Malcolm

A MACHINE FOR MAKING ERRORS

This structure of one of HIV's proteins, reverse transcriptase, could be called a machine for making errors. Reverse transcriptase is the primary source of the amazing number of mutations in HIV. After HIV has invaded a cell, reverse transcriptase copies the viral RNA into DNA that can be integrated in the cell's chromosomes. But reverse transcriptase lacks any proofreading ability and therefore makes a large number of errors—exactly what the virus needs to generate hundreds of variants, as diverse as snowflakes, which can evade both the immune system and most drugs.

In 1992, Thomas Steitz, an HHMI investigator at Yale University, deciphered the 3-D structure of reverse transcriptase bound to a drug, nevirapine. Reverse transcriptase consists of two subunits of very different shape and function. The larger unit (gold) does the work of copying the viral RNA strand (grey) into a strand of DNA (white). When the drug nevirapine (red) blocks the structure, reverse transcriptase becomes less flexible and less able to do its copying job; this prevents HIV from replicating. Being able to see this structure helped scientists to improve the design of the drug and make it more effective.

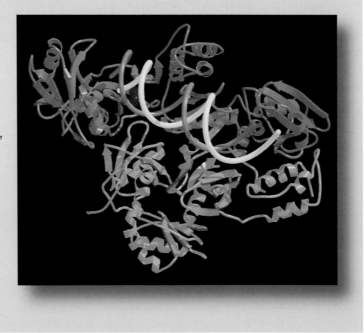

Martin, chief of NIAID's Laboratory of Molecular Microbiology, points out, "Ordinary mice cannot get infected with HIV. And while you can infect chimpanzees with HIV, usually they don't get AIDS. You can show that the virus replicates and makes new particles in the animal, but the chimp doesn't get the disease.

"What we need for the development of an HIV vaccine is a really good end point, a readout of disease," Martin explains. "Most vaccines don't prevent infection. Polio vaccine doesn't prevent infection by polio virus; it prevents the virus from getting to the nervous system and producing paralytic disease. When you give hepatitis vaccine, usually the infection is stopped in the gastrointestinal tract before the virus even gets to the liver, and therefore you don't get the disease. So for AIDS, we need a disease readout, like a loss of CD4 cells, which shows the beginning of the disease." The ability to prevent the loss of CD4 cells would then be a test of the vaccine's effectiveness.

Monkeys that are infected with a slightly different virus, SIV (simian immunodeficiency virus), do develop an AIDS-like disease. In 1991, Ronald Desrosiers of Harvard University's New England Regional Primate Research Center reported that monkeys could be protected against the disease by injecting them with a vaccine consisting of a live, but crippled, SIV from which the *nef* gene had been deleted.

In dealing with a virus as variable as HIV, however, scientists hesitate to use an attenuated live virus as a vaccine, for fear it might mutate, become more virulent, and lead to AIDS. Some of the experimental vaccines developed so far contain only subunits of HIV—certain proteins or parts of proteins. Others use weakened non-HIV viruses, such as cowpox (vaccinia) virus, to which an HIV gene has been added. Since 1988, more than 1,800 volunteers have taken part in studies of 13 investigational vaccines of this sort. "We have shown that these vaccines are safe, and we know they stimulate an immune response," says Robert Belshe, director of the Center for Vaccine Development at St. Louis University School of Medicine. But to date nobody knows whether the vaccines actually prevent disease.

People who believe that AIDS can be tamed place much weight on the fact that about 5 percent of all persons who are infected with HIV remain healthy, without any sign of the disease, for more than a decade. These long-term nonprogressors, as they are called, have stable, normal CD4 counts and far fewer viruses in their blood than people with AIDS.

"It means you can live very well with the virus if you can contain it," says Giuseppe Pantaleo of NIAID, who has studied a group of 15 nonprogressors. But opinions differ as to why the virus is contained. Do these people remain healthy because they happen to be infected with a particularly "wimpy" strain of the virus or because they have an unusually strong resistance to it?

When researchers analyzed the virus carried by one group of seven Australians who had received blood transfusions from the same infected donor 15 years ago and who did not develop AIDS (nor did the donor), they found that this particular virus is missing part of the *nef* gene. "It's a very important experiment of nature," says Anthony Fauci, director of NIAID. "It nails down the concept that one of the reasons people might be long-term nonprogressors is due to defective virus."

Another reason may be found in a human gene, CKR5, which normally produces a protein that allows HIV to enter CD4 cells. Scientists at New York's Aaron Diamond Center and the Free University of Brussels have discovered that mutations in the gene can block the entry of HIV into cells and protect some people from infection.

The jury is still out on this question, as it is on most issues involving AIDS. The only thing that all nonprogressors have in common is a relatively low number of viruses in their blood. So while the ultimate goals of researchers are still to develop a cure and a vaccine, a more immediate goal is to find ways to lower patients' viral loads so radically that their immune systems will be able to hold their own against the remaining viruses, giving the patients a nearly normal lifetime. •

SOME HARD-WON
VICTORIES

He is the last man in the world to have caught smallpox: Ali Maow Maalin, a cook in Merca, Somalia. Maalin came down with the disease in 1977, when he was 23. As soon as his illness was diagnosed, medical teams armed with smallpox vaccine and special jet injectors swept into town. They vaccinated more than 50,000 people against the disease. Since then, there has not been a single case of smallpox anywhere, except for a medical photographer who was accidentally infected in a lab in Birmingham, England, in 1978. She died of smallpox and transmitted the virus to her mother, who survived. But no one else has contracted this once terrifying disease, one of the deadliest in history.

As recently as in the 1950s, 8 million people a year died from smallpox. Millions more were blinded or permanently disfigured by it. Yet vaccination had been available since Edward Jenner developed the method in 1796. An immunizing scratch on the arm was a routine part of childhood in every industrialized nation, and travelers were not allowed back into their own countries unless they could show proof of vaccination, which required booster shots every decade.

In 1967, the World Health Organization started an intensive immunization campaign that reached

The last man in the world to catch smallpox, a young cook in Somalia, survived the disease in 1977.

into the remotest regions of the globe. Within 12 years it declared the first—and so far, the only—total victory over any human disease. Smallpox is now completely eradicated. Since the disease strikes only humans and there is no animal reservoir of the virus, vaccination is no longer necessary. The campaign broke the chain of person-to-person transmission and wiped the disease off the face of the earth.

Only two small stocks of the virus remain, under heavy guard—one in a maximum-security wing of the Centers for Disease Control and Prevention in Atlanta, Georgia, the other in a modern laboratory in Novosibirsk, Russia. The smallpox viruses in both places are slated for execution at a still undetermined date. They will be put into an autoclave and heated at 248° Fahrenheit; then the dead viruses will be incinerated.

Though smallpox is unlikely to reappear, scientists are planning to stockpile half a million doses of smallpox vaccine—just in case. In addition, they will keep stocks of the vaccinia virus that is used to make it. There is always the risk that someone has hidden away some smallpox virus in a secret freezer and that it might be used by terrorists or in biological warfare.

Researchers are also hanging on to a few segments of smallpox DNA that cannot cause infections but might contain valuable information about how viruses function; these will be grown in bacteria.

The next target for eradication is polio, another virus whose only reservoir is humans. In 1985, the Pan American Health Organization started a drive to eliminate polio from the Western Hemisphere. By 1994, after a vigorous immunization campaign, the Americas were declared free of polio for the first time. But the virus persists in Africa and Asia.

Measles, which has been called the most contagious of all human infections, is next in line. The existing measles vaccines are not perfect—they require constant refrigeration from factory to field, and they must be given right after a baby loses its maternal antibodies but before it is exposed to the virus, which is widespread in developing nations. Nevertheless, if vaccination teams could reach many of the 20 percent of children who are still not protected against measles, they might be able to eliminate the disease in most parts of the world.

No vaccines yet exist for the vast majority of diseases, however. Scientists hope that a new generation of vaccines will be developed when the full sequences of nucleotides making up the DNA of many disease-causing viruses and bacteria are deciphered.

In the summer of 1995, researchers worked out the first complete genetic blueprint of a free-living organism—a small bacterium known as *Haemophilus influenzae* (unrelated to the influenza virus, but a prime cause of ear infections and meningitis in children). James Watson, codiscoverer of the structure of DNA, declared this "a great moment in science." It gave biologists their first chance to study every one of the genes and proteins that a living cell needs to grow and reproduce. Many viruses have been sequenced, but their genomes are much smaller.

The *Haemophilus* bacterium was sequenced by a team led by Craig Venter of the Institute for Genomic Research in Gaithersburg, Maryland, and Hamilton Smith of the Johns Hopkins University School of Medicine in Baltimore. By now,

The first free-living organism to have its entire genetic blueprint revealed by scientists was a bacterium, H. influenzae. Each colored bar in the outer circle of this diagram represents a gene whose function is known or can be inferred. White bars indicate genes whose role is unknown.

Venter's group has published the sequence of another small bacterium, and more are coming. A really vicious bacterium, *Staphylococcus aureus*, a common cause of hospital infections and toxic shock syndrome, was almost completely sequenced by Human Genome Sciences, Inc., a biotechnology company in Rockville, Maryland, around the same time.

Comparing the sequences of various bacteria may reveal what makes some strains virulent while others are harmless. This should help in the design of new antibiotic drugs. Meanwhile, scientists are focusing on some of the proteins on bacterial coats—the proteins that are normally recognized and attacked by the immune system. Many bacteria have learned to alter these proteins in order to evade such attacks. But if researchers have access to a bacterium's full genome, they may be able to find some rare coat proteins that the bacterium cannot modify. These would be excellent candidates for vaccines.

A number of labs in various parts of the world have begun to sequence the parasites that cause malaria, sleeping sickness, Chagas' disease, and other major ills. These parasites' complex lifestyles have made them very difficult to study, and until the sequencing projects started, only a few of their genes had been identified.

Many researchers believe that this sequencing activity will lead to a new wave of victories in our never-ending fight against the microbes that threaten us. ●

The good news in the war against AIDS—at least in the United States—is that the death rate from this terrible disease has fallen dramatically as a result of the triple-drug therapy described in "Can AIDS Be Tamed?" Deaths declined for the first time in 1996 and have continued to drop. Though AIDS was the leading cause of death among Americans between the ages of 24 and 44 in 1995, it fell to number three on the list by the end of 1996 and was down to fifth place in 1997.

Good News/Bad News

The bad news, says Donna Shalala, secretary of health and human services, is that new cases of HIV infection in the United States have not declined. "Our ultimate goal is to prevent the estimated 40,000 new HIV infections that occur each year," she explains. Furthermore, there is still no safe and effective vaccine against HIV. And in large parts of Africa and Asia, the AIDS epidemic is still spreading disastrously.

While the new drug combinations can buy time for people who have been infected with HIV, the treatment works only when people are extremely con-scientious about taking all three drugs according to a very precise, demanding schedule. This is often diffi-cult because the drugs may have unpleasant side effects and because large numbers of pills must be taken at various times throughout the day. Besides, they are so expensive (up to $15,000 a year) that most of the world's HIV patients cannot afford them.

Patients who do obtain these drugs and take them but fail to stick to the prescribed regimen become walking time bombs. They risk not only developing full-blown AIDS more rapidly, but also encouraging the growth of a highly drug-resistant strain of HIV. This would limit their own options for future treat-ment and, in addition, might allow this extremely dangerous strain of virus to spread to other people.

Even when patients do so well on the drugs that they have no detectable levels of virus left in their blood, stopping the treatment brings a dramatic reemergence of active virus within a few weeks.

Fortunately, "more new drugs keep coming along," says Carl Dieffenbach, associate director of the Division of AIDS at the National Institute of Allergy and Infectious Diseases in Bethesda, Maryland. "We try to stay one or two drugs ahead."

The Lucky Few

One of the drugs that will enter clinical trials soon mimics the effects of a beneficial mutation in a human gene (CKR5, recently renamed CCR5) for a receptor on helper T cells. The mutation was discovered in a small number of people who remained uninfected with HIV despite years of high-risk behavior. These extraordinarily resistant people turned out to have two copies of the mutation, having inherited one from each parent. Yet even one copy of the mutation appears to be partly protective.

Another small group of people astonished researchers in a different way: They do get infected with HIV, but control the virus so well that they avoid developing AIDS for exceptionally long periods of time. A few years ago, Bruce Walker and his col-leagues at Massachusetts General Hospital in Boston heard about Bob Massie, who was infected with HIV through a transfusion of tainted blood-clotting factors (see p. 213) but remained free of AIDS 20 years after infection. Intrigued, the researchers started testing him regularly—"All I do is stick out my arm whenev-er they ask so that they can draw a syringe full of blood," Massie says. They also launched into a study of a dozen other long-term nonprogressors, people who had dodged the bullet of AIDS for at least 10 years after infection with HIV.

The team discovered that Massie produces remarkably potent helper T cells that specifically "recognize" his strain of HIV, as well as many killer T cells that aim straight for the virus. These helper T cells may be the keys to Massie's resistance, the researchers concluded, since helper T cells, the mas-ter cells of the immune system, are the primary tar-get of HIV. Other people's helper T cells generally start losing the ability to respond effectively to the virus within a few months after infection.

In an attempt to copy Bob Massie's success at containing HIV, Walker's team now prescribes a triple-drug regimen as early as a few days or weeks after infection—just after exposure, when people have only flu-like symptoms. The researchers hope this prompt treatment will reduce the patients' viral loads sufficiently to allow their own helper T cells to keep functioning at a high level and contain the remaining viruses. The 20 patients on whom they have tried this approach are doing well so far, says Eric Rosenberg of Massachusetts General Hospital; their helper T cells remain active against the virus, though not quite as active as Massie's.

A New Vaccine Against Malaria?

Parasites such as *Plasmodium falciparum*, which causes the most severe forms of malaria, are increasingly resistant to drugs. To make matters worse, the mosquitoes that transmit these parasites are increasingly resistant to insecticides. A vaccine is clearly needed. Yet anti-malaria vaccines have been largely ineffective until now—apparently because they target proteins that the parasites produce during only one stage of their complex life cycle.

In early 1999 a research team led by scientists at the Centers for Disease Control and Prevention announced it had developed a new "candidate" vaccine that combines 21 segments from nine immune-system-triggering proteins, or antigens, that *P. falciparum* produces during the four stages of its life cycle. Researchers led by Altaf Lal of CDC's Division of Parasitic Diseases chose the protein segments after analyzing the results of a large epidemiologic study in Kenya, where most people are repeatedly exposed to malaria and some of them acquire a natural immunity.

The new vaccine has been tried in rabbits, whose blood serum later showed high concentrations of antibodies that "recognized" the parasite at all stages of its life cycle. The antibodies also kept the parasite from invading the rabbits' liver cells and inhibited its growth in the rabbits' blood. If the vaccine proves equally safe and effective during tests now being carried out in monkeys at the CDC, it will go into human trials.

The TB Bacterium Reveals Its DNA Sequence

A major advance on the TB front: the entire genome of the tuberculosis bacterium has now been sequenced. A team led by Stewart Cole of the Pasteur Institute in Paris and Bart Barrell of the Sanger Center in Cambridge, England, won the race in 1998. This should open the door to a fuller understanding of how the bacterium becomes drug-resistant and how to deal with it, and also provide new targets for drug and vaccine development. According to HHMI investigator William Jacobs, the sequence "is greatly speeding up research on TB around the world."

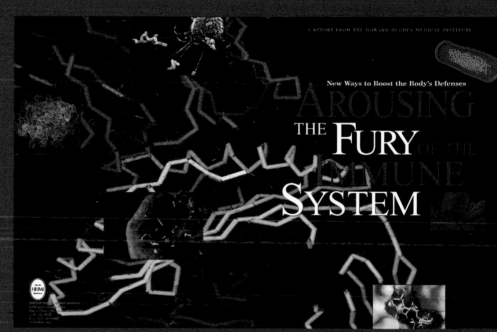

A REPORT FROM THE HOWARD HUGHES MEDICAL INSTITUTE

New Ways to Boost the Body's Defenses

AROUSING
THE FURY OF THE IMMUNE SYSTEM

FURY
THE IMMUNE SYSTEM

THE
SYSTEM

Images of 3-D structures such as these give clues to how molecules of the immune system function. Here a drug that suppresses immunity (rapamycin, yellow) is shown bringing together two human proteins (FKBP12, right, and FRAP, left). The areas of tightest binding are colored purple.

AROUSING THE FURY OF THE IMMUNE SYSTEM

New Ways to Boost the Body's Defenses

INTRODUCTION

· ·

I magine that a new kind of microbe lands on Earth from outer space. In a showdown with the human immune system, this alien would face at least a few weapons that seem made to order, ready and waiting to fight it.

Our immunological armories contain such a diverse array of warheads that one of these is bound to be capable of homing in on the invader. Humans can raise specific antibodies against a virtually limitless range of molecules, as the immunologist Karl Landsteiner showed early in this century; we can even make antibodies against synthetic chemicals that never existed before. In addition, our repertoire of T cells (specialized white blood cells) can recognize about 100 trillion different substances that would spur the T cells to sound an alarm, mobilize the troops, and launch a counterattack.

We need all these weapons because "every day is a battle against microorganisms," says Don Wiley, an HHMI investigator at Harvard University. "The immune system is really the reason that we're not dead."

But having the right kind of weapons isn't enough. In order to survive, we must be able to produce these weapons rapidly and abundantly enough to overwhelm the invaders before they overwhelm us.

"Microbes reproduce so quickly and mutate so quickly—how can a multicell organism that reproduces very, very slowly defend itself against them?" asks Mark Davis, a Hughes investigator at Stanford University School of Medicine. The answer, says Davis, is diversity. "You can compare the immune system to the Defense Department in terms of its massive duplication of effort and huge expense," he says. "You make billions of [the various] immune cells every day, just in case they're needed—and then you throw most of them away."

Furthermore, each person has a unique set of immunological weapons. This ensures that at least some members of the species will be able to respond to any particular threat.

"No two people, even if they're identical twins, have the same immune system," says Charles Janeway, an HHMI investigator at the Yale University School of Medicine, "because the genes for T and B cell receptors are defined by gene rearrangement, and that, in turn, has a degree of randomness."

The field of immunology has gone through an extraordinary flowering in recent years as molecular biology has provided new explanations for the complicated actions of immune-system cells. Although immunologists had previously described these actions in great detail, they also left many puzzles unsolved.

Some of the most important recent findings in immunology have actually come from crystallographers studying the 3-D structures of molecules, notes writer Stephen S. Hall in the article on "Shapes That Made a Revolution." In 1987, structural biologists at Harvard University provided key evidence that helped solve a central mystery of immunology: How the immune system's T cells distinguish "self" (the body's own cells or cell parts) from "non-self." The high-resolution image of a peptide bound inside an MHC Class I molecule produced by Wiley, Pamela Bjorkman (then a postdoctoral fellow in Wiley's laboratory, now an HHMI investigator at the California Institute of Technology), and Jack Strominger, a biochemist at Harvard, neatly answered many of the questions that immunologists had struggled with for decades.

"I don't know of any card-carrying immunologist who doesn't have, among his slides, that structure of MHC Class I," says Thomas Waldmann, a leading researcher at the National Cancer Institute. According to the citation of the 1995 Lasker Award, which Wiley shared with Strominger and three others, the structure "changed the ground rules of immunology...and made the rational design of specific molecules for antiviral vaccines a reality."

Then, in 1998, Wiley's laboratory uncovered the detailed structure of what immunologists viewed as their holy grail: the three key elements that must fit together to trigger a T cell's response. These consist of a receptor on a T cell, plus an MHC molecule on the surface of an infected cell (identifying the cell as "self"), and a peptide from the infecting virus (identifying the virus), as shown on page 322.

Explains Wiley: "We show everything between the two cell membranes—that is, the intercellular recognition complex that makes the decision, 'Do we kill, or don't we kill?' I think this is the first time that anybody has got both sides of any intercellular event in humans. It's what you'd see if you were standing between the two cells."

The T cell will not respond to such binding, however, unless the DNA in its nucleus "knows" that binding has occurred at the cell surface. This knowledge depends on intricate chains of signals that flash from cell surface to nucleus—Rube Goldberg-like sequences that used to be completely mysterious but are now coming under intense scrutiny as a hot topic of research.

A bicoastal collaboration between two HHMI scientists, Gerald Crabtree of Stanford and Stuart Schreiber of Harvard, is giving researchers an unprecedented level of control over such events. While certain drugs can prevent signals from reaching a T cell's DNA, the two scientists have shown that other drugs, such as the new, man-made "dimerizers," can induce a T cell to switch on specific genes.

The most efficient way to make the immune system work for us, however, is to prevent disease through vaccination. Vaccines prepare the system to respond swiftly to a specific attack before the invading microbes can multiply and produce damage. Many more vaccines are needed.

Maya Pines, *Editor*

Small but tough, killer T cells (white) surround a large cancer cell.

THE KILLERS THAT SAVE US

Scott Olson had dressed for the occasion as if he were ready to set off on a summer hike. A handsome young man of 24 years, with short brown hair and a granitic chin to go with his trim, athletic body, Olson was wearing a green T-shirt and shorts and high-topped Nike hiking boots. As he leafed through a copy of *Sports Illustrated* that Monday afternoon, his eyes lit up when he spotted an advertisement for shoes that purported to increase one's vertical leap—a skill that might come in handy, Scott explained to a visitor, in his job as a disk jockey performing at parties and weddings.

What made this all a little incongruous is that while he paged through the magazine, reclining in a chair with his feet propped up, he had an IV needle emerging from his left arm, with tubing that led up to a small, innocuous-looking plastic bag hanging from a pole. Scott Olson (not his real name) is infected with HIV, the AIDS virus, and the number of helper T cells—crucial components of the immune system—

in his blood had dropped from a normal level of about 1,000 per milliliter of blood to 350.

As he sat in an outpatient clinic at the Fred Hutchinson Cancer Research Center in Seattle, Washington, sipping from a large and frothy iced mocha coffee (this was, after all, Seattle: home of the drive-in espresso bar), Scott's immune system was receiving a pick-me-up of a different sort—a novel kind of medicine in the form of his own T cells. Selected by his doctors for their ability to target HIV, these cells were grown to prodigious numbers outside Scott's body and now returned to his bloodstream in a slow, milky drip through the IV tubing.

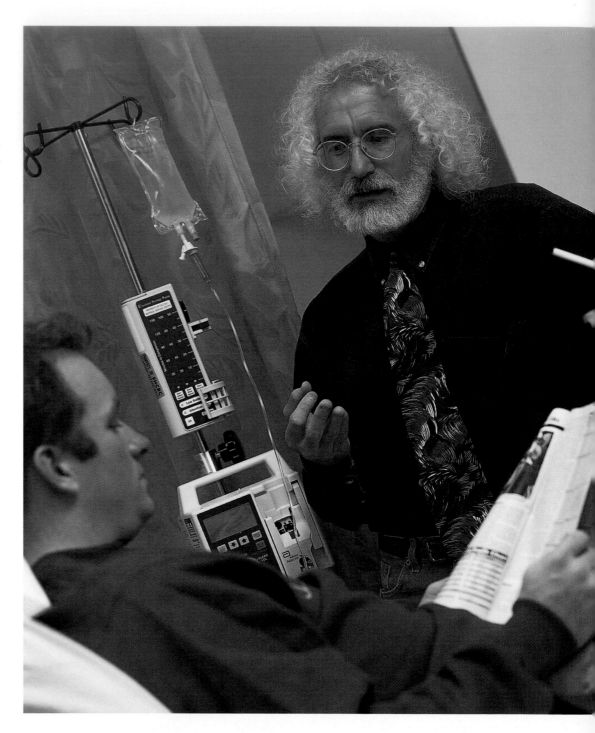

Philip Greenberg (standing) found that only 2 T cells out of 10 million T cells in a sample of Scott Olson's blood recognized HIV, the virus that causes AIDS. His lab made these two cells multiply to billions of identical cells. Here Olson receives an infusion of newly grown cells.

"It feels kind of cool going in," Olson said matter-of-factly, taking another sip of his coffee, "but that's about it."

"There are about 2.2 billion cells in that little bag," said Michael Wilson, the nurse in charge of administering this immunolog-ical transfusion. Stanley Riddell, Olson's doctor and one of the innovators of this new technique, added, "Your cells grow beauti-fully, so there's no problem there."

"They're still recognizing everything?" Olson asked.

"They're still doing beautifully," Riddell replied.

Recognition is a word pregnant with meaning—and health—in the world of immunology. Cells of the immune system are designed—genetically, biochemically, and even as 3-D shapes—to recognize an almost infinite variety of potential foes and eliminate them before they overrun the body.

As in any big production number, there are lots of players and many acts in the immune response, but a place of prominence is reserved for T cells, which mark the epicenter of the power that the immune system can unleash. Researchers have learned to pluck a few of these cells out of a patient's blood, cultivate them in laboratory flasks, and give them back as medicine—experimental medicine, to be sure, but a harbinger of potential miracles to come in the next decade in using the immune system as a tool of healing.

all of those cells to flow into Olson's arm.

"Compared to your average citizen, I'm probably an expert about the immune system," says Olson, who is an actor and performs in cabarets as well as working as a disk jockey. He understands, for example, the difference between two important subsets of T cells: the impresarios of the immune response, known as helper T cells (or CD4 cells), and the enforcers of that response, known as killer T cells (or CD8 cells). "What they're using here is CD8 cells," he explained. When it was suggested that he had become very familiar with immunology, he replied, "I wish I was less conversant."

Many of us have good reason to become more conversant with the immune system, in part because the mysteries of this splendid biological invention for self-protection have begun to yield to scientific exploration, and in part because our lives depend on it—every waking moment of

A NOVEL KIND OF MEDICINE: HIS OWN T CELLS

Philip Greenberg, who heads the laboratories at the University of Washington and the Fred Hutchinson Cancer Research Center where this novel approach has been developed, thinks of T cells as "single-minded" cells, an anthropomorphic adjective that captures the unique skills of these immunological avengers. In March 1996, Greenberg and his colleague Riddell drew a sample of Scott Olson's blood and identified two distinct T cells, among 10 million possible candidates, that recognized a tiny portion of a protein made by HIV. They cultivated these two cells in flasks that soon numbered 20 and then 40, and then expanded these cells to astronomical numbers, hundreds of millions of identical killer T cells, all genetically programmed to recognize HIV. It took barely 20 minutes for

every day. For some people, like Scott Olson, it may even represent one of the best hopes of diverting the fatal course of a disease. Using its cells, its secreted molecules, and sometimes combinations of the two, doctors have made slow but steady progress in turning the immune system into a tool that can be used to treat infectious diseases and malignancies.

T cells in particular (they are also known as T lymphocytes) have come a long way in the last 40 years. "In the 1950s, I think most biologists thought of the lymphocyte as a dead-end cell and very boring because it was this cell that floated around in the bloodstream and didn't have anything much to it except a nucleus," says Philippa Marrack, an HHMI investigator at the National Jewish Medical and Research

B Cell

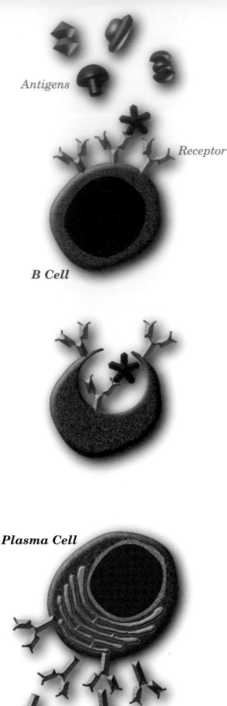

Antigens

Receptor

B Cell

Plasma Cell

Antibodies

Antigen

Receptors on the surface of a particular B cell will fit and bind with only one particular type of antigen (red).

After binding with an antigen, the B cell engulfs it.

The B cell then becomes a large plasma cell and starts to produce specific antibodies (yellow)—guided missiles that roam the bloodstream in search of specific invaders. Antibodies bind the same particular type of antigen (red) as the B cell receptor.

Center in Denver, Colorado, who, with her colleague John Kappler, has been a leading figure in T cell biology. But immunologists began to change their minds in the 1960s, when several researchers pointed out that if newborn mice—or humans, for that matter—lacked a thymus, they suffered impaired immune function and often died from infections. T cells are cells that mature in the thymus (hence the "T" in their name), a small triangular organ beneath the breastbone whose name derives from its resemblance to a thyme leaf.

It turned out that T cells not only kill microbes on their own but also send crucial signals that promote the formation of antibodies—potent weapons of the immune system produced by another class of lymphocyte, the B cell, which matures in the bone marrow.

All lymphocytes are studded with sensitive cell-surface receptors, organs of touch so subtle and precise that they can detect changes at the level of a single atom. The receptors on the surface of B lymphocytes are much like those on T cells, each set of receptors genetically programmed to recognize a different bit of bacteria or other pathogens. But B cells can also produce quantities of receptor-like molecules that are free to move out of the cell since they lack an anchor in the cell membrane. When a B cell grabs hold of an antigen and gets turned on, it metamorphoses into a kind of microscopic drug factory called a plasma cell, which churns out millions of these identical, foe-specific molecules and releases them into the bloodstream, where they can latch onto viruses, bacterial proteins, and toxins. These secreted, free-floating receptors go by the more common name of antibodies.

The combination of antibodies, which eliminate troublemakers outside the cells, and killer T cells, which eliminate troublemakers that hide inside infected cells, gives

IT FINDS
A WAY TO REACH MICROBES
WHEREVER THEY GO

the immune system a way to reach microbes wherever they go. And sitting in the director's chair, coordinating this fury by sending signals to both the antibody and killer cell arms of the response, is the helper T cell.

The T cell is a remarkable sleuth and slayer and, given a chance to perform, will enact wonders in the body. It is able to home in on fragments of a foreign invader in the form of small bits of protein known as peptides, which come nestled in trophy stands known as the major histocompatibility complex (or MHC) molecules. Greenberg and Riddell in fact have attempted to duplicate in the test tube the kind of process that occurs in the body when, for example, the flu virus pays a visit.

"A virus like flu," Marrack explains, "can probably give rise to maybe 20 or 30 different peptides which combine with the MHC molecules in one person, and a different 20 or 30 that combine with the MHC molecules in a different person. So now you've got 20 or 30 targets for T cell recognition. And the guess is that on the whole, about one in a million T cells will be able to recognize each of those targets when you start off. With help from their friends, the T cells will start to divide, in a good response, twice a day. They will be through 14 divisions by day seven, which is 2^{14}. So that's about 10,000 cells from one cell."

Researchers now have the ability to pick out the one-in-a-million T cells in a patient's blood sample that targets a particular virus, for example, and tinker with it in the laboratory to develop new medicines. Few labs

Philippa Marrack and John Kappler, who are married, have pioneered the study of how lymphocytes (B cells and T cells) recognize and attack invading organisms, as well as how the immune system can tell friend from foe. They are shown at the HHMI Holiday Lectures, which they presented in 1996.

in the world have become as proficient at this art (and it is so difficult that it is still as much an art as a science) as the Greenberg/Riddell lab in Seattle. In fact, it is probably fitting that this work has evolved at the Fred Hutchinson Cancer Research Center, named after the former baseball player and manager Fred Hutchinson, because Greenberg resorts to the dugout for his favorite metaphor to justify the stepwise progress made by his and other labs attempting immunological medicines: "There are no home runs."

There are, however, solid base hits. The Seattle group, for example, can identify T cells that attack certain herpesviruses and grow them to enormous numbers in the lab before infusing them into patients. They have now begun to genetically reengineer T cells so these cells fight infections better, and to modify viruses and tumor cells to create cancer vaccines. And they can avail themselves of the chemicals used by the immune system itself, many of them isolated and cloned and genetically engineered in the 1980s, to tweak the immune response even more. Their work started as an attempt to help patients who receive bone marrow transplants, and the preliminary success of this intervention is one of the most innovative and heartening stories to come out of applied immunology.

The story actually began in the mid-1980s, when Greenberg entertained hopes of using immune cells to treat cancer and recruited Stanley Riddell to assist in the effort. Greenberg hails originally from Brooklyn, New York. He speaks in a rapid boil of details and enthusiasms and looks a little like the late Grateful Dead guitarist Jerry Garcia—bearded, with rimless glasses and a curly nimbus of long, silvery locks. Riddell was born and raised in Swan River, Manitoba, a town of 5,000 people located about 800 miles north of Minneapolis; he speaks with the quiet authority of a minister and looks youthful enough still to pass

Helper T Cell

Antigens

Macrophage Swallowing Antigen

Macrophage Processing Antigen Into Fragments

Macrophage Displaying Antigen Fragment

Antigen Fragment

T Cell Receptor

MHC Molecule

Resting Helper T Cell

Activated Helper T Cell

The "helper" T cell is actually the immune system's master cell, giving orders to all other cells. But first it must be activated. It must recognize and bind to a particular fragment of antigen that is displayed by a less discriminating cell, such as a macrophage. Macrophages travel all over the body's tissues, swallowing up any antigens they can find. Here a macrophage (top) engulfs an antigen (red) and processes it into fragments.

The macrophage then displays one of the antigen's fragments (small red piece, lower left) together with an MHC Class II molecule, the macrophage's ID tag. A resting helper T cell (dark purple, with yellow receptor on the surface) recognizes this specific fragment and binds to it.

Now activated, the helper T cell is ready to arouse the entire immune system to fight the invader it has recognized. It can boost the number of B cells or increase the number of killer T cells.

as the president of the high school 4-H Club. In a field where other clinicians sometimes leap before looking at all the consequences of their interventions, Greenberg and Riddell have established a reputation for learning about T cells step by step.

Greenberg's interest, in particular, focused on the killer T cell. When helper T cells recognize an antigen—anything that provokes an immune response—they send signals to fellow immunological cells that shape and organize the overall strategy of containment and elimination. But when killer T cells recognize an antigen, something remarkable happens: premeditated murder of a very precise sort.

A killer T cell that recognizes a target on

memory as a killer T cell, which is why Greenberg and Riddell thought killer T cells would make a particularly adept weapon against disease, especially cancer.

In order to learn more about how T cells "see" a malignant cell, the University of Washington group began by studying the interaction between T cells and a more familiar adversary: a common viral pathogen known as cytomegalovirus. This relationship is a microcosm for understanding how the immune system deals with enemies that lurk inside cells.

Cytomegalovirus—usually abbreviated as CMV—belongs to the herpes family of viruses. It has infected many people without causing conspicuous illness and is

SWIFT AND STUNNING ASSASSINATIONS

a cell—either a cell that harbors a virus or a cancer cell—more or less attaches itself to the tainted cell. Biologists spying on this fatal embrace with the help of high-powered microscopes have witnessed the swift and stunning assassinations that follow. Little bubbles of preformed toxins inside the killer T cell, known as granules, begin to gravitate toward the side of the T cell that has made contact with the target. Within 5 or 10 minutes, all of the bubbles have migrated to the region of the T cell where its surface has made contact with the virus-infected cell. In less than an hour, these little bubbles erupt against the membrane—or outer skin—of the target cell.

One might think of the killer T cell as throwing water balloons against the outer wall of the target cell, but since the bubbles contain a potent weapon called perforin—which literally perforates, or pokes holes in, the target cell—this cell soon bursts apart. No medicine known to man can find a virus-infected cell with as much precision, doggedness, mortal intent, and long-term

generally considered harmless. In people whose immune systems are impaired, however, including people with AIDS and leukemia patients who have undergone bone marrow transplants, it can cause life-threatening ills. Bone marrow transplant patients are especially vulnerable for about three months after the procedure, before their new immune system is fully up and running. So Riddell and Greenberg mounted a search for killer T cells that specifically target CMV. They looked for the cells in blood from the donors of the bone marrow transplants, because those cells would be compatible with the new immune systems of the recipients.

They began this effort around 1985. They studied CMV intensively. They identified viral "epitopes," the little bumps and bits of viral protein that a killer T cell's receptors may recognize on an antigen (an antigen may have a dozen different epitopes, each of which can provoke a distinct response). Most important, starting with a test tube of blood from the bone marrow donors, which contained roughly 40 million white blood cells, and using new technologies of identification, they managed to pluck out several killer T cells that were "single-minded" about eliminating cytomegalovirus, and only CMV. Over the next several years, Riddell developed the cellular and molecular strategies that allowed him to begin with a solitary T cell and, nudging it through a succession of proliferative divisions with various inducements, to cultivate hundreds of millions of

HAPPILY, T CELLS DON'T SLEEP ON THE JOB

copies, or clones, of this one CMV-specific immune cell.

The first test of the strategy came in connection with bone marrow transplants, which was fortuitous because the Hutchinson Center performs more such transplants than any other medical center in the country. Patients with leukemia have rogue blood cells that run amok, replicating incessantly until they literally overrun the bloodstream. In an attempt to eradicate all possible cancer cells, the patients undergo full-body radiation, which essentially kills all their blood cells including, it is hoped, the cancerous ones. Then they receive a transplant from their donor, who may be an identical twin, another relative, or, increasingly, an unrelated person with compatible blood. In 1991, the Hutchinson group performed its first infusion of lab-cultivated T cells as a form of medicine in a bone marrow transplant patient. Greenberg gathered his entire team—researchers, technicians, and nurses—to witness the treatment. The patient didn't share the momentousness of the occasion;

he promptly fell asleep.

Happily, T cells don't sleep on the job. By cleverly marking their custom-grown T cells, the Seattle doctors could follow the fate of the cells they had transfused, seeing where they went, how long they lasted in the body of patients, and whether they had any effect. They published a progress report in *Science* on the first 14 patients, none of whom succumbed to CMV pneumonia. Although still preliminary, the results have been, in Riddell's words, a "proof of principle."

By the summer of 1998, 26 patients who received bone marrow transplants at the Hutchinson Center from their siblings had been treated with killer T cells. While half of them would have been expected to suffer from a serious CMV infection, only one developed any evidence of virus in his blood during the critical period of vulnerability, and even he did not progress to CMV disease. In other words, transfusions of customized T cells seem to protect vulnerable patients from potentially fatal complications of bone marrow transplants.

Although these T cells did not sleep on the job, they did eventually die on the job, and that unusual wrinkle led to the next step in T cell transfer.

To nonexperts, the elaborate ritual of immune signaling can sometimes seem to require as many steps as the secret handshake of a fraternal lodge. A killer T cell must receive a set of signals—not just from an antigen on the target cell but also from helper T cells—in order to remain alive and replicate. Otherwise, the killer T cells sense that something is wrong and, rather than wreak cell-killing havoc, they commit suicide. The cells grown by Riddell and his colleague Kathe Watanabe did their job for about 12 weeks, which was enough time to get bone marrow transplant patients over the hump; then they disappeared. So Riddell and Greenberg went back to the drawing board and came up with a plan to more closely emulate a natural immune response by invoking the help of the helper T cells as well.

Typically, when a virus like CMV infects the body, the two sets of T cells act in concert. The killer T cells attack any cells in the body that contain the invading virus; this attack is short-lived, however, unless the helper cells continue to send signals to the killer cells. And the helper cells send their signals only when they have been alerted to the viral infection by other cells of the immune system known as antigen-presenting cells. So now the Seattle group is cultivating both helper and killer T cells in the lab and giving both of them back to bone marrow transplant patients, compressing the treatment period to nine days. "If you give them a helper response and a killer T cell response, you've fully reconstituted the natural immune response," Greenberg said, explaining that helper T cells also tell B cells to make antibodies. "We're just looking now, but we feel this will provide long-standing immunity."

There is no question that the technique is labor-intensive and expensive, but Greenberg argues that in the long run, it may be cheaper and healthier for bone marrow transplant patients than the use of an antiviral drug known as gancyclovir. The drug works against CMV, but it can cause dangerous suppression of the immune system. Once the drug is discontinued, a number of bone marrow transplant patients develop late-onset viral pneumonias caused by CMV that ultimately prove fatal. T cell immunotherapy, while not exactly "natural," nonetheless offers an alternative.

"It's clearly less toxic, it doesn't appear to be more expensive, and it may ultimately prove to be substantially more effective," Greenberg said. "If that's the case, then there's lots of reasons to think that this ought to be broadened to a large population."

Killer T Cell

The "killer" T cell is a ferocious fighter that can destroy infected cells. But first it must be activated—it must recognize and bind to a particular fragment of antigen that is displayed by another cell. Here a cell (top) becomes infected with a virus (yellow). The virus replicates inside the cell and some of its proteins are processed into fragments.

The infected cell then displays one of the antigen fragments (small yellow piece, lower left) together with an MHC Class I molecule, the cell's ID tag. A resting killer cell (dark purple, with green receptor on the surface) recognizes this specific fragment and binds to it.

Now activated, the killer T cell attacks an infected cell (red) that bears the viral fragment. As a result, the infected cell dies.

Antigens

Virus Infecting
a Cell

Cell Processing
Viral Protein
Into Fragments

MHC
Molecule

Cell
Displaying
Viral
Fragment

T Cell
Receptor

Viral Fragment

**Resting
Killer T Cell**

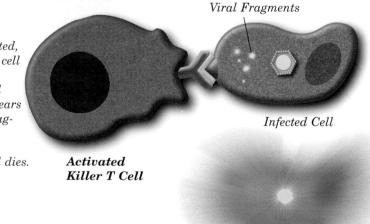

Viral Fragments

Infected Cell

**Activated
Killer T Cell**

Dead Cell

Like all good clinical experiments, the Hutchinson group's experience with anti-CMV T cell immunotherapy is leading to other, more general medical applications that will soon be tested in the clinic. Because the immune system is so versatile and potent, so too are the variations on this therapeutic theme that can be applied to different medical situations.

First of all, having raised customized T cells against CMV, the group has tried to do the same thing with T cells against HIV, the AIDS virus. Scott Olson was among the first dozen patients to receive killer T cells against an antigen of HIV.

Secondly, they have begun to extend the CMV T cell therapy to a larger group of bone marrow transplant patients. The approach is now being expanded to include not only transplants between siblings but transplants from unrelated donors, who are not as closely matched to the patients.

What makes this approach especially promising is that doctors may be able to turn an old problem into a spectacular advantage. Many transplant patients suffer an immunological side effect known as graft-versus-host (GVH) disease, in which transplanted T cells from the donor attack the tissues of the recipient. As terrible as this side effect can be (and sometimes it is fatal), doctors have noticed that leukemia patients who survive GVH have a much lower incidence of relapse, apparently because the transplanted T cells also attack any of the patients' leukemic cells that remain after the radiation treatment. This collateral effect is known as graft-versus-leukemia, or GVL.

Using the same technology that has proved so successful against CMV, Greenberg and Riddell now plan to search the blood of transplant recipients for donor T cells that recognize antigen expressed only by the patients' blood cells, including the leukemic cells, expand them in the lab,

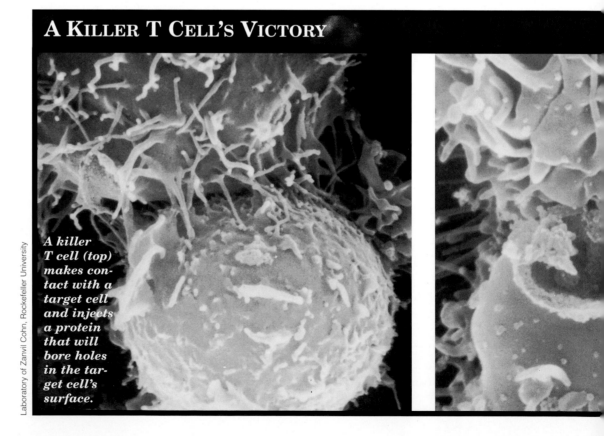

A KILLER T CELL'S VICTORY

A killer T cell (top) makes contact with a target cell and injects a protein that will bore holes in the target cell's surface.

Laboratory of Zanvil Cohn, Rockefeller University

and give them back as a cancer-specific medicine.

"Since all the normal blood cells are killed during the transplant, the only target cells remaining in the patient are the leukemic cells. So we may be able to get the GVL reaction without causing GVH disease," Riddell explains. "It's a well-defined circumstance where you can really show whether or not T cells can work against a malignant disease. And one in which almost certainly they will work."

Finally, the Seattle group is spearheading an effort to unleash these educated, single-minded T cells against other cancers as well. Certain cancer cells—melanocytes in the skin cancer melanoma and Reed-Sternberg cells in Hodgkin's disease —display antigens on their surface that are typical only of the cancer cells; this makes them excellent potential targets for killer T cells. The group has already begun treating melanoma patients with this approach, and Greenberg expects to start testing the idea in Hodgkin's disease patients soon.

Perhaps the most promising aspect of all is that the Seattle work represents but one of many exciting new approaches to immune system manipulation that are reaching the point of clinical testing. Not all of them will work, of course—Greenberg still doesn't believe in home runs when it comes to novel treatments—but there is palpable excitement that the enormous power of the immune system can be harnessed to predictable, and positive, effect.

"The immune system," Greenberg says, "has had millions of years to evolve and refine itself and become effective. So what we can do now is take advantage of all those refinements and use what the immune system is good for—its remarkable specificity and efficiency—but expand it to make it more effective in doing what it's supposed to do." ●

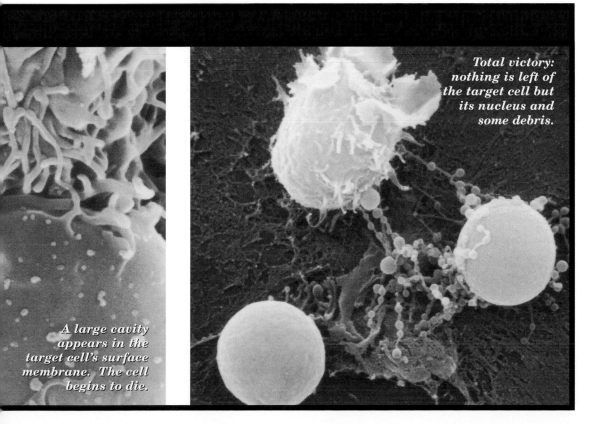

Total victory: nothing is left of the target cell but its nucleus and some debris.

A large cavity appears in the target cell's surface membrane. The cell begins to die.

INNATE DEFENSES THAT HOLD THE FORT

Recent experiments on animals have revealed just how sophisticated and important the earliest stages of the immune response truly are.

From the time an infectious agent enters the body to the moment that the big guns of the immune system, namely B cells and T cells, restore the peace, several unheralded cells prepare everything that follows. These early cells perform two essential tasks: they blunt the initial spread of infection until reinforcements can arrive, and they present information—in the form of a specific antigen—to the later-arriving cells.

No less a champion of the T cell than Philippa Marrack makes the point that crucial containment strategies are well under way long before T cells arrive on the scene. "T cell division takes seven days to get into a full flowering (in humans)," she says. "That's a long time. And if you've got a bacterial infection and the bacteria are dividing once every 40 minutes, you're going to be dead long before these T cells ever get off the ground. So we have these innate mechanisms that help hold the fort."

Emil Unanue, chairman of pathology at Washington University School of Medicine in St. Louis, has parsed out these innate mechanisms in exquisite detail using *Listeria* bacteria in mice. *Listeria monocytogenes* provides a particularly good window on these events. Known as an intracellular parasite, it burrows inside the cells it infects and sets up housekeeping there. The immune system must not only pick off the parasite as it floats through the body but also destroy the cells in which it hides. Unanue's group has shown how the

A macrophage (red) prepares to swallow up E. coli bacteria (green) that are caught in its long, thin pseudopods.

SOLDIERS OF THE IMMUNE SYSTEM

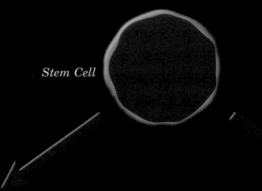

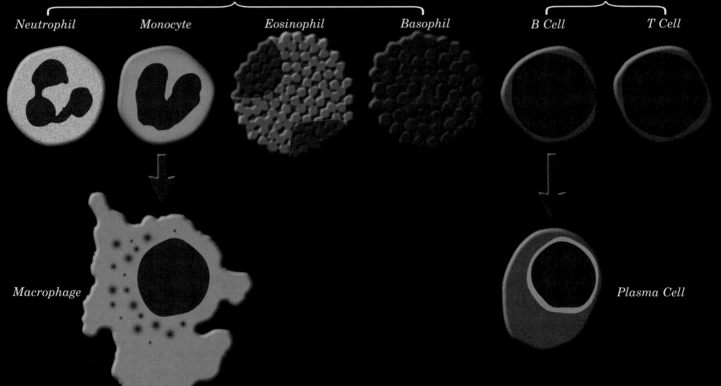

Stem Cell

Instant Defense

Later Defense (lymphocytes)

Neutrophil   *Monocyte*   *Eosinophil*   *Basophil*   *B Cell*   *T Cell*

Macrophage

Plasma Cell

The soldiers of the immune system are six kinds of white blood cells that all come from a single stem cell in the bone marrow (top).

The body's first, nonspecific response to infection involves four of these cells: neutrophils, which engulf bacteria upon contact and send out early-warning signals; monocytes, which turn into macrophages that swallow up invaders; eosinophils, which attack parasites; and basophils, whose granules contain histamine and other compounds related to allergies.

The lymphocytes—B cells and T cells, which are more specifically tailored to the invader—arrive later at the field of battle. The B cells turn into plasma cells that manufacture thousands of highly specific antibodies, which are released into the bloodstream. The T cells coordinate the entire immune response and eliminate viruses that hide inside infected cells.

microbe sets off alarms that alert the entire immune system.

The first cells to show up at the site of inflammation are the neutrophils. A microbe or a splinter, it makes no difference—if it's foreign, it attracts the immediate notice of these nonspecific cells. Like sandbaggers at a cresting river, neutrophils wade into action with the crudest of tools and simply try to contain the impending damage by walling off the infective agent before it floods into adjoining tissues.

Besides holding off the invasion, the neutrophils send out a generalized distress signal that tells the body something is wrong. It is an early warning that a threat has been spotted but does not reveal the exact nature of the threat.

Responding to that signal is a second wave of immunological cells, the macrophages, which generally show up within 24 hours. The name literally means "big eaters"; Unanue sometimes refers to them more informally as "macs." Macrophages simply engulf foreign material, surrounding and swallowing it whole. Once they've made a meal of something unfriendly, these large cells become "activated": they unleash chemical messenger molecules known as cytokines, particularly tumor necrosis factor and interleukin-12, which alert still more cells of the immune system that the threat is real and trigger an ever more complex cascade of signaling molecules.

At the same time, the macrophages become more aggressive and attack bacteria at the site of infection. As Philippa Marrack puts it, "The macrophages start chewing up bacteria and keeping them under control while the lymphocytes are still busy thinking about whether or not they should be dividing."

"The macrophages take up the *Listeria* and turn on the system to make these cascades of early cytokines," Unanue explains. "If the macrophages are not there, the mouse dies. The cytokines have many effects on tissues, on the vessels in the tissues, and on the cells—they sort of poise the immune system to react."

MACROPHAGES SIMPLY ENGULF FOREIGN MATERIAL, SURROUNDING AND SWALLOWING IT WHOLE

Once they've ingested a microbe, macrophages also prepare a biological identikit of the invader for the benefit of the later-arriving lymphocytes. They entomb the microbe in a kind of bubble called a vacuole and begin chopping it to pieces; these pieces then get repackaged and presented on the cell surface, like evidence at a trial, before the ultimate judge and jury in an immune response: the helper T cell. To offer up this evidence, the macrophages use a packaging and shipping system called MHC (major histocompatibility complex) Class II (see "Showing the Flag").

Even at the microbial level, it's a dog-eat-dog world. Because *Listeria* is an intracellular parasite, it can bite back at the very cell that has eaten it. It possesses the ability in some cases to escape the vacuole by punching holes in the bubble and slipping into the guts of the macrophage. This is a short-lived escape, however, like falling through a trapdoor into an even worse predicament, since the vertebrate immune system has evolved an additional trick: a complementary presentation mechanism that enables a cell to alert T cells when the cell's interior has been breached, especially by a virus.

This mechanism cuts up whatever happens to have landed in the cell cytoplasm and ships short segments of chopped-up protein, bound to MHC Class I molecules, to the cell surface for inspection by the killer T cells. If a bit of *Listeria*'s flag is waving from the surface of a cell, for instance, killer T cells will spot it and destroy the cell, at the same time signaling for the creation of more like-minded T cells that are biologically programmed to attack *Listeria*.

These newly made T cells will always remember a foe, and they play a major role in the so-called specific response of the vertebrate immune system. But as Unanue points out, none of this would be possible without the original innate response. "Really, it is the nonspecific immunity part of the response that sets the tone of the reaction," he says. ●

by Stephen S. Hall

In the past 20 years, scientists have learned two fundamental rules about T cells: the cells like to see small bits of a foreign invader, rather than the whole beast; and they will recognize such bits only when the foreign pieces are presented by larger molecules known as MHC.

MHC stands for major histocompatibility complex and refers to a set of markings on the surface of every cell in the body, flagging it as uniquely "self." Indeed, when transplant surgeons speak of organ rejection, they are in fact referring to cases where the immune system of a recipient recognizes the donated organ as foreign because it possesses non-self MHC molecules—which prompts the immune system to attack this organ furiously.

MHC molecules are as distinct in each person as blood type and come in two categories: MHC Class I (of which each human possesses 6, out of at least 200 possible variations) and MHC Class II (of which each human possesses 8, out of about 230 possibilities). Class I molecules are expressed on all cells, providing global surveillance; Class II molecules can be found only on certain specialized cells. The two classes of MHC molecules present evidence to two different populations of T cells: killer T cells recognize antigens gripped by Class I MHC molecules, while helper T cells recognize antigens in the grasp of Class II MHC molecules.

Oddly enough, this complex process is a by-product of the massive but normal recycling process that occurs in each cell in the body. In the shrewd, energy-saving ecology of cellular metabolism, proteins are broken down into constituent parts, or peptides, before being reused.

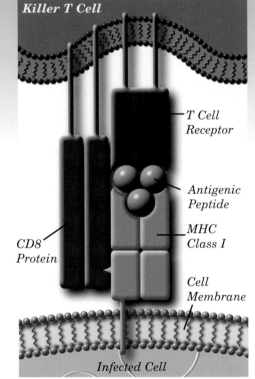

Killer T Cell

T Cell Receptor

Antigenic Peptide

MHC Class I

CD8 Protein

Cell Membrane

Infected Cell

MHC Class I molecules serve as ID tags on every cell in the body, identifying it as "self." A cell that is infected with a virus or other invader will process the alien proteins into fragments (peptides). These peptides are then scooped up by MHC Class I molecules and carried through the cell membrane to its surface, where they are displayed. When a killer T cell recognizes both the peptide and the MHC molecule, it binds to the infected cell and destroys it.

"Proteins are continually being synthesized and continually degraded," says Peter Cresswell, an HHMI investigator at Yale University School of Medicine, who has studied the process of peptide presentation by MHC molecules. If the peptides are recognized as "self," other cells in the body remain tolerant of them.

One might even liken this recycling to an industrial demolition process—tearing proteins apart, hauling off the debris with a fleet of small forklift trucks, and heaping the scraps into a long line of railroad cars that then snake their way to the cell surface. The action occurs around the clock, rain or shine, in every cell, and usually attracts no notice—until there is an infection.

When the flu virus infects epithelial cells lining the respiratory tract, for instance, the virus gets into the cytoplasm of those cells and begins to make viral proteins. "You're now introducing into that mix new proteins, which come from the virus," Cresswell says. Once peptides are generated by the recycling process, they cannot go anywhere until they are scooped up by local transport devices called TAP (for transport associated with antigen processing) proteins and escorted to a kind of assembly line snaking through the cell—the endoplasmic reticulum, or ER.

Class I MHC molecules creep along the ER like empty coal cars in an endless freight train. Not every peptide fits into every MHC molecule—there are six differently shaped "cars" in a Class I train. If nothing fits into these open-topped hoppers, they simply fall apart. Nature designed MHC molecules to be held together by the right kind of cargo, and those that remain empty never make it out of the recycling plant.

Once a Class I MHC molecule wraps the flanges of its groove around a peptide, the TAP escort falls off. The Class I MHC molecule, with the peptide firmly in its grip, then gets routed into a cellular corridor called the Golgi apparatus, a separate passageway that leads to the cell surface. Ultimately, the MHC molecule and its peptide are delivered to the cell surface and simply parked there, where they can be spotted by killer T cells.

"In an uninfected cell, you have these MHC Class I molecules with peptides that are representative of the pool of proteins always being turned over in the cytoplasm," says Cresswell. "But when you get an infection, new proteins are introduced. So in the middle of this sea of familiar peptides, now you have new ones floating around. And that's what the T cells recognize."

Class II MHC molecules function in much the same way, but they operate only in cells that are specialized to present antigens, such as macrophages and dendritic cells. These cells exhibit their catch to helper T cells.

Suppose a macrophage has gobbled up some intruding bacteria, digested them, and displayed scraps of the invaders' unique protein clothing on its surface. The macrophage then takes its bounty to the nearest lymph node.

The lymph nodes are like crowded convention centers for lymphocytes, where a disease-fighting lottery is constantly being held. Macrophages that have scavenged tissues around the throat, for example, drain to lymph nodes located under the jaw bone. These small glands teem with B cells and T cells,

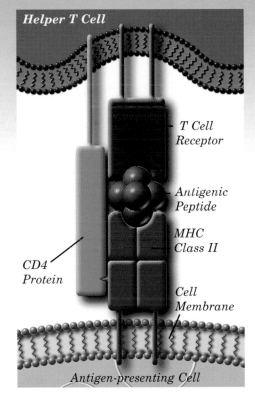

MHC Class II molecules are found only in macrophages and other antigen-presenting cells. When a macrophage detects bacteria in the bloodstream, it engulfs the bacteria and breaks down their proteins into peptides. The MHC Class II molecules then trap these peptides and display them on the macrophage surface. There they are recognized by helper T cells, which alert the entire immune system to fight the invaders.

the major weapons of the immune response, and what happens next might irreverently but accurately be described as a "group grope" at the atomic level.

B cells, each of which has a unique and distinct antibody acting as a receptor on its surface, mill and jostle about and bump into the antigens that have been brought to the lymph node by the macrophages. In a variation on the old Cinderella theme, the macrophage visits every corner of the kingdom, looking for the perfect fit for the antigen displayed on its surface, and only one or a few B cells will possess the antibody that fits the antigen perfectly. When the antigen finds a match, however, a remarkable transformation occurs.

The B cell that locks onto the antigen suddenly convulses to life. The interaction turns a key in its cellular ignition, and its protein-synthesizing engine throbs to life as it becomes what is known as a plasma cell. A plasma cell churns out antibody molecules at an astonishing pace—up to 10 million an hour.

The lymph node expands with the cells' activity—the routine "swollen glands" of a sore throat are in fact lymph nodes choked with immune cells and proteins responding to the incursion. It may take three or four days for a B cell to bump into its antigen, but once the connection is made, millions of antibodies pour into the bloodstream, all of them targeted to attack a particular microbe. This, too, is part of the vertebrate immune system's "specific" response. It takes a little longer to develop than the nonspecific response, but is much more efficient at finding and eliminating a troublesome pathogen.

When immigration authorities forced a Japanese-born molecular biologist named Susumu Tonegawa to leave the United States in the early 1970s, they lent an unwitting hand to the solution of one of immunology's most long-standing mysteries. In effect, they placed a brilliant mind in close proximity to an important scientific puzzle. The result was that Tonegawa received the Nobel Prize in Physiology or Medicine in 1987 for solving the genetic problem known as antibody diversity.

Even today, Tonegawa—an HHMI investigator at the Massachusetts Institute of Technology in Cambridge—has a hard time explaining how he got interested in immunology. "I got interested in—well, it's by default!" he admitted in an interview. "You know, I was not trained as an immunologist."

Indeed, far from it. Born in Nagoya, Japan, on the eve of World War II, Tonegawa obtained his Ph.D. in molecular biology at the University of California, San Diego, in the late 1960s and was working on a monkey virus known as SV40 (simian virus number 40) in the laboratory of Renato Dulbecco at the Salk Institute when immigration authorities came calling. They informed him that his visa had expired and that he had no choice but to leave the country.

"I was quite fascinated by the SV40 work, but I had to leave," Tonegawa recalled. "After looking around at several possibilities, I ended up in Switzerland. The Basel Institute of Immunology was just being created at the time, and I knew nothing about immunology, but Dulbecco said it might be a good idea for molecular biologists to get involved in the fundamental issues of immunology. I didn't exactly know what he meant by that, but more or less by default, I went to Basel."

Tonegawa arrived in Basel in February 1971. The new institute encouraged collaboration and independence but did not offer unlimited resources. Like other researchers, Tonegawa had a very small laboratory that took up half a room, which he shared with one technician. Over the next 18 months, he learned immunology "by osmosis," and that is how he became aware of what one immunologist called the "preeminent mystery" in the field over the previous three decades. It went by the deceptively simple name of antibody diversity.

The problem was straightforward, yet profound. At birth, human beings possess a phenomenally large repertoire of antibody molecules—a different immunological tool, as it were, to fit every possible loose screw in the body. When Tonegawa started his work, researchers did not know the full extent of

MIXING

AND

MATCHING

PIECES OF

GENES

Susumu Tonegawa discovered how we produce billions of different antibodies with a limited number of genes.

THE
SECRET
OF OUR

SUCCESS

the paradox. We now know that each human is born with the capability of generating upward of 10^{12}, or 1 trillion, antibody molecules, each with a different shape. Yet we also know that each person possesses far fewer genes, the latest estimates putting the total number at between 50,000 and 100,000 genes for all the functions of the body. If, as molecular biologists insisted, one gene produced one protein, how could the immune system manufacture up to 1,000,000,000,000 different antibody proteins from fewer than 100,000 genes?

This puzzle had intrigued and attracted prominent immunologists, one of the most visionary being an Australian scientist named F. Macfarlane Burnet, a circumspect, bespectacled, soft-spoken man who always gravitated to important problems in immunology. In the 1950s, Burnet and David Talmage (then at the University of Chicago, now at the University of Colorado) independently advanced a radical theory known as clonal selection to explain antibody diversity. They argued that each individual has to be born with the full repertoire of potential antibodies, impossible as

that sounds, and that it is the germs invading the body that, in effect, randomly intersect with—or "select"—the one-in-a-trillion cell whose antibody matches the bug.

This wonderfully astute hypothesis nonetheless did not answer the fundamental question: How does the body manage to generate an unlimited number of antibodies with a limited number of genes? That was the problem that Tonegawa began to attack.

Toward the end of 1972, Tonegawa started working with B cells from a mouse cancer known as myeloma. These cells manufacture too much of one particular antibody, and Tonegawa hoped their genetic material might explain how antibody genes function.

At the time, only a little was known about the structure of antibodies. The molecules are typically depicted as a Y-shaped protein, like a tree with two main branches. The trunk is the same from molecule to molecule and is therefore called the constant, or C, region. Variable (V) regions at the tips of the two branches—the business ends of the antibody—fit into different antigens. The junction of the two major domains of the molecule is known as the J region. If there

Making an Antibody

The tips of antibodies—their business ends, where they bind with antigens—are amazingly varied. To achieve this variety, each antibody molecule is made up of two separate chains, a heavy chain and a light chain, which are derived from DNA in two different chromosomes. Each chain is encoded by a very rare

combination of DNA fragments, and when the two chains assemble they form a unique molecule.

The formula for a heavy chain is as follows: Randomly select 1 out of 400 variable (V—orange) segments, 1 out of 15 diversity (D—yellow) segments, and 1 out of 4 joining (J—green) segments.

Splice together with a constant (C—blue) segment. This leads to as many as 24,000 possible combinations for the DNA encoding the heavy chain alone. (The chromosome segments that encode the light chain are not shown here.)

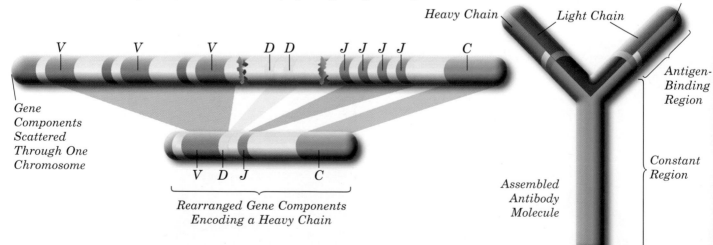

Gene Components Scattered Through One Chromosome

Rearranged Gene Components Encoding a Heavy Chain

Heavy Chain   Light Chain   Heavy Chain

Antigen-Binding Region

Constant Region

Assembled Antibody Molecule

were one gene for each different antibody, a molecular biologist would expect the DNA containing that gene to be a single, continuous stretch. It was precisely that point that Tonegawa sought to address.

A debate was then going on about where all the variation occurs: Is it in the germline cells, which give rise to sperm or egg, or in somatic cells, cells of the tissues? If the site was the germline, it meant that an improbably large number of genes would have to be devoted to antibody production. But the somatic theory was even more heretical; it suggested that the body somehow mixes—or rearranges—modules of antibody genes to generate different combinations. "I did have a bias," said Tonegawa, who thought the answer resided in the somatic cells. "Although I felt that this idea of genes being altered during one's lifetime was drastic, I was more tolerant of it."

Tonegawa brought less immunological expertise to the experiments he performed in Switzerland over the next three years than did most of the scientists who explored this problem, but he had an expertise in the technologies that would strip away the mystery. "My personal training and the timing were perfect for this project," he admitted. He knew how to use biochemical scissors, known as restriction enzymes, to cut up DNA, and he knew how to use a procedure known as hybridization to identify active genes in a given cell.

This expertise was brought to bear on the mouse myeloma cells. These cancerous B cells provided lots of the raw material needed for the experiments: messenger RNA that indicated which genes were active in the cell. Working with Nobumichi Hozumi, Tonegawa conducted hybridization experiments comparing the DNA of malignant B cells to that of normal liver cells. The turning point, Tonegawa recalls, took place in the winter of 1972.

"The previous night we set a large number of samples in the scintillation counter," he said, referring to a device that measures radioactive traces in experimental samples. "We went home around midnight, very tired, and the next morning I came in very early, around 6 a.m., to look at the data.

When I saw the printouts, I could tell the pattern of hybridization was very different [from that of normal genes]. I thought, 'My goodness, what's going on?' I got very excited and called Nobu, and we plotted the data and it looked...quite *unusual*."

To a molecular biologist like Tonegawa, the pattern in the data was unambiguous. It told him that the gene that instructed a B cell to make an antibody was, shockingly, a patchwork gene—a gene stitched together from the DNA of 1 out of roughly 400 V genes and 1 out of 4 J genes for the antibody's heavy chain, plus the DNA of 1 out of 250 V genes and 1 out of 4 J genes for the antibody's light chain.

These gene components were randomly selected within each group; the combinations then generated the enormous diversity in antibody molecules. Tonegawa and Hozumi published their results in 1976 in *Proceedings of the National Academy of Sciences*, and the report revolutionized thinking about how antibody genes can be spliced together to achieve healthy variety. Indeed, Tonegawa's lab was the first to notice that additional antibody diversity arose simply in the process of DNA rearrangement: imprecise joining of the V, J, and related D (diversity) genes often added a few letters of DNA here and there, subtly affecting the antibody's specificity.

That was not the only surprise. As other researchers, particularly Leroy Hood, who was then at the California Institute of Technology (he is now at the University of Washington, Seattle), focused on this fascinating patch of DNA that controlled antibody formation, they discovered something even more startling. Once a B cell that fit a microbial antigen had been selected, the cell underwent mutations in its antibody gene to improve the tightness of the fit—to increase the antibody's "binding affinity," in the words of biologists. These mutations occurred in so-called hot spots in the gene. In other words, the immune system had a way of taking a loose-fitting antibody and refining it by way of mutation to evolve a more effective, tighter fit.

This was of more than academic interest, because it was these more evolved, later-generation B cells that retained immunological memory of an enemy. A few of these specialized B cells, grizzled veterans of an earlier war, are always present in the body; they possess the biological memory to recognize an old foe with greater speed the second time around, and they possess the biological ability to replicate prodigiously whenever that foe pays a return visit, in effect intercepting second infections.

In 1974, just as Susumu Tonegawa was uncovering the first hints of antibody gene diversity, a young fencing enthusiast at The Johns Hopkins University in Baltimore named Mark Davis began to learn some tricks of molecular biology that would result, 10 years later, in the first solution to the other great riddle in immunology: how the receptors on the surface of T cells achieve their diversity.

Compared to antibodies, T cell receptors belonged to what seemed in the early 1970s to be "a much more mysterious category," according to Davis. One Japanese scientist had gone so far as to refer to the T cell receptor as "an imaginary monster." The problem haunted the sleep of many immunologists, not least because of a bolt of lightning out of Australia—the discovery by Rolf Zinkernagel and Peter Doherty that T cells recognize foreign antigens and MHC molecules simultaneously (for which they won a Nobel Prize in 1996). This news caused great puzzlement. Did the T cell receptor bind to both antigen and MHC at one time? Or was the T cell a two-headed beast, possessing two receptors that acted in concert, one grabbing onto antigen while the other grasped the MHC molecule? The smart money at the time was on the T cell as a two-headed beast.

"Through the mid-to-late 1970s and through the early 1980s, arguments raged back and forth in the noisy and confused fashion that seems to characterize many of these early immunological debates," Davis later recalled. There were also many attempts, by Hood, Tonegawa, and others, to see if T cells use the same genes as B cells do when they make antibodies. That, alas, would have been much too simple. "Thus," Davis continued, "by 1982–1983, a sense of desperation began sinking into the immunology community regarding the nature of the T cell receptor. It seemed that the slate had to be wiped clean and some completely new approaches taken."

Davis, who learned molecular biology at the California Institute of Technology as a graduate student and then went to the National Institutes of Health, decided to attack the problem with some clever molecular genetics. He reasoned that since both T cells and B cells are lymphocytes, the genes they use were likely to be very much the same. By comparing them side by side, however, it might be possible to identify a few genes that are uniquely active in (or "expressed" by) T cells and not B cells. Those genes, Davis realized, would probably include those of the T cell receptor.

To make a long and molecularly complex story a little simpler, Davis and his NIH colleagues found that only 100 to 200 genes are expressed by T cells but not B cells (out of a total of 10,000). As they refined the search over several years, they narrowed it down to a gene that encodes the genetic information for a subunit of the T cell receptor known as the beta chain.

When Davis disclosed these results at an immunology conference in Kyoto, Japan, in September 1983, it was Susumu Tonegawa, sitting in the audience, who alerted Japanese journalists to the fact that this was a major development. Davis, now an HHMI investigator at Stanford, and his colleagues published their results on the mouse gene in *Nature* in March 1984, and Tak Mak of the Ontario Cancer Institute in Toronto reported in that same issue on the human gene for the beta chain.

As it turns out, the T cell receptor does indeed follow the same pattern as the antibody receptor. Humans, for example, possess about 50 versions of the variable region (V) and 50 joining (J) segments for the alpha chain of the T cell, so there are roughly 2,500 different possible combinations of this chain alone. In addition, the beta chain has 30 possible V segments, 2 D (diversity) segments, and 12 J segments. John Kappler of HHMI in Denver likens these gene rearrangements to ordering from

LIKE ORDERING FROM A PRIX FIXE MENU WITH DIFFERENT CHOICES FOR APPETIZER, MAIN COURSE, AND DESSERT.

a prix fixe menu with different choices for appetizer, main course, and dessert.

The tremendous diversity of T cell receptors is generated by the enormous number of possible combinations of these elements. When additional changes that may result from imprecise joining of the segments are factored in, the total number of possibilities could reach a staggering 10^{16}.

This picture is even further complicated by the fact that T cells come with two different kinds of receptors. Some cells, like the well-known killer and helper T cells, possess receptors made up of alpha and beta chains that grab onto antigens and MHC molecules. But Susumu Tonegawa accidentally discovered that other T cells have a different kind of receptor. Though he originally believed his group had identified the T cell receptor's alpha chain, it turned out that they had found an entirely different chain, now called gamma, which is part of a previously unrecognized species of T cell receptor. This gamma chain is joined to a delta chain, and just to show how small the world of molecular immunology was in the 1980s, the delta chain was finally identified in 1987 at Stanford by Yueh-hsiu Chien, Mark Davis's colleague, to whom he is married. To this day, immunologists are still struggling to understand the role of these gamma-delta T cells, which account for 10 percent of the average person's T cell supply. "Their function is still very mysterious," says Davis.

Genetic rearrangements of antibody and T cell receptor genes explain the enormous diversity in the repertoire of immunological cells. These cells are literally ready for all comers from the world of microbes, because their receptors are genetically programmed to recognize a mind-bogglingly broad array of potential targets. What is perhaps most remarkable is that evolution provided organisms like ourselves with a proactive form of defense. The immune system does not build from scratch the T cells and antibodies that will eliminate the chicken pox virus, for example, or the streptococci that cause a strep throat. The specific agents are already on the shelf, warehoused in immunological organs like the spleen and lymph nodes. Once an infection occurs, the

bugs in a sense select their own poison. Given that cue, all the immune system has to do is cause the appropriate B cell or T cell to proliferate and make sure it has access to the infection.

As awesome and wondrous as this preparedness is, Susumu Tonegawa likes to point out one of the greatest wonders in the entire story: the fact that the human mind could intuit such an immense, complicated, and utterly nonobvious system even before genes or DNA or T cells were well understood.

"As you know, the clonal selection theory is an *incredible* theory in terms of its insightfulness," Tonegawa says. "I often use this as a rare example in biological science where the systems handled are very complex, but nevertheless, purely by insight, the basic principle of the immune system was laid out, and correctly. With virtually no molecular or cellular information. It's tremendously remarkable. There are lots of ideas around, but to be so correct, for something which is so counterintuitive, is amazing." ●

Mark Davis uses a video-fluorescent microscope to visualize how a T cell recognizes antigens during an immune response.

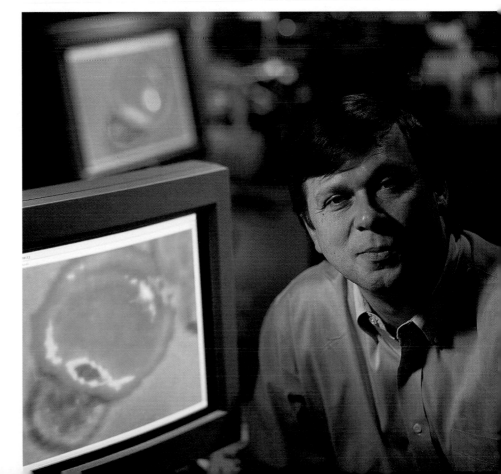

Dave Garboczi and Partho Ghosh were sitting in front of a video screen, Garboczi narrating a close encounter of the fatal kind while Ghosh twisted knobs and manipulated a dazzling 3-D electronic sword in a downward arc.

To the untrained eye, they might have looked like a couple of college kids—Garboczi with a scruffy beard and husky voice, Ghosh neat in both appearance and remarks—deeply absorbed in a video game in which they maneuvered an attacking six-pronged probe in order to deliver a

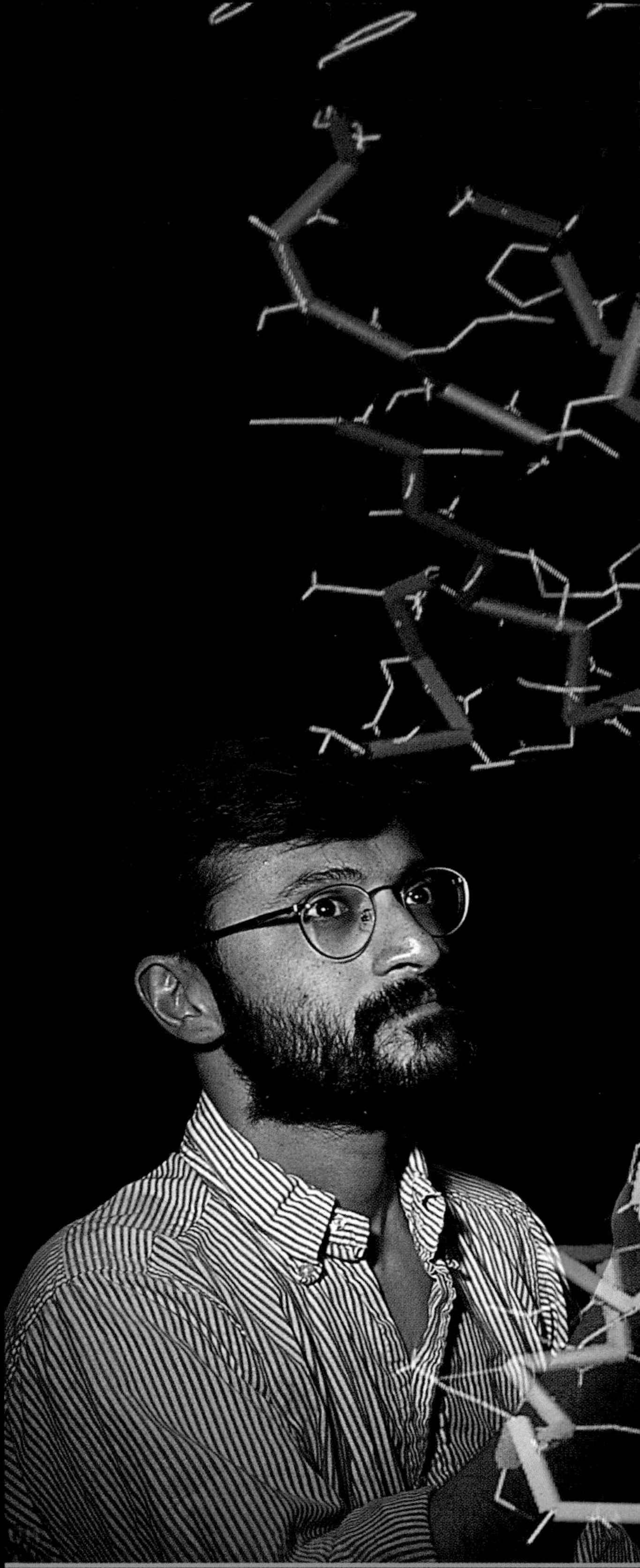

SHAPES THAT MADE A REVO

BY STEPHEN S. HALL

JTION

Partho Ghosh (left) and Dave Garboczi admire their new structure, which represents the "Holy Grail" of immunology: the three-part interaction of human T cell receptor (light and dark purple), antigenic peptide (yellow), and MHC molecule (blue). It shows that almost the entire peptide is buried in the interface between T cell receptor and MHC.

lethal blow to a target below. In reality, they are highly skilled Ph.D.'s in biology, and the interaction on the screen represented a "game" the human body plays for keeps every waking and sleeping moment of life: the battle between cells of the immune system and viruses, bacteria, and other pathogens.

What they were actually doing on this warm and rainy July afternoon in Cambridge, Massachusetts, sitting at a $10,000 Silicon Graphics workstation in the laboratory of Don Wiley at Harvard University, was studying an atomic interaction that occurs within only 600 square angstroms—a field of battle that could be hidden in a wavelength of light. In the real estate of immunology—where "location, location, location" can be a matter of life and death—they were attempting to discover the ultimate "view" in all of immunology: what a T cell "sees" when it encounters a foreign protein, usually the first sign that the body's defenses have been breached. With a wondrously precise technology known as x-ray crystallography, which allows scientists to identify the position of thousands of atoms in the crystals of proteins and nucleic acids, Garboczi and Ghosh were racing to determine the 3-D architecture of this crucial interaction.

They weren't quite there yet. "We've been crunching numbers for months," Garboczi said at one point, "and we haven't done one lick of biology yet. Now it's biology. Now it's immunology. This is really going to explain what happens in immunology."

By twisting a few knobs, Ghosh brought into relief first one, then two, three, four, five fingers of antigen reaching up and the double mitts of the T cell receptor reaching down. The exercise is a little like holding two mountain ranges, one in each hand, and trying to get the teeth of all the peaks to fit snugly together. By the time they finished their 3-D jigsaw puzzle in mid-October, they had science's first glimpse, at the level of atoms, of how the human immune system draws a bead on the enemy.

"This is a recognition event between two cells," Don Wiley explained later. "A target cell, which in this case will be some cell that needs to be killed, and a cell of the immune system, a cytotoxic T cell, a so-called killer cell. And what we've got is a crystal of everything between the two cell membranes—an intercellular recognition complex of T cell, viral antigen, and MHC molecule that contains enough information to allow the T cell to make the decision, 'Do we kill or don't we kill?'"

The picture—even the out-of-focus, uncertain picture of July—was spectacular. Yet it was not just a pretty image conjured up by a powerful computer. This T cell and this MHC molecule and this viral antigen all came from a patient at the National Institutes of Health who is battling a deadly neurological disorder.

This three-part interaction—T cell, antigen, and MHC molecule—has been referred to by some as the Holy Grail of immunology, and the autumn of 1996 was a very good season for structures of this sort. In late October, Ian Wilson's group at the Scripps Institute in La Jolla, California, provided the first view of a T cell-antigen–MHC interaction in a mouse. Then Garboczi and Ghosh published the human version of this interaction, providing the latest in a veritable gallery of immunological structures that have been turned out by the joint Harvard laboratories of Wiley and his colleague Stephen Harrison, also an HHMI investigator.

When you ask Don Wiley how he became interested in immunological molecules, he blames the flu. Not the illness per se, but the cunning virus that causes influenza. In fact, he gestures toward a three-foot-high sculpture in the corner of his office as a way of explaining how he got involved in immunology in the first place. The sculptor is Nature, and the work is a masterpiece of molecular mischief. It shows the exact structure, atom by atom, of hemagglutinin, a molecule on the surface of influenza viruses that mutates with extreme rapidity to evade the immune system.

"The real motivation for studying that molecule in the first place was that it beats the human immune system," recalled Wiley, who sometimes speaks with the irascible inflections of a teenager, yet sports a few grey hairs as well. "That's why you get the flu more than once. And so we started out in immunology."

Even as he spoke, colleagues in the lab pursued a broad range of immunological targets: in addition to Garboczi and Ghosh's work on the T cell receptor, graduate student Qing R. Fan was attempting to capture a picture of the natural killer (NK) cell receptor, which allows NK cells to decide whether to kill or not, and Theodore Jardetzky was working on the structure of a superantigen. Meanwhile Stephen Harrison was working on molecules that are involved in signaling inside the T cell. All this work, in a sense, aspired to let us see what the immune system sees.

Structural biologists focus on exactly what their name implies: structures. They produce atom-by-atom, 3-D models of biological molecules. Among the first, and no doubt the most famous, of models to emerge

Don Wiley became interested in molecules of the immune system through his work on the structure of hemagglutinin (below, right), a molecule on the surface of flu viruses that mutates rapidly to evade antibodies.

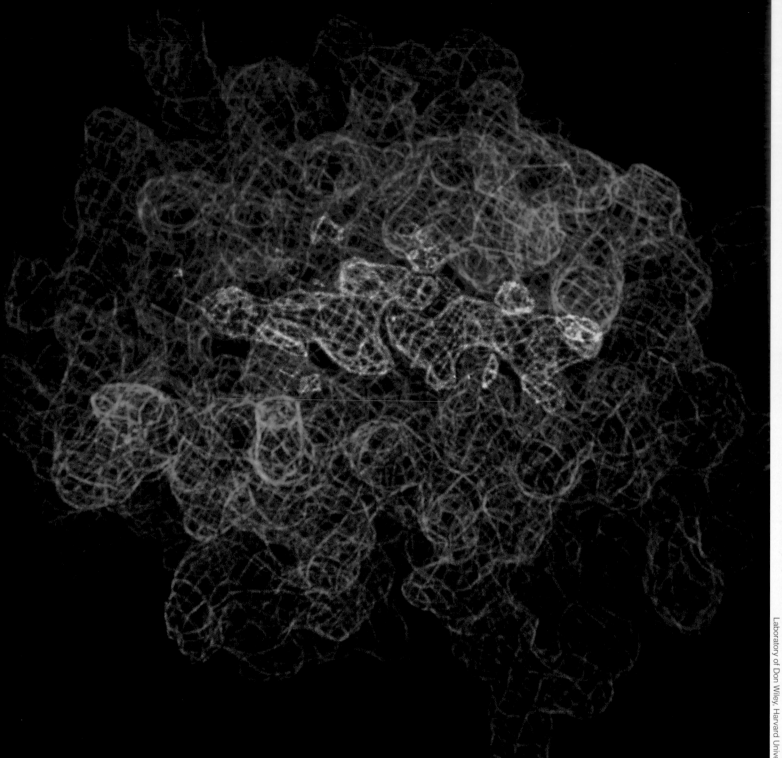

"All of a sudden it made sense"—the first view of how a peptide fits inside an MHC molecule's binding groove.

from this sort of work was the double-helix structure of DNA reported by James D. Watson and Francis Crick in 1953. Their structure covered a tabletop and was held together with wire and clamps; today's structures can be viewed on the screen of a laptop, rotated with a knob, and colored with every hue of the rainbow.

X-ray crystallography proceeds in three stages. First, you must grow crystals of the purified molecule—or molecules—to be studied, so that all the atoms that make up the molecule are arrayed in a precise, repetitive order. Recombinant DNA technology revolutionized this step, because it allows even rare proteins to be reproduced abundantly by genetic engineering once the gene has been identified—and in order to make crystals, you need lots of protein. Second, you bombard these crystals with x-rays, whose wavelengths are small enough to be scattered by the atoms in the crystal, creating what is called a diffraction pattern. Finally, you need to interpret the resulting diffraction patterns and infer the exact location of the atoms doing the deflecting, which will lead to an overall structure.

When Wiley started doing this work, it was not unusual for a structural biologist to spend 10 years or more on a single structure; now, with the combination of genetic engineering, more powerful x-ray sources, speedier number crunching by computers, and better software for modeling, structures can be "solved" in a year or two, sometimes less. The slowest step is the first one: growing the crystals.

Like many other scientists, Wiley and his Harvard colleagues were intrigued by the reports emerging from immunology labs in the 1970s and 1980s. "I am interested in the interaction between infectious agents and the immune system," Wiley admitted, "but then, you are also opportunistic." Opportunity knocked first with the report from Zinkernagel and Doherty, which suggested that MHC molecules somehow play a role in immune recognition, and then with the discovery by four researchers in Denver—R. Shimonkevitz, John Kappler, Philippa Marrack, and Howard Grey—that T cells recognize not whole proteins as antigens, but merely small, partially digested bits of them known as peptides.

"It was a mystery how the T cell receptor could be recognizing the MHC molecule and the antigen simultaneously," Wiley says, picking up the story, "so a lot of people were drawn to studying it, from all different points of view, including the structural point of view. Pamela Bjorkman in our lab, in collaboration with Jack Strominger, got going on trying to get the structure of this Class I MHC molecule, and you know how one thing leads to another."

Bjorkman, now an HHMI investigator at the California Institute of Technology, spent years leading the effort to solve this structure as a graduate student and then a postdoctoral fellow in Wiley's lab. Its publication in *Nature* in 1987 left the immunological community happily dumbstruck. An x-ray picture is worth more than a thousand words because it provides such an astonishing view of a landscape so minuscule and so otherwise inaccessible to human vision.

This famous picture revealed that a Class I MHC molecule has a deep groove down the middle and, surprisingly, that a peptide nestles inside this groove, its thorny side chains of amino acids sticking out like outcrops on a mountainside. The picture of the entire complex, showing a hot pink tube of peptide inside the wraparound groove of a neon blue MHC molecule, captivated the scientific community. Among less fastidious biologists, the structure was said to resemble a hot dog (peptide) sitting in an MHC bun.

Upon seeing the structure for the first time, Bjorkman immediately realized that the peptide would have to be bound in the groove. Ironically, this caused more consternation than joy at first. "If we were truly stupid, we would have been unbelievably happy," Wiley recalled, "because, you know, it's obvious! The trouble is, you had to ask yourself: How could a peptide still be in there after all this crystallization procedure? I mean, it is staring you in the face, but how could it be?"

The picture invited experimentation, and it turns out that MHC molecules hold onto peptide exceptionally tightly; the grip is so firm that, even when the complex appears on the surface of the cell, the MHC clutches a peptide for roughly a hundred hours or so before letting go. There is an excellent scientific reason for the MHC molecules' maintaining such a tight grip. "If the peptides are on the surface of a cell and they fall off," Wiley explained, "they will never get back on because the concentration is so low." And if the peptides fall off, T cells will not recognize the intrusion. End of immune response.

It is difficult to exaggerate the impact of this one structure on the field of immunology. "I think the first picture, the original Bjorkman and Wiley 1987 paper, just opened up everyone's eyes to what was going on," Peter Cresswell recalled. "When the structure came out and you saw this fantastic binding groove and the way that individual side chains in the peptides fitted

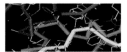

It is difficult to exaggerate the impact of this one structure on the field of immunology.

into pockets [of the MHC molecule]—all of a sudden it made sense. It explained visually, I think, something that we'd all been having a very hard time grasping from an intellectual level."

The thing that impressed John Kappler, a T cell maven in Denver, was the way one picture supplanted dozens of papers and years of difficult experiments by scores of researchers in one fell swoop. "That was such a powerful paper," Kappler said nearly a decade later. "There were 50 papers, maybe even more, listed in the citations, all trying to get at this question of how the molecule worked by mutational analysis and antibody binding and so on. And, you know, with just a sweep of the hand, those 50 papers were all collapsed into this one picture. And it all made sense. Everything that you had been pondering made sense just by looking at this structure." (Kappler took the lesson especially to heart; he has become a convert to structural biology, believing it offers the best hope of explaining the function of the immune system.)

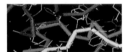

"That's it, folks..."

"Looking is particularly evocative," Wiley acknowledged. "You look and you see, whereas other types of experiments may be equally informative, but to only a few people. And you can sometimes settle issues [of uncertainty in a field] by seeing something and getting everybody to agree. When everybody looks at it, it's 'Aaah, we actually stand *here*! Let's go forward from here.' It's an amazingly powerful effect, and I think structure sometimes has that effect."

Wiley then bent over his desk, searching for an archival piece of white scrap paper. "The day we had that picture," he said, referring to an image of the MHC structure on his bulletin board, "we actually did this...," shuffling papers on his desk. "I don't think it's still lying around here...." He finally produced a facsimile of the missing piece of paper—his prediction of the T cell receptor's shape. It measured about two inches by three inches, the expected size of a T cell receptor at the scale of the photo on the bulletin board, and matched the footprint of the MHC-peptide complex. "And we went, 'That's it, folks, that's how a T cell could bind.'" The piece

of paper, he neglected to add, was blown up about 10 million times from the actual size of the MHC surface.

"And now," he said, turning to a large computer screen in his office, "we are at the next stage, which is, How do T cells actually recognize this? And what happens to T cell receptors when they bind to MHC molecules? How do signals get sent?"

A tragic opportunity to look at this even bigger picture presented itself in the early 1990s when a patient with a severe neurological disorder was referred to doctors in the Neuroimmunology Branch of the National Institute of Neurological Disorders and Stroke in Bethesda, Maryland. At some point in his life, this gentleman had become infected with a human retrovirus called HTLV-1 (human T cell lymphotropic virus). This virus and its siblings, distant relatives of HIV, were originally discovered because they cause T cell leukemia, a rare cancer in white blood cells. But in similarly rare instances, they also seem to be associated with a degenerative neurological disorder called tropical spastic paraparesis, or TSP. However, "many, many people infected with HTLV-1 don't have any disease," according to William E. Biddison of the NIH. "Most develop nothing."

Biddison's patient first sensed that something was wrong when he had difficulty walking. "Most patients initially present with problems in their gait," Biddison explained. "The symptoms of the disease are much like those of multiple sclerosis (MS). In fact, frequently there's a question of whether the patient has MS or TSP." As in MS, the nerve axons of TSP patients lose their protective outer sheath of myelin, and the patients slowly lose control of muscular function. "Ultimately," Biddison said, "they end up in a wheelchair."

Biddison and his colleague Ursula Utz observed a very unusual situation in their patient, however. When they analyzed his blood, they found an abundance of a type of T cell that recognized HTLV-1; specifically, it recognized a peptide made only by the virus. Biddison's group identified the peptide and began to collaborate with Wiley's group around 1991 to help their analysis.

What made this situation especially unusual, and especially promising to researchers, is that an invading virus whose protein is chopped into a variety of peptides by immune cells normally would provoke many different T cells to recognize it—so many, in fact, that it would be difficult to isolate only one. By finding one T cell in such conspicuous abundance, the NIH researchers were able to clone its genes, and this led to a collaboration with the HHMI team at Harvard.

Ultimately, they managed to isolate the three critical components of a T cell response: the peptide from HTLV-1, the MHC molecule into which it fit, and, from the NIH patient, the exact T cell receptor conformation—the 1-in-100-trillion shape—that recognized this peptide. In September 1995, Biddison recalled, "things started to take off"—specifically, Dave Garboczi could genetically engineer bacteria to make the T cell receptor—the business end of the T cell—in the copious amounts that were needed to grow crystals.

Even more ambitiously, Garboczi and Ghosh grew three-part crystals containing the T cell receptor, the viral peptide, and the MHC molecule in which it nestled. They had such crystals by the fall of 1995, and they traveled to Ithaca, New York, in February and May of 1996 to gather x-ray diffraction data, using the high-energy synchrotron at Cornell University—"the difference between a slow drip and a fire hose," says Garboczi—as a source of x-rays. From that point, it took another five months to solve the structure, which appeared in the November 14, 1996, issue of *Nature*.

Once again, the Wiley group produced a remarkable picture. Garboczi and Ghosh had kept refining their image of the interaction until it assumed definitive shape and great explanatory power. The size of the interaction turns out to be larger than the MHC surface alone would suggest: the HHMI researchers estimate the real estate in this case takes up approximately 1,000 square angstroms (to put this in perspective, a wavelength of red light is about 4,000 angstroms).

If one imagines an MHC molecule as a horizontal rectangle with the groove running through the middle, from left to right, the HHMI crystallographers have shown that the molecule has a forbidding topography that makes access difficult. There are "peaks" in the upper left and lower right corners of the rectangle, so in order to get a good 3-D grip on the peptide in the groove, the T cell receptor has to come in diagonally, squeezing into the valley between the two MHC peaks and athwart the main groove. "The top of the molecule isn't always flat," Garboczi explained. "It's undulating, which surprised us. So the T cell receptor has to go between the hills and into the valley to get to the peptide."

By nestling into this second, diagonal groove, the T cell receptor maneuvers its business end directly over the heart of the peptide that it is "interrogating," as Stanford immunologist Peter Parham puts it. Parsing this interaction down to the level of contacts between atoms, the HHMI researchers determined that the T cell receptor buries the MHC-peptide complex, touching it in 20 different places. It is these 20 different, wavering hydrogen bonds, barely the diameter of an atom in length, that tell an immune cell that something is wrong. "There's a lot of contact," Garboczi said. From this twitch of recognition, where atoms of friend and foe graze each other, the T cell becomes activated, and from that derives all the commotion that occurs in the blood during an immune response.

Don Wiley stresses that the structural biologists who are providing pictures of the shapes that make the immune response possible bring no special visionary insight to the effort. "If Doherty and Zinkernagel hadn't done what they had done, there is no way we'd have done what we did," he says. "We just sort of roll along...And that's a key thing, I think, about the value of the work that you ever do. It's much more likely that somebody else will read about it and see the next step more clearly as a result of it. And good science is when lots of people see new steps more clearly." ●

Perhaps this has happened to you—you are out on a summer picnic, and the chicken salad seems a little...*off*, but you wolf down your sandwich nonetheless. Several hours later, you start feeling nauseated, and before you know it, you are vomiting or suffering from diarrhea. It's a classic case of food poisoning. What you probably don't know, according to John Kappler, an HHMI researcher at the National Jewish Medical and Research Center in Denver, is that this common malady is sometimes caused by your immune system. More specifically, it is caused by T cells run amok.

A small but significant portion of food poisoning cases can be traced to the family of bacteria known as *Staphylococcus aureus*. Kappler and his colleague Philippa Marrack have played leading roles in demonstrating that the real root of the problem is not the microbes themselves, but tiny freefloating bits of potent poison that they produce, called enterotoxins. Because certain of these toxins trigger a wildly exaggerated T cell response, Kappler and Marrack decided to call them superantigens.

"In modern times," Kappler explains, "*Staphylococcus* poisoning usually comes from food on which the bug has grown

and made some toxin, but often the bug isn't even there. You're not eating staph; you're eating what's been left behind."

Superantigens possess remarkable properties, beginning with their ability to rile large numbers of T cells. The average viral antigen that is nestled in the groove of an MHC molecule during antigen presentation may activate at most 1 in 100,000 T cells, and more probably 1 in a million T cells. Superantigens can sometimes arouse 5 percent of all T cells—tens of millions of different T cells, all frothing with cytokines because of this activation.

Staph enterotoxins provide an excellent illustration of how superantigens stir up so much trouble. Once ingested in contaminated food, these bacterial toxins are hearty enough proteins to survive the highly acidic environment of the stomach and the equally lethal enzymes of the intestines, which normally break down proteins into peptides that can be used by the body. When the toxins cross the surface tissue of the gut, known as the epithelium, and get into the body tissues and bloodstream, the real trouble begins.

Normally, foreign proteins attract the notice of cells such as macrophages, which capture the alien

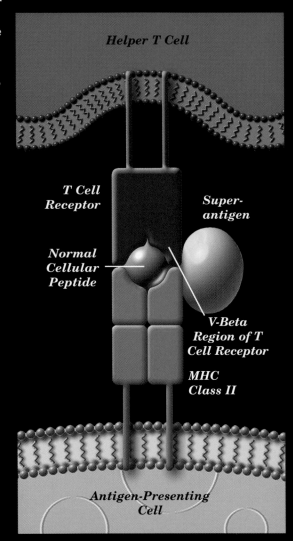

Helper T Cell

T Cell Receptor

Super-antigen

Normal Cellular Peptide

V-Beta Region of T Cell Receptor

MHC Class II

Antigen-Presenting Cell

Rather than snuggling into the groove of MHC molecules, superantigens grab hold of the outside of the molecules. This enables them to present much larger targets for T cell receptors and to activate many more T cells.

proteins, break them down into peptides, repackage them in the groove of MHC molecules, and then hold them up for inspection by helper T cells. This interaction, detailed in the main story, is extraordinarily precise: the T cell receptor recognizes only a few atomic bumps in the peptides as they project up and out from the MHC groove. The fit must be nearly perfect, down to the level of atoms, and thus only a few T cells interact with antigens presented in such a restrictive fashion.

Not so with superantigens. For some reason, they resist the normally neat packaging of the immune system; rather than snuggling into the groove of MHC molecules, they grab hold of the outside of the molecule, like a child clinging to the leg of a parent instead of being cradled in the parent's arms, and thus present a much larger target to the T cell receptor.

In the case of staph toxins, the superantigen binds to—that is, interlocks with—a more generic part of the T cell receptor known as the V region of the beta chain (V-beta). Each person possesses, roughly speaking, about 30 different V-beta segments that combine with D and J segments, as well as with a separate alpha chain, to produce the staggeringly large number—tens of trillions—of different, highly specific T cell receptors. If by chance an antigen clings to a more generic, invariant part of the receptor, however, it will activate many more T cells. That is exactly what happens with superantigens, each of which binds to about three different V-beta segments.

"Paradoxically, the effect of that is extremely pathological," says Kappler. Activated T cells manufacture very powerful, often nasty chemicals known as cytokines. "Usually they only do those

nasty things in a very confined area," Kappler continues. "But when we encounter superantigens, so many T cells get activated that instead of having this occur in one little tiny place, it occurs throughout the body, and they do very nasty things on a global basis." In the case of food poisoning, they create the agitation in the gut that leads to vomiting and diarrhea. Usually that only lasts about 24 to 48 hours, but it can be a lot worse.

Toxic shock syndrome, which has been linked to bacterial infections hastened by superabsorbent tampons, is also a superantigen disease. The bacterial toxins make their way into the bloodstream, where they trigger a broad T cell response. These activated T cells become little pharmaceutical factories, churning out huge amounts of an inflammatory chemical called tumor necrosis factor (or TNF). TNF has the unhappy effect of causing blood vessels to become leaky and porous, leading to the sharp and sometimes deadly drops in blood pressure known in cases of fatal shock.

There may even be a superantigen connection to AIDS. It turns out that a retrovirus in mice called mouse mammary tumor virus can function as a superantigen. This finding inspired researchers to investigate the possibility that the human immunodeficiency virus (HIV), a retrovirus, might also do that. Since HIV replicates in T cells, a superantigen would provide the virus with that many more targets for infection.

The HIV connection remains theoretical at this point, but there is nothing theoretical about the problems superantigens can cause at a picnic. "Superantigens are very powerful things, not to be trifled with," Kappler says. "You can't go putting them into humans without worrying about dire consequences."

Too Little:

Life in a Germ-Free Bubble

David was not allowed to touch anyone or anything outside of his bubble.

David the "Bubble Boy," as he came to be called, was born without a working immune system, which meant that even the mildest infection could kill him. To protect him, his family built a germ-free bubble in their Houston home and placed David inside it. He remained there, in isolation, nearly all 12 years of his short life.

His parents had already lost a son at 6 months of age from infections that could not be cured by antibiotics. They were told he suffered from some form of severe combined immunodeficiency (SCID), a condition that affects 1 baby in every 100,000, but there was no way to determine exactly what type of genetic error caused it—nor whether it was likely to recur in a sibling.

The odds against its happening again in their family were very low—only 1 in 10,000—if this was a new genetic mutation, the parents were told. On the other hand, if the defect was inherited from one of the mother's X chromosomes, any male child stood a 50-50 chance of being born with SCID. (The mother and her daughters would remain healthy despite this error, since females have two X chromosomes, one of which usually compensates for a defect on the other; males, who have one X and one Y chromosome, lack this margin for error.)

As they had no history of SCID in their families, the parents decided to try again. This time, however, the doctors took special precautions. Immediately after David's birth in 1972, they put him in a plastic isolator to protect him from microbes, in case he had the same immune deficiency as his sibling. The nurses held, fed, and changed him through built-in gloves. When his doctors discovered, to their dismay, that David also had SCID, they said that his life could be saved—as long as he had no contact with the outside world.

Nobody knew how long this condition would last, since no child with X-linked SCID had lived much beyond infancy. Perhaps David's immune system was just "immature" and would spontaneously begin to function as he grew up, his doctors suggested. If so, he might need to be in a protective bubble for only a couple of years. In any event, they hoped to keep him alive long enough for a successful treatment to become available.

The parents took David home and set up a large plastic bubble for him in their living room. For the next 12 years, David breathed filtered air, ate sterilized food, and remained in the bubble, except for brief trips in a miniature space suit. He played with sterilized toys and read sterilized books. He talked to his family through the plastic barrier. He even did his homework with the aid of large rubber gloves that were sealed into the bubble. But his body still failed to make T cells, and in addition, his B cells were defective. In some ways David resembled patients with AIDS (*acquired* immunodeficiency syndrome), who cannot fight infections because the AIDS virus has killed off their T cells.

In 1984, in an attempt to stimulate his immune system and free him from the bubble before he became a teenager, David underwent a bone marrow transplant. But complications set in. He grew very ill, then died, leaving behind distraught parents—

as well as cultures of cells that researchers had taken from his blood.

Ever since then, David's blood cells have continued to grow in the lab. And for a decade after his death, scientists studied them in an effort to find the precise cause of his disease. Finally, four years ago, Warren Leonard and his associates at the National Heart, Lung and Blood Institute in Bethesda, Maryland, discovered the genetic error involved. It was a mutation in a subunit (the gamma chain) of the interleukin-2 (IL-2) receptor—a receptor on the surface of T cells that receives the signal that it is time for the cells to grow and divide.

As a result of this discovery, it is now possible to tell families whether or not they are at risk of having a child with this type of SCID, which represents about half of the total number of cases. This information can be of great value in family planning and may lead to new treatments for the defect.

One by one, the causes of previously mysterious immune deficiency diseases are being uncovered, thanks to recent advances in immunology. Warren Leonard's team was not actually looking for the cause of X-linked SCID but was studying the IL-2 receptor and its subunits. In the course of this research, he and others mapped the gamma chain gene to a position on the X chromosome, and when they realized that this region had been implicated in SCID, they rushed to test the Bubble Boy's cells. Sure enough, the DNA from these cells had a mutation in the gamma-chain gene. So did the DNA from two other SCID patients. The importance of the gamma chain goes far beyond this defect, however. The chain

> **Patients with immune deficiencies will eventually be cured through gene therapy...**

is also a critical component of four other growth-factor receptors on T cells and plays a major role in the complex signaling that leads T cells to grow and differentiate.

Other breakthroughs have come from the study of B cells, the cells that produce antibodies. That is how Owen Witte, a developmental immunologist and an HHMI investigator at UCLA, discovered the cause of the very first immunodeficiency disease ever described—Bruton's agammaglobulinemia, the inability to make antibodies.

Also called XLA, the disease had been known since 1952. Witte ascribes his finding to "a stroke of very good luck." He was examining what regulates the growth of B cells so that just the right number of cells are made—neither too many, which could produce leukemia, nor too few, which might result in immunodeficiency—when Satoshi Sukata, a postdoctoral fellow in his lab, discovered a previously unknown regulatory gene. The gene was soon mapped to a particular part of the X chromosome.

"When we looked in the various textbooks to see what X-linked diseases there might be in that region of the X chromosome that had anything to do with B cell development," Witte recalls, "there, staring us in the face, was probably the best-studied primary immunodeficiency of all, which is XLA."

Witte at once began to collaborate with Max Cooper, a pediatrician who is a leader in the study of primary immunodeficiencies and a Hughes investigator at the University of Alabama, Birmingham. After gathering specimens from XLA patients, the researchers demonstrated that the patients' cells failed to express the new gene normally. Meanwhile, a group of European scientists who had been using a more traditional genetic mapping technique also found a gene whose mutations caused XLA. It turned out to be the identical gene, now named Btk (for Bruton's tyrosine kinase), and the two reports were published in 1993, only a week apart.

To create an animal model of the disease, Witte and others knocked out the Btk gene in mice. Now they are trying to cure the disease in descendants of this mouse by inserting a normal gene into the defective mice. Witte's group has already succeeded in correcting the defect in cells that were taken from XLA patients by adding the Btk gene to these cells.

Patients with XLA and other immune deficiencies will eventually be cured through gene therapy, Witte hopes. Even more people may benefit from the fact that "Btk provides a very good target for intervention in several diseases that involve too many B cells or too much B cell activity," he says—for example, autoimmune diseases such as systemic lupus erythematosus, or perhaps even certain cancers that are derived from B cells.

"It's the flip side," Witte says. The same need for regulation and balance applies to other immune system cells, he emphasizes. In every case, there can be too little immunity or too much. ●

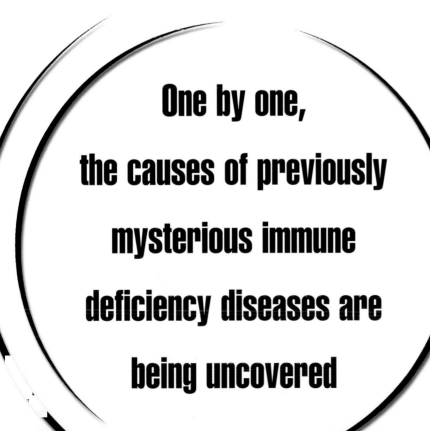

One by one, the causes of previously mysterious immune deficiency diseases are being uncovered

Autoimmune Diseases

Always prepared for emergencies—attacks by bacteria or viruses, damage from radiation, wounds—the body makes billions of immune-system cells every day, far more than are generally needed. Most of these cells die without ever being called to duty.

While some of the cells just wither away from old age, a surprisingly large number commit suicide right after birth. They sacrifice themselves after they prove incapable of telling friend from foe, "self" from "non-self," which indicates they are extremely dangerous. Such cells would have no built-in restrictions against attacking the body's own tissues. The body then orders them to self-destruct, and they obey.

"Inevitably, because of the random gene rearrangement process that brings about their diversity, some lymphocyte receptors will be generated that react to the body's own proteins," explains

Charles Janeway, a professor of immunology at Yale University and an HHMI investigator, who has written a leading textbook in the field. To prevent cells with such receptors from causing havoc, the body gets rid of them early in development.

"Before T cells are allowed out to go on patrol in the rest of the body, they're checked for whether or not they can react with anything in the thymus, where they grow up," says Philippa Marrack, one of the researchers who discovered how such T cells are identified and destroyed. "If their receptors react with some bit of self, this

"It is caused by T cells that migrate toward the islets of Langerhans and kill the beta cells of the pancreas that make insulin. Then that individual, for life, is deprived of insulin and has to depend on daily injections to metabolize his sugar."

Once the beta cells are gone, it is too late to treat the disease by clamping down on the T cell response. But other autoimmune diseases might be controlled in this way. "A lot of people hope to treat such disorders by deflecting the T cell response one way or the other—to interrupt the reaction," Max Cooper says. "On the other hand, you

The body then orders the cells to self-destruct, and they obey.

causes the cell bearing these receptors to die." First the cell's nucleus condenses, then the cell disintegrates in an orderly process called apoptosis, or programmed cell death. Only those lymphocytes that are tolerant of self are allowed to mature.

Any cells that escape this dragnet become outlaws. Sometimes they go on a rampage, attacking their own mother and sister cells and producing severe autoimmune diseases such as multiple sclerosis (MS), rheumatoid arthritis, systemic lupus erythematosus, or juvenile diabetes.

Autoimmune diseases used to seem very mysterious, but they are increasingly well understood. "Look at juvenile diabetes. We now know definitely that it is a T cell-mediated disease," says Emil Unanue, of Washington University School of Medicine.

don't want to go too far. You don't want to create a generalized immunodeficiency, or a significant gap in immunity, while you turn off an attack on your joint in rheumatoid arthritis or on your brain in multiple sclerosis."

Researchers who want to find the right balance are now turning to animal models, where they can study a variety of treatments. Molecular biology has made it possible "to create almost any autoimmune disease in mice just by knocking out some element of the immune system," Cooper points out. He hopes that studies of these mice will help scientists answer a fundamental question about diseases that are caused by either too little or too much immunity: "How do you adjust the thermostat in a really precise way?" ●

Twelve years ago, Gerald Crabtree, a lithe, easygoing biologist and HHMI investigator at Stanford University in California, set out to discover how T cells "know" when it is time to send out an alarm that will turn on the immune system.

A T cell sounds a molecular alarm about 45 minutes after molecules on its surface have recognized an intruder. Its signals then tell other immune system cells to become fruitful and multiply. Dozens of cells become hundreds, hundreds become thousands, and before long, the lone T cell sentinel has switched on millions of other cells, all specifically attuned to that particular intruder. This powerful wave of immune defenders then fights off infection.

But how does this first T cell get activated? When the T cell's surface has detected an enemy in its vicinity, how does the knowledge of that enemy reach the cell's nucleus and switch on genes whose products will unleash the entire immune system?

If he figured this out, Crabtree reasoned, he would have a new way to study what he was really interested in: how the tissues of a developing embryo take shape. During early development, many cells communicate with one another, but in 1985 no one knew how cells interpret the signals they receive. At that time, "lots of things about embryonic development were hazy," recalls Crabtree. So he decided to seek a solution to the general problem of signaling inside cells by looking inside T cells, where at least the activation of a single cell led to a well-known response.

At about the same time, Stuart Schreiber, a chemist who is now an HHMI investigator at Harvard University, became interested in the question of how to turn the immune system off. Although the immune system protects us from infection, there are times when turning it off with immunosuppressive drugs is a matter of

Scientists Design "Double-Stick

URNING OFF
THE IMMUNE SYSTEM   by Steven Dickman

life and death. Following a kidney transplant, for instance, the body protects itself from the perceived "attack" of the donor organ with an overpowering immune response.

All that immune firepower could shrivel a donated kidney, and with it the patient's chances of survival. Saving the patient's life requires that physicians use potent immunosuppressive drugs.

Schreiber had studied two of these drugs. He knew that though they are frequently effective, they are extremely toxic and can leave the patient just as dead as when a transplant is rejected. So he was eager to uncover the mechanism by which the drugs acted: What exactly were they doing to prevent the T cells from attacking the grafted organ? Could the drugs be rejiggered to be less toxic?

Coming from opposite directions, Schreiber and Crabtree "bumped heads," as Schreiber describes it, in 1989, one looking for the secrets of immune activation, the other for the basis of the immune system's shutdown. It turned into a natural, if transcontinental, collaboration.

But the fruits of their labors went far beyond either researcher's wildest dreams. In seeking answers to these two problems inside the T cell, Schreiber and Crabtree learned how to flip a switch that determines the outcome of many vital cel-

Gerald Crabtree (left) and Stuart Schreiber discuss the next step in their research on how signals from outside a cell turn genes on or off in the cell nucleus. The two scientists, who live on opposite coasts, are shown working in Crabtree's parents' home, one of the many meeting places for their transcontinental collaboration.

Molecules to Switch Genes On at Will

lular functions. Their findings were of such importance that they both plan to spend the rest of their careers working out the consequences. "It was a delightful bump," says Crabtree.

When he started on this road, Crabtree recalls, T cell activation was a black box. Only one thing was known: after an alien substance attaches itself to a receptor on the T cell's surface, the cell turns on a gene that produces a protein called interleukin-2 (IL-2). This protein then alerts other T cells. But "there was not a clue, not a single molecule identified between the outside of the T cell and its nucleus" that could account for switching on this IL-2, he says.

Crabtree's approach to scientific work is perhaps best exemplified by his leisure-time pursuits: After he designed an addition to his ridgeline, ocean-view stone house, he built it himself, one rock at a time. He attacked the problem of T cell activation with the same kind of quiet determination.

By 1987, with the help of some post-doctoral researchers in his laboratory, Crabtree had identified the molecules that activate the IL-2 gene in the T cell's nucleus. This gave him the far end of the pathway. To analyze earlier steps in the pathway, he started to work with two drugs that suppress the immune system, cyclosporin A and FK506, the same two molecules that, unbeknownst to Crabtree, had also interested Schreiber.

Cyclosporin A had been discovered in Norway, whereas FK506 had been found halfway around the globe on the slopes of Mt. Fuji in Japan. This geographic spread, together with their utterly unrelated chemical structures, implied that the drugs might act on cells in very different ways. Nevertheless, both drugs could be used to bring the immune system to a halt in time to stop graft rejection. And inside the T cell, Crabtree's team discovered, the drugs seemed to act in a similar fashion: they both blocked the activity of certain transcription factors (proteins that bind to DNA and regulate gene transcription) and thus stopped the IL-2 gene from being turned on.

Crabtree and his postdoctoral fellow,

Michael Flanagan, soon found that a transcription factor called NFAT provided a link between the cell surface and the nucleus, and that cyclosporin blocked communication between them. Their report caught the immediate attention of Schreiber, who was working (across the continent at Harvard University) on how FK506 turns off the immune system.

If Crabtree is a scientific rock-piler, then Schreiber is the academic equivalent of an Olympic athlete—one who switches from, say, the discus to the pole vault in mid-Games and wins gold medals in both events. What sets Schreiber apart, according to cell biologist Tim Mitchison of the University of California, San Francisco, who recently spent a few months in Schreiber's lab, is his gargantuan self-confidence. "He picks projects that are new and pushes the envelope with them even though their success is not guaranteed," says Mitchison.

Schreiber started out as a chemist, receiving a Ph.D. at Harvard and then taking a faculty position at Yale in 1981. He returned to Harvard seven years later, but he also returned transformed. He had been bitten by the bug of biology.

The turning point came when Schreiber completed the synthesis of a molecule that happened to have a striking biological effect. The molecule, periplanone B, is the sex attractant of the American cockroach. After synthesizing it, Schreiber took some of it down into the basement of the Yale chemistry building to run a field test. For the male cockroaches there, the smell of the chemical made it seem as if the cockroach equivalent of Marilyn Monroe had walked in. All Schreiber had to do was puff a tiny amount of his compound into the air and the insects stood on their back legs and flapped their wings frantically. Some nearly destroyed themselves in the process, collapsing in a tangled heap of broken antennae and twisted wings.

Schreiber's discovery earned him an *Esquire* magazine Dubious Achievement Award for starting a "cockroach dating service," but it also launched him upon a new career. "Seeing the physiologic response of the insect seemed almost magical," he

T cell activation was a black box.

recalls. "It was fascinating, mind-boggling. But it was not particularly accessible to me as a chemist." It was only when a Yale colleague showed Schreiber how to gingerly put a spring-loaded clip on the end of an insect antenna, puff the compound on, and see an electric signal on a tiny voltmeter that magic became science. "I thought to myself, that's just chemistry. There must be a receptor in the cell membrane, communicating with something inside the cell."

Once he began to think about following the trail of a molecular signal when it touched the outside of a cell and caused changes inside, he was hooked. Biology offered Schreiber something entirely new. "Chemical synthesis," Schreiber continues, "is an activity you engage in to make very complicated molecules. Those molecules typically exist in nature, and you're finding a route to make them in the laboratory."

By contrast, finding the cellular target of a drug made him "the first person on the planet to know that this protein binds that particular molecule," Schreiber says. "I was just filled with excitement and curiosity about how that could occur." Making the switch from chemistry to biology was "like jumping into a pool of water without knowing how to swim," he says. But the beguiling nature of the problems at hand led him on.

Now Schreiber's new mission became clear: to explore the reactions inside cells to signals coming in from outside. In particular, Schreiber set out to find the targets of cyclosporin A and FK506. This posed a similar problem to the one Crabtree had faced in determining how T cells are activated. Schreiber's group had to determine how the two drugs prevent the normal signal (the transcription factor NFAT) from reaching the T cell's nucleus, thereby blocking the cell's activation.

Not long before Schreiber became interested in immune suppressants, his Yale colleague, Robert Handschumacher, had identified cyclophilin, a protein to which cyclosporin attaches itself inside T cells. Schreiber then identified a similar protein target for FK506, which he called FK506-binding protein, or FKBP. Before anyone

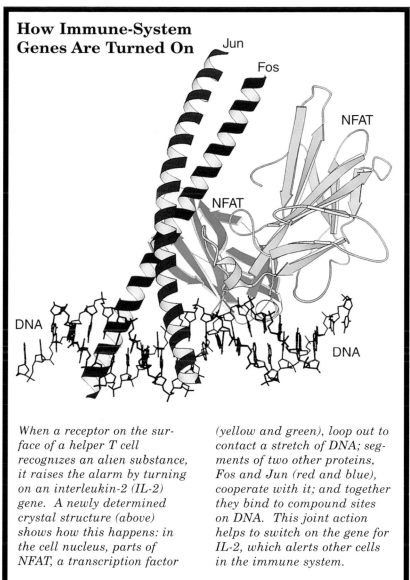

How Immune-System Genes Are Turned On

Jun

Fos

NFAT

NFAT

DNA

DNA

Laboratory of Stephan Harrison, Harvard University

When a receptor on the surface of a helper T cell recognizes an alien substance, it raises the alarm by turning on an interleukin-2 (IL-2) gene. A newly determined crystal structure (above) shows how this happens: in the cell nucleus, parts of NFAT, a transcription factor *(yellow and green), loop out to contact a stretch of DNA; segments of two other proteins, Fos and Jun (red and blue), cooperate with it; and together they bind to compound sites on DNA. This joint action helps to switch on the gene for IL-2, which alerts other cells in the immune system.*

could start to understand how the immune system was turned on and off, however, two questions needed to be answered: How do the drugs interact with their docking sites, and where are these docking sites located?

Both were simple questions with deceptively complicated answers.

The cell's exterior is so studded with receptors and channels of various kinds that in electron micrographs it looks somewhat like a fruitcake. Some drugs act by attaching themselves to docking molecules located at the cell's surface membrane and then use the cell's own machinery to exert

Forcing a Marriage

A natural "double-stick" molecule, rapamycin (green) links two proteins that normally ignore each other, FKBP12 (right) and FRAP (left). As soon as these proteins are bound together, they act as an immuno-suppressant, blocking the proliferation of T cells.

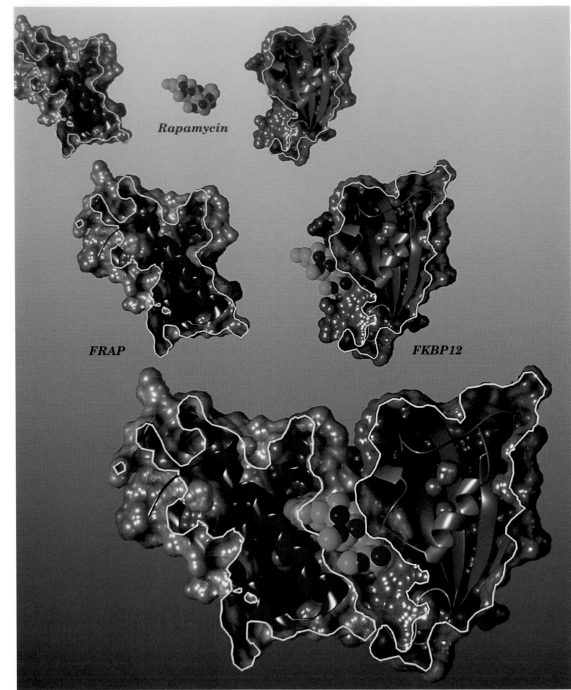

Rapamycin

FRAP

FKBP12

their effects. The conventional wisdom dictated that cyclosporin A and FK506 would act in the same way. Indeed, by the late 1980s, researchers were publishing "a thousand papers a year" based on the assumption that the drugs were docking at the cell surface, recalls Crabtree.

But in 1991, postdoctoral fellows Jun Liu (in Schreiber's lab) and Neil Clipstone (in Crabtree's lab) discovered that both drugs were actually binding to a protein called calcineurin, which sits inside the cell. The drugs did this in a surprising way: they seemed to melt through the cell's surface membrane to reach their target. According to this upstart model, cyclosporin slips into the cell, interacts with cyclophilin, and then binds to calcineurin. Similarly, FK506 first interacts with FKBP and then binds to calcineurin.

This news came as a shock to the wider research community. "A major review article appeared in 1992 saying that this could not possibly be the mechanism," recalls Crabtree. By now, though, the consensus has shifted, and the opposition has "given way to a nearly unanimous view that calcineurin is the functional target" of both drugs.

Unfortunately for the pharmaceutical industry and for transplant patients, this implied that it would be nearly impossible to reduce the drugs' toxicity. Every cell has calcineurin, Liu explains—not just T cells. Kidney cells have it too. "We now know," says Liu, "that the calcineurin inhibition these two drugs cause is responsible for the major side effect of the drugs, kidney toxicity."

On top of that, the drugs' extreme specificity was a problem for researchers. Chemists like Schreiber routinely tweak drugs into slightly more useful configurations by nudging a few atoms around. In this case, that would have been about as useful as trying to fine-tune a race car engine with a sledgehammer.

Consequently, knowing the identity of the protein to which the immunosuppressive drugs bind did not lead to an improvement in the drugs themselves. Their effect is too powerful and too hard to modify.

As happens so often in science, however, a blind alley suddenly turned into a road to success. While Schreiber and Crabtree pondered about how FK506 turns off the immune system, they had an inspiration. The mechanism that they envisioned—and later proved—was this: FK506 is sticky, like Scotch tape. After FK506 slips inside a T cell or other cell, it sticks to FKBP, the target Schreiber had identified initially. But—and here was the subtle and unexpected part of their proposed mechanism—the other side of FK506 was sticky too, albeit in a different way. And that was the side that stuck to calcineurin.

FK506 was like double-stick tape. The side that sticks to FKBP is different from the side that sticks to calcineurin. Once both proteins are brought together by the immunosuppressive drug, Schreiber and Crabtree reasoned, something startling happens: calcineurin is blocked from performing its normal function inside the T cell and, *voilà*, the T cell comes to a sudden halt.

In other words, FK506 acts as a molecular matchmaker, bringing together two molecules that would otherwise be extremely unlikely to get near each other. Once they are together, though, powerful things occur, such as the shutting down of T cells.

Schreiber and Crabtree began to wonder whether they could activate cellular processes—as tiny as the activation of a single gene, as large as the switching on of the entire immune system—simply by bringing two cellular components closer together.

As simple as the concept of proximity sounds, it runs counter to the established wisdom of cell biology and biochemistry, observes Crabtree. "At first, I had a terrible time persuading anyone to work on this project," he recalls. "Every postdoc said, 'Proximity is not going to help you very much because the two proteins could come together anyway.'" They were justified in their skepticism, he agrees. "It just seems so counterintuitive."

Nevertheless, according to Crabtree, these postdoctoral fellows were forgetting two important principles. First, the energy that powers chemical reactions is extremely sensitive to distance. "Think about stoking a fire," says Crabtree. "The logs are scattered, the embers are dying. But if you just push them together, you get a quantitative change. The fire starts burning again."

Second, the inside of a cell is more like a gumbo than a consommé. "It's hard for a protein to get from one place to another without bumping into something," says Crabtree. That dramatically raises the odds against two particular molecules coming together and staying together long enough for something biologically interesting to happen. Therefore, if you overcome

FK506 was like double-stick tape.

these odds and nudge two components so that they draw close to each other—dimerize them, in chemistry lingo—then you could make strange and beautiful things happen.

Schreiber and Crabtree decided to take advantage of the property of FK506 that made it so effective inside both T cells and kidney cells: its sticky and specific attachments, first to FKBP and then to calcineurin. When a molecule of FK506 stuck to FKBP, one section of the FK506 molecule was involved. When the complex then stuck to the larger calcineurin molecule, a second region of FK506 became involved. If the researchers could chemically modify the second region, then FK506 could no longer bind calcineurin. It would be unable to affect immune function.

Schreiber the chemistry whiz sprang into action. He altered FK506 so that it would still be sticky on one "side"—the side that normally binds FKBP—and added a second FK506 molecule to the other "side." The result was a molecule that resembled two pieces of double-stick tape stuck together: the outside "sides" were still sticky, while the inner ones were stuck to each other. The new molecule was dubbed FK1012.

FK1012 belongs to an entirely new category of biological molecules: "dimerizers." A dimerizer can create a connected twosome out of two disparate ingredients. The resulting dimer does not have to be long-lived; all that matters in this case is temporary proximity.

The first big test of dimerizers in living cells came in 1993. Schreiber and Crabtree elected to work with the immune system—specifically, the T cell. Could they turn on the immune system by increasing the proximity of a pair of targets on the T cell? they asked themselves. David Spencer, a postdoctoral fellow in Crabtree's lab, turned his attention to the first protein in the signaling cascade that activates the T cell, the receptor. This receptor spans the outer membrane of the cell the way an iceberg spans the surface of the ocean.

One thing Schreiber and Crabtree knew for sure about antigen receptors: they cannot leave the membrane in which they are located. Strong chemical forces prevent the receptors from sinking through the membrane and submerging themselves completely inside the cell. Equally strong forces keep them from escaping the membrane completely and floating away from the cell. That leaves only one way for these hulking, iceberg-like molecules to move: sideways.

Perhaps, Schreiber and Crabtree said to themselves, pushing the receptors together would cause something to happen inside the cell. Some other kinds of receptors work in a similar way: singly, they have little effect on a cell, but as soon as they pair up they can send strong signals to the cell's nucleus.

David Spencer then altered a T cell's antigen receptors at their submerged base and made them look very much like FKBP. The researchers hoped that FK1012 would stick first to one of these receptors, then to a second one, and thereby create a pair that would cause a signal to be sent to the cell's nucleus. That is exactly what happened. When Schreiber and Crabtree slipped FK1012 inside such modified T cells, it stuck to the FKBP-like regions of the antigen receptors and paired up the receptors as if it were rounding up icebergs, two by two.

The result was a powerful confirmation of their theory. Just by bringing the antigen receptors together, FK1012 activated the T cell. With a drug they had designed, and in a roundabout and utterly unexpected way, Schreiber and Crabtree had turned on the immune system.

Just as important, they had done so in a controlled manner. By adding, say, 5,000 doubly sticky molecules to the modified T cells, they were able to turn up the T cells' response, getting the cells to churn out more IL-2, which can activate other parts of the immune system. Adding more molecules—say 10,000—cranks up the cells' reaction even further. By the same token, giving the modified T cells a dose of just 2,000 molecules of FK1012 would cause the T cells to produce less IL-2, leading to a more subdued immune response.

The importance of such control cannot be

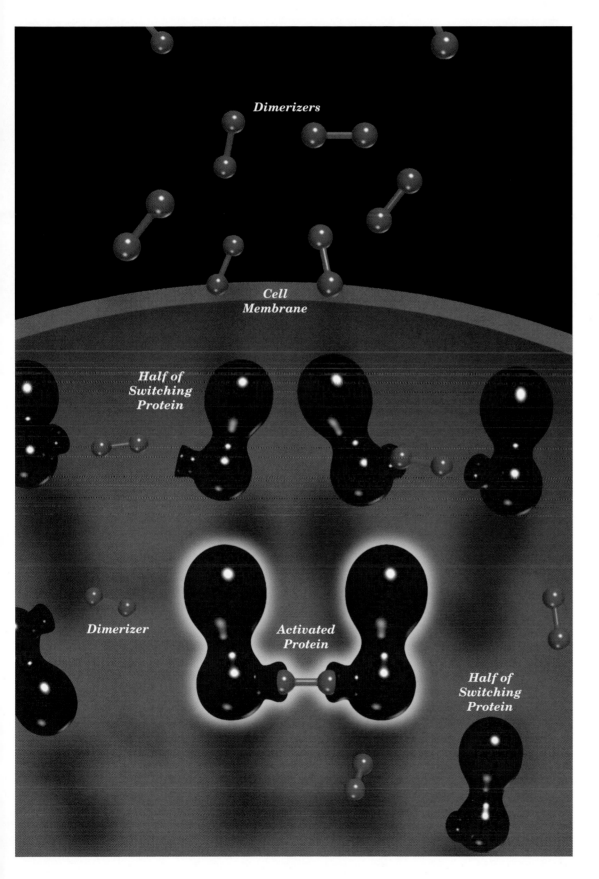

Here Come the "Dimerizers"

Man-made dimerizers (dumbbell shapes, tan) are used as a drug to control an inserted gene's activity. Once the dimerizers (in this case, rapamycin molecules) are injected into genetically modified mice, they bring together two halves of a switching protein (red). The activated protein (yellow outline) then switches on human growth hormone genes that the scientists introduced into the mice. The more dimerizer is injected, the more hormone is produced by the mice.

Dimerizers

Cell
Membrane

Half of
Switching
Protein

Dimerizer

Activated
Protein

Half of
Switching
Protein

overstated. The dosage for most drugs is still determined by a patient's body weight, not by the expected cell-by-cell response. The exquisite control that dimerizers offer represented remarkable progress—if Schreiber and Crabtree could harness the technique.

They could indeed. In fact, FK1012 turned out to be just the beginning. FK1012 and other, similar molecules provided Schreiber and Crabtree with a set of tools they might use to crack open any pathway inside the cell, not just the one they had studied originally.

Signaling pathways are everywhere in cells, telling them when to be born, when to die, and whether to churn out hormones and other powerful products. "We have taken nearly every step of a signaling pathway and shown we can regulate it," says Crabtree. "We can deliberately turn these molecules on and off, one by one." As a tool for understanding how cells work, dimerizers have no peer.

Though he cautions that "we don't yet have a way to apply this to everyday medicine," Crabtree believes that "the route to making an activating drug is clear: you screen libraries of chemicals for small, harmless molecules that bind to one side of a protein, and then you simply chemically link the two small molecules and presto! you have a new drug." For instance, pancreatic cells might be made to generate more insulin by this method. Cells near a burn or wound might be made to generate larger quantities of molecules that promote healing. "We can interfere or activate wherever we like," says Crabtree. "It's like a biological rheostat."

And what better target for deliberate switching on and off than genes themselves? he asks. Gene therapy might yet become the most striking medical application of this approach. When Schreiber and Crabtree used FK1012 to activate the T cell, they were switching on genes in an indirect way, he points out. The pairing up of antigen receptors in the T cell membrane led to T cell activation via genes in the nucleus. But these were normal genes that were being activated in an unusual fashion. It

would be much more appealing, says Crabtree, to put in a therapeutic gene and then use a dimerizer to activate it directly and turn its level of activity up and down as needed.

A team from ARIAD Pharmaceuticals, a small biotechnology company in Cambridge, Massachusetts, has now done just that in genetically modified mice. "If gene therapy is ever going to work, it has to be dosage controlled," says Michael Gilman, the team leader at ARIAD. Dimerizers gave Gilman's team a switch to control the dosage.

Gilman and his colleagues used a molecule that nature has made doubly sticky: rapamycin. By coincidence, rapamycin is also an immunosuppressant. Gilman injected rapamycin into the mice, and, once inside their cells, it performed its matchmaking feat. What it joined together were two parts of a protein that, in turn, activated the gene that produces human growth hormone.

The scientists hope that similar systems will allow them to sidestep one of the toughest barriers to making gene therapy work: the problem of delivering therapeutic genes in the right quantities and at the right time. The way these scientists envision it, gene therapy will someday be a two-step procedure. First, physicians will introduce the therapeutic genes in dormant form into the patient. Then, at exactly the right moment, they will give the patient a dose of a dimerizer to activate the genes.

The patients who might benefit range from dialysis patients who need blood-boosting proteins to children who need human growth hormone. "Someday, we might be able to use dimerizers to deliver these proteins in just the right amounts," says Crabtree. A half-dozen dimerizers, both natural and man-made, are known to exist so far.

For Schreiber, the chemist-turned-biologist, and Crabtree, the stalwart rockpiler, the immune system was a great launching pad. Now they can rise to an even bigger challenge: using the power of proximity to turn on and off the stuff of life itself. ●

"We can interfere or activate wherever we like," says Crabtree. "It's like a biological rheostat."

A Signaling Pathway Reveals Its Secrets

As in a game of volleyball, the proteins that bring messages from outside a cell to the genes in its nucleus must pass the ball to one another. But in a cell the players must follow a particular order, as shown here by arrows.

Until recently, even the key players in major signaling pathways were unknown. Now scientists have identified all the proteins in certain pathways, and in some cases, such as the "RAS pathway" (right), they have solved many of the proteins' 3-D structures, with more being added rapidly.

The signals in this pathway tell a skin cell it is time to grow and divide. They start with two molecules of fibroblast growth factor (FGF), a growth hormone released from another cell, and two FGF receptors (FGFR1), which span the membrane of the skin cell. When the enzymatically active domains of these two receptors (red—the only parts of the receptors whose structure is known) bind together, forming a dimer, they activate one another and trigger the signaling cascade.

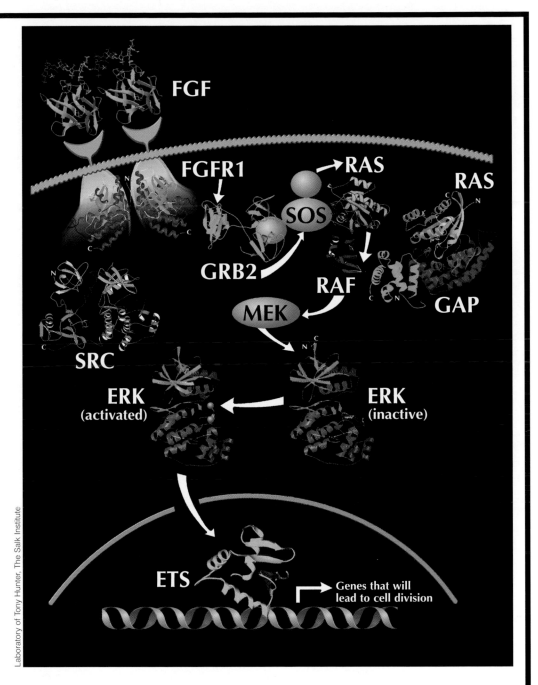

Laboratory of Tony Hunter, The Salk Institute

The pathway goes from these receptors to GRB2 (growth factor receptor binding protein-2), SOS (son of sevenless), RAS (an oncogene), RAF (another oncogene), and MEK (mitogen-activated protein, or MAP kinase kinase). MEK then adds two phosphate molecules (red balls) to ERK (extracellular-signal regulated kinase), activating it. ERK moves to the nucleus and activates a transcription factor

called ETS (E twenty-six). ETS binds to DNA in the nucleus and switches on genes that will lead to cell division.

At each stage of the cascade, the original signal can be amplified 10 times, producing a 10,000-fold amplification at the end. Mutations in some of the signaling proteins, particularly RAS, can make the cell divide too rapidly and may lead to cancer.

A BULL'S-EY

As a deer tick feeds, its
head and mouthparts
are buried in flesh.

Aspeck the size of a poppy seed, the deer tick holds on tight, its harpoon-tipped mouthparts sunk deep into the skin of its host. Hour after hour it sits there, motionless—a particle of life, it could seem, scarcely worthy of the name.

But the tick harbors a microscopic corkscrew-shaped bacterium, a "spirochete," which gives Lyme disease to thousands of Americans a year. In most cases, a

E ON LYME DISEASE

New Vaccine Will Prevent Ticks from Transmitting Lyme Parasite

by Robert Kanigel

Sam Telford drags a flannel blanket through the underbrush on Nantucket Island to collect ticks for study

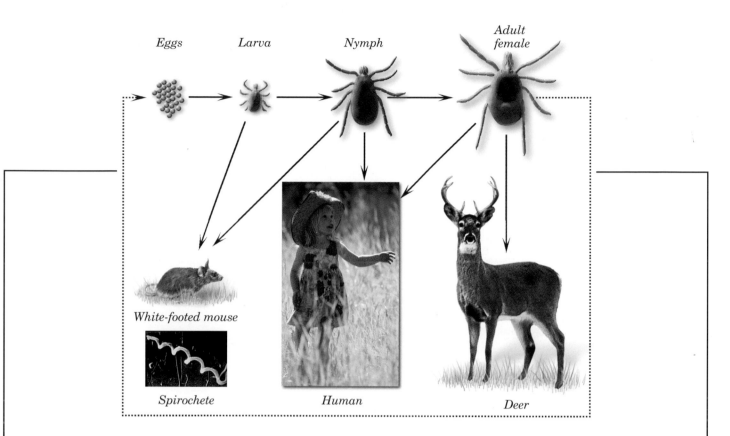

Eggs　　　Larva　　　Nymph　　　Adult female

White-footed mouse

Spirochete

Human

Deer

The Tick's Life Cycle

The eggs of a deer tick are not infected with the spirochete that causes Lyme disease. Nor are the larvae that emerge from them. But in certain parts of the Northeast more than half of the wild field mice called "white-footed" carry these parasites, and when a deer tick larva bites an infected mouse to drink its blood, the larva becomes infected.

After this blood meal, the deer tick larva molts into a nymph. Nymphs, too, like to feed on white-footed mice

(thus giving ticks another chance to become infected with the Lyme disease spirochete). But they also feed on other small mammals, birds, or—all too often—humans. And if the nymphs carry spirochetes, they can pass them along. Because of the nymphs' small size there is a good chance they may not be discovered or removed until they have transmitted the disease.

The nymphs then grow into adults. That is when they start feeding on deer, which supply the large quantities

of blood a female tick needs to enable it to produce eggs. Adult ticks may also bite other large mammals or humans, but they are more obvious and therefore more likely to be noticed before they have done much damage.

While feeding on deer, the female tick mates with a male. Shortly afterwards, the female lays its eggs. Then it dies. Another two-year life cycle begins with the new crop of eggs.

red bull's-eye rash spreads across the skin. Fever develops, joints ache. Untreated, the disease causes shooting pains, numbness, muscle weakness, cardiac problems, disabling arthritis. In great swathes of New England, in Pennsylvania and New York, in the north woods of Minnesota and Wisconsin, fear of Lyme—or else just the thought of having the disgusting, blood-sucking thing upon them—keeps woods-loving vacationers, hikers, and backyard barbecuers indoors.

As an infected tick leisurely sucks up its blood meal, it becomes engorged; the almost flat ellipse of its body swells like a

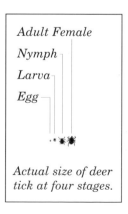

Adult Female

Nymph

Larva

Egg

Actual size of deer tick at four stages.

toy store balloon. The blood serves as a rich nutrient stew for the spirochetes within the tick's gut, which multiply over the next two days from a few hundred to perhaps a million. Moreover, responding in part to the blood's warmth, they assume a biochemically different form. In time they migrate from the tick's gut to its mouthparts, and thence to the flesh of deer, human, mouse, dog, or other mammal, infecting it.

Not long ago, no such account of the intricate biological dance culminating in Lyme disease could have been written; or had it been written, it would have been wrong. Knowledge gleaned and premises revised only in the last few years help explain how a new Lyme vaccine, now proven effective in a trial involving 11,000 volunteers, really works.

And explain, too, those moments of profound bafflement and confusion, verging on panic, that Sam Telford felt one day in 1991. Vials of ticks had arrived at his lab at Harvard University for routine analysis —merely to confirm the obvious, he thought. Telford squished the ticks onto slides and slipped them under the lens of a microscope, only to find that, as a group, they looked nothing like what they were supposed to look like.

Sam Telford III, lecturer in tropical medicine at Harvard's School of Public Health, had literally grown up with parasites; his father was himself a parasitologist. Before 1990, when he was first drawn into the search for a Lyme vaccine, he had for years haunted the low scrub of Nantucket Island, 30 miles off the coast of Cape Cod, dragging his flannel blanket through the underbrush to collect ticks. Stop him today as he barrels across the foggy, windswept island's dirt roads in a pickup, and he'll point to one stretch of landscape as virtually barren of ticks and another nearby, all scrub oak, bayberry, poison ivy, and greenbriar, as tick heaven. Telford knows his ticks.

So when, back in 1990, he and his Harvard mentor, veteran entomologist Andrew Spielman, learned that a team of Yale researchers claimed to have demonstrated a vaccine against Lyme disease, they'd been skeptical. The Yale group had immunized mice, injected them with a syringe full of spirochetes, noted that they remained disease free and—blare of trumpets—decided they had a vaccine. Or that was how, a little too hastily, you could interpret their account in the pages of *Science*.

Didn't they know, for God's sake, that a syringe is not a tick?

Lyme disease got its name back in 1975 when two mothers in Lyme, Connecticut, whose children had come down with what seemed to be rheumatoid arthritis, contacted the state health department. Soon Allen

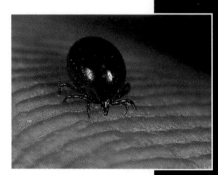

Filled with blood after a two-day meal, a deer tick nymph is engorged to more than five times its usual size.

When this corkscrew-shaped bacterium or spirochete, Borrelia burgdorferi, **passes from an infected tick to a human, it causes Lyme disease.**

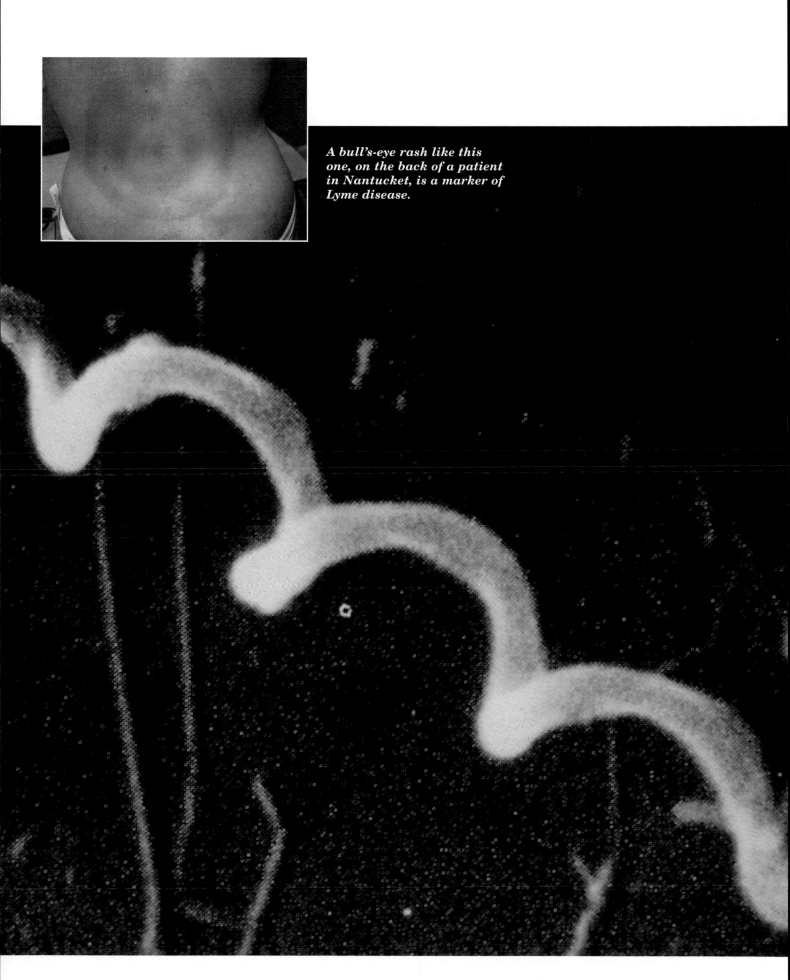

A bull's-eye rash like this one, on the back of a patient in Nantucket, is a marker of Lyme disease.

Steere, then a research fellow with Yale rheumatologist Stephen Malawista, was on the case, embarking on medical detective work worthy of Paul de Kruif's *Microbe Hunters*. Recalling the bull's-eye rash one of the mothers remembered seeing on her child, learning of a 1909 European study implicating ticks in similar rashes, and correlating the prevalence of cases with that of ticks on either side of the Connecticut River, he tracked the disease to the deer tick, *Ixodes dammini*, endemic around Lyme. Later, physicians came to see other conditions—Bell's palsy, aseptic meningitis, Montauk knee, various mysterious rheumatisms—as coming under Lyme disease's reach.

Within a few years, a researcher at the Rocky Mountain Laboratory of the National Institutes of Health had identified the parasite passed along by the tick. His name was Willy Burgdorfer, and the spirochete, kin to the bacterium that causes syphilis, is today known as *Borrelia burgdorferi*. Meanwhile, through much hard scientific slogging, researchers were teasing out the complex life cycle through which ticks (in three distinct stages of growth), deer (upon whose bodies ticks breed), white-footed mice, humans, and other mammals pass the spirochete back and forth among them, keeping it in circulation and spreading the disease. And Alan Barbour, of the University of Texas Health Sciences Center in San Antonio, cloned the genes for several of the spirochete's surface proteins.

Much ado about not much, a skeptic might argue. Lyme disease, after all, conjures up neither death, nor horrific wasting, nor Ebola-like bloody vomit. Rarely, a case can veer off and, the spirochete reaching the heart, claim a life. By one reckoning, there have been perhaps 10 such cases—ever. Much more often, Lyme disease is more like a bad flu. Identified early, it can usually be treated by penicillin-like antibiotics in every physician's kit bag.

Yet Lyme disease, the most common vector-borne illness in the United States, is *not* always identified early. Left untreated—as it often was in the early days—the spirochete can migrate through the body, work its way into the joints, and cause debilitating arthritis; even today 1 in 10 of its victims contracts arthritis that persists across the years. Lyme isn't so bad? It is if you see the big red rash on your child and—as with polio among parents of an earlier generation—imagine her crippled for life.

The disease's early notoriety as a victimizer of innocent children in a leafy, prosperous suburb left a deep imprint. Seven in ten Lyme victims, it has been estimated, get it in their own yards. Idyllic, semi-wild areas at the fringes of town are often most afflicted; so campers don't camp, hikers don't hike, and tourists stay away. On the ferry to Nantucket, bright posters warn travelers: "FAILURE TO TREAT may lead to chronic arthritis...DO NOT DELAY TREATMENT!" Millions of people in Lyme-endemic areas flock to the doctor's office each year at the first sign of a summer flu. Other millions warily heed the public health warnings: wear light clothing, tuck pants legs into socks, examine yourself at night—all to detect the tiny tick, loathsome and disease-bearing.

YALE PUSHES AHEAD

As their Harvard critics suspected, Yale researchers on the trail of a Lyme vaccine in 1990 could not have said precisely how, as a transmitter of parasites, a tick differed from a syringe—or for that matter, anything much about ticks at all. But they did know plenty about Lyme disease.

Yale had been at the forefront of Lyme research from the start. Yale researchers had first identified the disease. Yale veterinarian Stephen W. Barthold had, over a span of several years, developed the first good animal model—he found that a particular strain of mouse known as C3H is susceptible to Lyme disease and that its symptoms are very similar to those of humans. Infect one of these mice and, during autopsy, you can make straight for the

Seven in ten Lyme victims, it has been estimated, get infected in their own yards.

joints or the heart and see the sort of inflammation you'd see in a human victim; test a vaccine on such a mouse and you'd be a leg up on predicting its effectiveness in humans. By around 1989, recalls Fred Kantor, one of a group of Yale Lyme researchers who began meeting periodically in the library next door to his ninth floor office, "a lot of people were feeding at the Lyme trough." The time was ripe. One doctoral student, whose dogs were routinely getting Lyme disease, fairly pleaded with Kantor to push ahead on a vaccine. When Richard Flavell arrived at Yale, they did.

Professor and chief of immunobiology,

When Erol Fikrig (left) showed Richard Flavell the promising results of an experiment, Flavell said, "Repeat it."

head of Yale's effort to develop a Lyme vaccine, and a Howard Hughes investigator, Flavell came to Yale in 1988 from Biogen, a big biotechnology firm whose research he had overseen for seven years. There, he'd first learned of Lyme disease, which almost from the start had struck his immunologist's heart as a promising area of study. A vaccine against Lyme would do a great deal of good, of course, but in Flavell's thinking, that was only part of its appeal. He knew that while most inflammatory diseases, such as rheumatoid arthritis, have no clear cause, Lyme disease is different. It *does* have a clearcut mechanism—a particular species of tick, *I. dammini*, which transmits a particular organism, *B. burgdorferi*. Vaccine research might thus lead to deeper knowledge of the immunological roots of arthritis.

So when Kantor suggested that maybe it was time to go after a vaccine, Flavell replied he'd been thinking along identical lines. "I was convinced it didn't take 300 people to develop this vaccine," he recalls, referring to the resources a private biotech company like Biogen could lavish on a hot project. "Still, I expected it to be really difficult and interesting."

He was half wrong. "It was interesting," says Flavell, "but not difficult," at least not as difficult as he'd anticipated.

They call it Ospay, which is less of a mouthful than its proper name, Outer Surface Protein A, or OspA.

Early on, Lyme researchers had identified several proteins not tucked away within the recesses of the spirochete but, as electron microscopy made clear, inhabiting its outer surface. That, at least in principle, made them candidate targets for a vaccine; the immune system can't attack what it can't "see," but these were parts of the disease organism that it probably could see. And most abundant among these surface proteins were two called OspA and OspB.

A strategy was hatched: From, say, the OspA gene cloned by Alan Barbour in Texas, synthesize OspA protein. Inject this

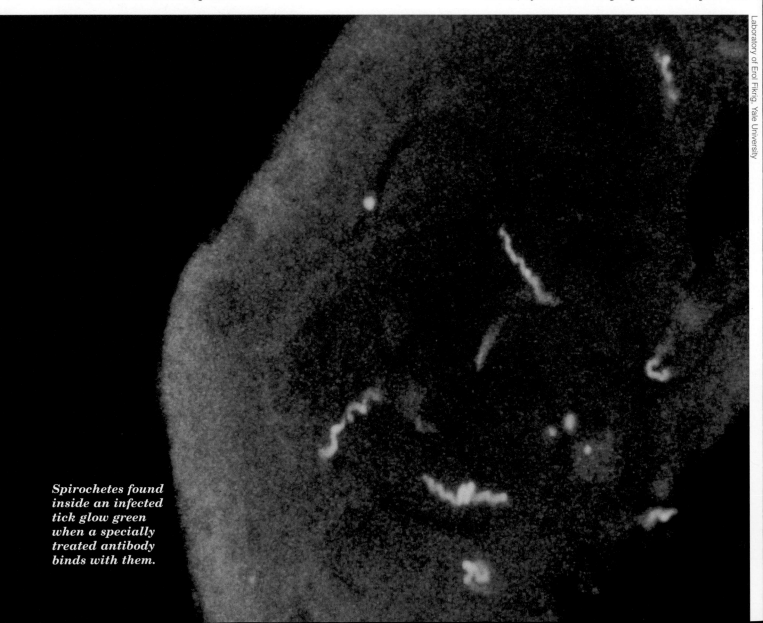

Spirochetes found inside an infected tick glow green when a specially treated antibody binds with them.

protein into the body as a vaccine to raise antibodies. Then, when a tick disgorges its spirochetes, the antibodies should recognize their outer surface proteins and launch the biochemical mechanisms that should culminate in the bacteria's destruction.

Forget it, other Lyme researchers as much as told him, says Flavell; it won't work, since humans mount only a feeble antibody response to OspA. But maybe just give it a nudge, he reasoned. During infection, the host doesn't normally kill the pathogen, nor the pathogen the host; they coexist in uneasy balance. Enhance immune response just a little, through a vaccine, and you might shift the balance in favor of the host. Besides, it seemed to Flavell that something beyond cool reason lay behind the naysaying. Fifteen years had passed since Lyme disease had surfaced in 1975 and no vaccine was in sight. By now, the Lyme spirochete had taken on some of the mystique of its infamous syphilis-causing cousin, renowned for lying low in the body and eluding destruction.

Believing his strategy sound, Flavell pressed on. What made OspA and B particularly promising was that they were plentiful and large, the better to raise antibodies. "Those two stupid, boring facts," says Erol Fikrig, assistant professor of medicine at Yale, "made them good targets."

Fikrig, a new postdoctoral fellow in Flavell's lab, whose experience had been almost wholly medical, down to treating AIDS patients in Haiti, came armed with an M.D. rather than a Ph.D. "Their minds are super, but their hands are useless," says Flavell of the breed. Fikrig, who by Flavell's reckoning had scarcely ever touched a pipette, now hoped to do basic research, applying the powerful tools of molecular biology to immunology. Unaccountably, everything he touched turned to gold.

Fikrig immunized Barthold's chosen mice, some with OspA and some with OspB; gave them weekly booster shots; "challenged" them with exposure to *B. burgdorferi*; killed them after a couple of

weeks; cultured their blood and microscopically searched it for spirochetes; and examined their joints and hearts for evidence of inflammation.

The climax of the experiment, capping two months of work, came one Friday afternoon. Earlier studies had led Fred Kantor to suspect OspB might be the better vaccine candidate. So when, around six that evening, the first results showed OspB granting no apparent protection, he went home, leaving Fikrig to carry on with two technicians.

However, OspA *did* protect. Fikrig remembers reading them off, "First mouse, negative.... Second, negative...." The results were open-and-shut, free of ambiguity: the mice immunized with OspA, all five of them, were clean, whereas those not immunized bore both parasites and the physiological markers of disease. "We were very excited," Fikrig recalls.

Richard Flavell's response?

"Repeat it."

They did so, and the experiment worked once more. Soon their results were in the pages of *Science* as "Protection of Mice Against the Lyme Disease Agent by Immunizing with Recombinant OspA."

In its promise of a vaccine that might end the summertime scourge, this was big news. But when an enterprising reporter asked Andrew Spielman, Sam Telford's boss at Harvard's department of tropical public health, what he thought of the Yale work, Spielman replied that it proved nothing. What could you learn by infecting mice with organisms from a syringe? After all, he reminded them, a syringe is not a tick.

Spielman, with a long line of research papers going back to his early days at Harvard in the late 1950s, had once gotten Lyme disease himself, though he'd not realized it at first. He was just back from China when it started: sweats, chills, high fever. He drove to the hospital and told the doctors he had malaria—reasonable enough, given that he was just back from a malaria-ridden part of the world.

Soon he was in bed, with a temperature

Unaccountably, everything he touched turned to gold.

BEWARE OF OTHER TICK-BORNE INFECTIONS

Timothy Lepore, shown here with stacks of reports about the volunteers on whom he tried out the new vaccine against Lyme disease (including himself and his family), feels happy that the trials worked. But he's still worried about a few other diseases that are carried by the same ugly deer ticks.

"The ticks can't give you Lyme if you've been vaccinated," he says. "But they can give you babesiosis, a malaria-like infection, or ehrlichiosis, which is even worse than Lyme—it kills 7 percent of the people who get it."

Lepore points out that while these infections are much rarer than Lyme, they may become a serious problem if people who have had the Lyme vaccine feel so safe that they neglect taking precautions against ticks.

"Children between the ages of 5 and 10 appear to be at high risk for tick-borne disease," he says. "So we're about to start testing kids in elementary school every year, to get a handle on the incidence of all these diseases." Working with the Harvard School of Public Health, Lepore has sent out letters to parents, asking their permission to take small samples of their children's blood for this purpose. The blood will be analyzed from a few drops on a piece of filter paper.

He warns parents that dogs and cats are frequent carriers of deer ticks. "Ticks will feed on mice, dogs, cats, or people wandering through the brush," he says. "They are typically in knee-high areas, where it's moist, waiting for something warm to go by."

At least one fourth of the dogs on Nantucket have been infected. But cats are an even greater menace, Lepore believes, because "if a cat is in bed with you, it may lick itself, feel a tick, and flick it off. So now the tick is half fed and more dangerous, because it will infect you sooner. Cats are better groomers than dogs. Dogs may not even notice the tick." – *M.P.*

of 104, looking at a wet photomicrograph of his own blood. It wasn't malaria, but the doctors didn't know what it was. So he was treated for his symptoms and released. Propped up on pillows, he managed to give a scheduled lecture on Lyme disease, but was soon in the hospital again. There, as he sat naked in the examining room, a nurse spotted in his groin area the great bull's-eye rash that is Lyme disease's most distinctive marker. He had probably got it before China, Spielman figured, in Cape Cod, when he'd wandered down to the beach clad only in a bathing suit. Within a day, antibiotics left him feeling better.

Spielman had worked on Lyme disease before there *was* a Lyme disease, or before it was called that, and knew ticks and how they did their dirty work better than anyone. He knew that the spirochetes migrate through the skin itself and only rarely through the blood, so a vaccine might prove ineffective because the blood that bore it might fail to reach the sites of infection. He knew that the bite of some insects, like the sand fly that transmits leishmania, bears substances that inhibit an immune response. The same might be true of the deer tick, which, unlike a mosquito, feeds slowly over the course of days. If it produced an itch or inflammation, the tick might get scratched away, so evolution had likely adapted it to secrete anti-inflammatory agents.

The Yale team, it seemed to him, knew none of this. "There was much whoop-de-do about how wonderful this vaccine was," Spielman recalls, but it struck him as "almost laughable" that they thought it would work with a real tick rather than a sterilized syringe of lab-grown organisms.

Admits Erol Fikrig, thinking back to those early days: "We were very naive."

Flavell took the Harvard objections seriously and soon was on the phone with Spielman. The upshot? They would dispense the spirochete by tick, not syringe. And they would enlist their biggest skeptics, Spielman and Telford themselves, as collaborators.

Over the phone and through faxes,

"There was much whoop-de-do about how wonderful this vaccine was," Spielman recalls, but it struck him as "almost laughable."

Telford found himself explaining to Fikrig—"very bright, very ambitious, with lots of energy," Telford calls him—all about ticks. "I had to explain everything from square one" about dormancy and reactivation and much of the rest of an entomologist's stock of knowledge. Soon they'd embarked upon a key experiment—one that Fikrig and Flavell, to listen to them today, never doubted would work and that Spielman and Telford were just as sure would not.

First, mice were injected with the spirochetes, infecting them. Then, ticks drawn from a Harvard laboratory colony were allowed to feed on them and grow to the nymphal stage in pools of 100 to 200. Samples drawn from each pool showed that at least 70 percent of the ticks, and typically more, were infected, usually with 100 or more spirochetes each.

Then the ticks were allowed to "feed to repletion" on two groups of uninfected mice—Barthold's mouse model again—half of them immunized with OspA and half not. Each mouse had three or eight nymphal ticks feasting upon it. A few days later, having had their fill, the ticks fell off into a pool of water over which the mouse cages were suspended.

The absence of spirochetes in the blood being only an imperfect marker of protection, Fikrig and his technicians looked in their autopsied mice for visible evidence of arthritis and carditis. In one group of five unvaccinated mice, just two had spirochetes in their blood—but every one showed signs of joint and heart inflammation. Of the immunized mice, not one showed a single spirochete or even a hint of inflammation. Telford had been sure the vaccine wouldn't work. It had. "Uh oh," he remembers thinking, "I may have to eat my words."

Routinely, as a nod to method and procedure, he had the ticks sent back to Harvard for examination. What for? one might ask. The scientists knew that at least 70 percent of the ticks had started out infected. And though a single tick was enough to infect a mouse, they had placed several on each. Upon falling off into the water, the ticks had been collected in vials, one vial to a mouse.

What difference would it make now if a particular vial proved to have only one or two, as opposed to five or six, infected ticks?

Still, Telford stuck to the protocol. He'd rip open each tick, squeeze its insides onto a slide, fix it with acetone, and add an antibody molecularly linked to a fluorescing compound known as fluorescein. If the antibody "found" spirochetes with which to bind, its linked fluorescein would light up under a mercury vapor lamp equipped with a special filter. In a crowded little room with a big compound microscope, Telford examined remains from each tick. Under the microscope, infected ticks looked like dense felt—threads of matted fiber, each a spirochete, glowing green against a black background.

When Telford performed this test on the ticks sent back from Yale, something went wrong. He didn't get the 80 percent infection rate he'd expected. He didn't get the 70 percent their sampling had virtually guaranteed. Instead, he got about 40 percent—way, way below anything that made sense. The results, recalls Telford, were "unmistakable"—yet couldn't be. "How did I screw up?" he asked himself. "I was scratching my head and wondering, 'How am I going to tell Fikrig?'"

When he did, Fikrig reports, "It was like you could hear Sam shaking his head over the phone. 'What a bummer,' I said."

Stoking their bewilderment was that impossibly low raw infection rate, that damning 40 percent. They had not actually broken the number down into its elements to see which mouse each tick had fed upon, however. When they did at last take this step, the mystery abruptly—and astonishingly—cleared up.

As they cross-checked the vials over the phone, they realized that ticks from the nonimmunized mice had just the infection rate they'd expected all along, around 80 percent. It was only the ticks feeding on immunized mice that pulled down the average. *Pulled it down?* Why, their infec-

tion rate was down near *zero*. It wasn't just the mice that were being immunized; by the time they'd finished feasting on mouse blood, the ticks were immunized too.

Fact: if you remove a tick from your skin that's been attached for less than about two days, you probably don't have Lyme disease. That 48 hours, as Andrew Spielman and others showed in the late 1980s, is how long it takes the spirochetes in the tick's gut, bathed in warm blood from the host, to divide, multiply, and make their way through the tick's blood-like hemocoel to its mouthparts; only then can they can pass to the host and transmit the disease. It was these two days that the vaccine exploited: Swimming in immunized blood, the spirochetes faced not a nourishing environment but a toxic one. Far from multiplying, they died before ever reaching their mammalian host.

Nothing like such a mechanism of vaccine protection had ever been seen before. "Whoa," Fikrig recalls Telford yelling into the phone as the mystery cleared up before them, "we've got to get writing fast." By mid-March they and their colleagues had gotten their paper to the *Proceedings of the National Academy of Sciences*. By June 1992, it was in print, as "Elimination of *Borrelia burgdorferi* from vector ticks feeding on OspA-immunized mice."

Harvard, now, as well as Yale, saw a Lyme vaccine as promising. So did several drug companies, including SmithKline Beecham, to whom Yale licensed its prospective vaccine, and Pasteur Merieux Connaught, which was developing a similar, competing vaccine.

VACCINE OR PLACEBO?

It's a simple three-by-five-inch postcard, with Yes and No columns and nine straightforward questions: Since the patient's last contact with the clinic, has he or she newly experienced an expanding rash? Joint pain or swelling? Numbness or tingling of a limb? Memory difficulty? Over a year and a half, scores of Nantucket residents got

"How did I screw up?" he asked himself.

such postcards in the mail each month. Recruited by Timothy Lepore, medical director of Nantucket Hospital, a 19-bed shake-shingled facility that fits right in with island architecture but boasts its own helicopter pad, they were volunteers in a large, Phase III study to test the Lyme vaccine that had emerged from the Yale-Harvard collaboration.

A preliminary Phase I had involved 30 patients. Phase II, designed to establish the vaccine's safety, hinted that it worked, but 250 patients were far too few to say for sure, since even in Lyme-endemic areas only about 1 to 4 percent of the population comes down sick with it each season. The Phase III trial, sponsored by SmithKline Beecham, involved 11,000 patients at 31 sites across the Northeast and Midwest. A hundred or so came under Dr. Lepore's care on Nantucket.

For each new volunteer, Lepore made a call to a central registry. The patient got an initial shot and later two boosters. Neither doctor, nor patient, nor SmithKline Beecham knew whether these immunizations were the real thing—based on OspA—or a placebo; a computer in Pennsylvania randomly assigned each volunteer to one group or the other.

Over the months, the volunteers—each of whom received about $200 for travel expenses and time off from work—went about their business and sent in their postcards. Some got sick; most, of course, didn't wait to send in their postcards but went straight to the doctor.

"On this island," Lepore cautions, "if people feel sick they figure it's Lyme disease," which was notoriously hard to diagnose. *Was it Lyme, or wasn't it?* They had to know for sure. So for Phase III, Lepore and his counterparts photographed the rashes. They took skin biopsies. They collected blood, urine, and sometimes joint fluid samples. All these went to Boston for analysis, where Allen Steere, who'd helped identify Lyme disease more than two decades before, coordinated the study from his laboratory at Tufts University School of Medicine. Month by month, the cases reaching him in Boston accumulated. By late 1996, he estimated,

he had records of some 200—enough to statistically ensure a verdict one way or the other.

Some months into the trials, a tremor jostled the scientific underpinnings of the vaccine. In their first joint paper, the Yale-Harvard team had observed that the vaccine might work dually—first, by killing spirochetes within the tick before they were transmitted, and second, in the more usual way, within the host. But in 1995, Tom Schwan and his colleagues at Rocky Mountain Laboratories, where *B. burgdorferi* had been identified 14 years before, furnished evidence suggesting that only the first way worked; and Yale's Fikrig supplied more.

During the two days it takes to migrate to the tick's mouthparts, they found, the spirochete changes form. It sheds the OspA outer coat that Flavell and Co. had designed their vaccine to attack and dons another, OspC. Deer or mouse or human, it seemed, never even see OspA. If the vaccine worked, perhaps it did so only within the tick itself, during the two days in which it sucks blood laden with OspA antibodies from its host.

The vaccine, in any case, did work. For a year and a half, the Phase III trial could be likened to a high school lab report's Table of Results—rows and columns all neatly ruled across the page, comparisons between this group and that set up, statistical tests prepared—without, however, so much as a morsel of data to compare, all the blanks still empty. Then, late in 1996, a computer command was issued that linked each patient's medical history—ill with Lyme, or not—to whether he or she had received placebo or vaccine. With the code thus broken, data flowed through the computer, blanks filled in, statistical calculations were performed.

In early February 1997, SmithKline Beecham issued its first comment on the Phase III trial—one that, for all its brevity and caution, spoke volumes: The company intended to file a product license application with the Food and Drug Administration. And an independent safety monitoring board was recommending that volunteers who had received the placebo now be vaccinated. ●

Tick Watch on the Nantucket Ferry

Many of the people who visited Nantucket Island last summer had their first look at deer ticks right on the ferry—from a gifted juggler called Brian Smith.

"Deer ticks can't jump, they can only crawl," he demonstrated with a set of colored balls. "Each day, please, you should check your body for ticks. They like the warm, moist areas of your body: the armpits, the back of your knees."

"If you find a tick, you don't need to panic," he explained. "Simply pull it out with tweezers! The worst thing you can do with a deer tick is to burn it or to put vaseline on it—that makes the tick regurgitate into the wound and deposit its spirochetes."

After his juggling act and monologue, which were part of a "Summer Health Study" run by Brigham and Women's Hospital in Boston in an effort to prevent Lyme disease, Brian Smith showed his audience an artificial arm with real ticks on it. It made quite an impression.

An old lady stared at the three sizes of ticks stuck on the arm and asked, "Which one is most dangerous?" "This one," said Brian Smith, pointing to the nymph. "Oh, my God—so tiny!" said the woman.

"Are they real?" asked a child warily. "Yes," Smith replied, "but they're dead and they're glued on, they won't come off."

Smith and Charlotte Phillips, director of the study, also distributed questionnaires. "Everyone who fills them out gets a goodie bag," Smith announced, "and in the goodie bag there's a lollipop!" The goodies included a coupon for 20 percent off on tweezers anywhere in Nantucket, a card showing how to recognize and get rid of ticks, and information about the disease. The bright green lollipops carried the study's slogan: "Lick Lyme." — *M.P.*

Juggler Brian Smith uses colored balls to show what deer ticks cannot do: jump.

An artificial arm bearing real (but dead) ticks attracts the passengers' attention.

Filling out question-naires about their exposure to ticks, these passengers qualify for a reward.

A small passenger with a green lollipop carries out the Summer Health Study's slogan, "Lick Lyme."

The immune system has a lust for vengeance—and seldom forgets an insult. A microbe that threatened us once had better not mess with us again, or the immune system will pounce on it and batter it. Even a speck of matter that merely *looks* like an offending microbe could trigger a major assault.

Taking advantage of this vengeful nature, vaccines simply mimic a dangerous bacterium or virus. This primes the immune system for action, so if it is ever attacked by the real thing it will respond swiftly. At that point the microbe will usually be defeated and the vaccinated person will remain unharmed—immune to the disease.

"Vaccines are the most cost-effective medical intervention to prevent death and disease," declares Barry R. Bloom, an HHMI investigator at the Albert Einstein College of Medicine in the Bronx, New York. But substances that are challenging enough to provoke full immunity and also safe enough for widespread use have proved very difficult to find.

The Chinese appear to have been the first to use some form of vaccination, as early as the 11th century. Realizing that people who had recovered from smallpox were somehow protected from another bout of the same disease, they figured out a way to transfer this protection to healthy people. First they collected scabs from the pus-filled blisters of persons who were recovering from smallpox. Then they ground these crusts to a powder and made uninfected persons inhale it; or else they put some of the powder on a needle and pricked people's skin with it.

Most of the people who were treated in this fashion became resistant to smallpox after a mild illness. However some fell seriously ill or died. It seemed a chance worth taking, since in those days smallpox

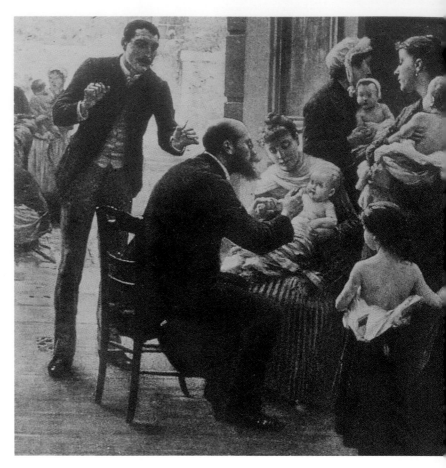

was the most dreaded of all diseases. Everyone expected to catch it sooner or later, and a quarter of the people who caught it died.

The Chinese method of preventing smallpox spread to various parts of the world. When Lady Mary Wortley Montagu, the wife of the British ambassador to Turkey, saw it in action there in the early 1700s, she became so enthusiastic that she told all her friends in England about it and made the procedure fashionable. That is how an English country doctor named Edward Jenner came to be immunized in the Chinese way. Some years later, in 1796, Jenner heard that milkmaids—famed for their clear, unblemished skin—believed they were protected from smallpox because

Edward Jenner inoculates English children with cowpox in 1798 in an attempt to prevent smallpox.

Vaccines

they generally caught a much milder disease, cowpox, from their cows. This gave Jenner an idea for an experiment. He could not experiment on himself, since he had been immunized. So he decided to experiment on an 8-year-old boy, James Phipps.

After taking some pus from the blisters of a milkmaid who had cowpox, he scratched James' arm with a pus-covered needle. As expected, the boy developed cowpox and promptly recovered. Jenner then scratched his skin again—this time, using pus from the blisters of a patient who had real smallpox. Fortunately, James remained healthy and a new type of disease prevention was born: vaccination, named after "vacca," the Latin word for cow.

During the next two centuries, smallpox was completely eradicated with the help of vaccines. Additional vaccines were developed against other viral diseases, such as polio, measles, yellow fever, and influenza, as well as bacterial diseases—whooping cough, diphtheria, tetanus, and others—saving millions of lives each year.

The early vaccine era is sometimes called the "stumbling-along period" because scientists worked mostly by trial and error, without quite understanding what they were doing. The vaccines were generally produced from whole microbes—live viruses or bacteria that were weakened to make them safer, or killed microbes. Starting in the 1900s, inactivated microbial toxins, such as tetanus toxoid, also were used as vaccines.

In the 1960s, scientists began to create more selective and sophisticated vaccines. These often consisted just of subunits, such as parts of a microbe's envelope, rather than the entire microbe; this ensured that the microbe itself could not reproduce in the body and cause disease. The first vaccine against hepatitis B was developed in this way in 1980. Since it

was derived from human blood, however, it had certain drawbacks: it was in short supply and in theory carried some risk of infection with other bloodborne viruses.

The birth of recombinant DNA technology in the 1980s gave scientists the power to make vaccines out of totally synthetic ingredients. They soon isolated the genes that code for hepatitis B coat protein and introduced these genes into yeast cells. The yeast cells then churned out large quantities of virus particles, which could be used as a vaccine because they aroused the immune system. The resulting hepatitis B vaccine carries no bloodborne viruses and is available in unlimited quantities. Many other vaccines were designed or improved with similar methods.

Amid all these successes, one huge challenge remains: AIDS. "I always get asked, 'When will we have a vaccine for AIDS?'" remarks Barry Bloom. "My answer is, 'The easy vaccines have all been made.' If we could do AIDS, we could do cancer, because it's the same mechanisms."

Bloom points out that HIV, the virus that causes AIDS, behaves quite differently from polio and other viruses for which we have effective vaccines. "Polio comes into the body as a free virus, as far as we know," Bloom says. "If it encounters an antibody to the right antigen, it's deader than a doornail. But much of HIV is inside cells, and antibodies don't do anything to a virus that's in a cell.

"You really need helper T cells and killer cells to deal with things that live inside cells," Bloom explains, "and we don't yet understand how to induce cell-mediated immune responses very well. That's why vaccines against TB, parasitic diseases, cancer, and AIDS are not going to be that simple."

There is no lack of ideas, however, about how the next generation of vaccines might be developed.

Naked DNA and Other Vaccines of the Future

One new approach to vaccines is to inject some "naked" DNA from a microbe right into a muscle, or actually shoot it into the skin. Then the cells of the host will carry out the DNA's instructions and produce small quantities of the microbial antigen, stimulating an immune response.

This has worked in experiments with animals, and "it's very likely...that over the next 10 to 20 years many of our vaccines are going to be replaced by a new generation of DNA vaccines," according to Anthony Fauci, director of the National Institute of Allergy and Infectious Diseases.

DNA vaccines seem to arouse not only antibodies, but also T cells. As a result, they could be very effective against microbes that hide inside human cells, such as herpesvirus or HIV, as well as TB bacteria and a variety of parasites.

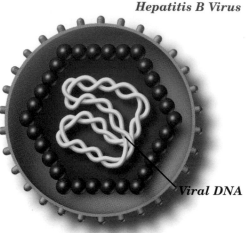

Hepatitis B Virus

Viral DNA

Plasmid

Viral Promoter

Viral Gene Spliced Into Plasmid

Plasmid

Viral Gene

Plasmids Containing a Gene From the Hepatitis B Virus

To make naked DNA vaccine, scientists use plasmids—tiny rings of DNA, normally found in bacteria, that reproduce separately from the chromosome. They pluck a gene from a disease-causing microbe (such as the hepatitis B virus, above), splice this gene into the plasmids, and choose a method of delivering the vaccine: either the plasmids are dissolved in saline solution and injected into muscle tissue with an ordinary hypodermic needle, or they are sprayed onto tiny gold beads and shot into the skin with a specially designed gene gun.

The gold beads used in these vaccines are so small—only 2 micrometers, or .000078 of an inch in diameter—that researchers shoot 10 million beads into a person's arm at one time. As the gene gun is fired, these microscopic beads are propelled through a disposable, transparent nozzle by a small burst of helium gas. The beads then form a golden cloud and painlessly enter the outer layers of the skin.

"It feels just like a puff of air," says William Swain, one of the developers of this method at PowderJect Vaccines, Inc., of Madison, Wisconsin, who experimented on himself before proceeding with clinical tests.

After penetrating a skin cell's nucleus, the plasmids separate from the gold beads and begin to direct the manufacture of microbial proteins that will act as antigens.

Gold Beads Covered With Plasmids

A Gun That Sho DNA-Covered Gold Pellets

Recent research on fruit flies is spurring scientists to focus ever more sharply on the very first stages of the immune system's response to invaders—the crucial events that limit infection in the early hours after exposure to microorganisms, as described in our article on "Innate Defenses That Hold the Fort."

It turns out that *Drosophila's* innate defenses are very similar to those of humans, pointing to a common ancestry of these systems. While analyzing various signaling pathways in fruit flies, researchers have discovered large numbers of peptides that have specific antimicrobial activity both in insects and in mammals.

Choosing the Right Peptide

About 400 antimicrobial peptides have been identified so far. They have turned up in every multicellular organism in which researchers have sought them—including plants. Even the tiny fruit fly uses them selectively, choosing different peptides to attack bacteria and fungi, for instance, so as to wreak maximum damage on each type of invader. If scientists were able to produce such peptides in large quantities, they might provide new classes of antibiotics, suggest four leading American and European immunologists in a review article in the May 21, 1999 *Science*.

Studies of the signaling pathways that lead to innate immunity in *Drosophila* may also bring new treatments for humans whose innate immunity is impaired, the *Science* reviewers add. Many people suffer from such impairments. For instance, patients with severe burns are at great risk of infection not only because they have lost the physical barrier provided by the skin, but also because healthy skin supplies several components of innate immunity—including antimicrobial peptides as well as certain cells, such as macrophages, that can swallow up invaders. Likewise, people with cystic fibrosis develop frequent infections because the antimicrobial peptides produced by the normal respiratory epithelium are unable to function in their airways. As a result of recent work on fruit flies, medical researchers may eventually find ways to supply replacements for certain parts of the signaling pathways involved.

Seeking a Microbe's Tracks in the Immune System

In the decade since researchers showed that *Helicobacter pylori* causes peptic ulcers, microbes have come under increasing suspicion in other diseases not originally linked to infections. The suspicion has proved true in certain illnesses, such as in Kaposi's sarcoma, a cancer affecting people with HIV. But all too often, by the time researchers look for microbes, the perpetrators have vanished. Infections may trigger an immune response that produces disease long after the microbes have been cleared from the body.

This problem has led David Relman of Stanford University Medical School to propose a new strategy. Rather than seeking traces of the microbes themselves, Relman asks, why not try to identify the body's specific reactions to a wide variety of infectious agents? In collaboration with HHMI investigator Patrick Brown, also at Stanford, and Lou Staudt of the National Cancer Institute, Relman is now doing just that, using microarrays that contain complementary DNA for more than 15,000 human genes.

The group's hypothesis is that each type of microbe will provoke a particular pattern of genetic responses, "producing profiles that can be used to distinguish among different infectious agents," writes Relman in the same issue of *Science*. These profiles would be far more specific than traditional markers of inflammation, and thus might help physicians uncover the source of many unexplained illnesses.

In addition, the profiles might show that certain diseases are caused by the interactions of several microbes. In Relman's view, these conditions might include several rheumatic or autoimmune diseases, as well as malignancies.

Onward and Upward with Vaccines

The Lyme disease vaccine is doing well. So are various trials of vaccines using naked DNA. Researchers are also making some headway in dealing with staph infections. Every year, some 500,000 patients in American hospitals contract staph infections during their stays. These infections can lead to pneumonia, meningitis, or infections of the heart and bloodstream. They used to succumb rapidly to drugs, but *Staphylococcus aureus* has become increasingly resistant to antibiotics.

Now researchers at the National Institute of Allergy and Infectious Diseases have produced an experimental vaccine against this stubborn bacterium. The vaccine protects mice against many strains of staph. It is expected to go into human trials soon.

"By putting the DNA molecules right into cells, we can get away with a much lower amount of DNA," explains Swain, who says that the DNA covering 10 million gold beads adds up to only half a milligram for each dose and would be relatively inexpensive. Delivering the vaccine with a hypodermic needle into muscle would require 1,000 times more DNA, he says.

Hypodermic Needle

Disposable Nozzle

Although the gene gun is applied to the arm, a one-inch "spacer" leaves a gap between the nozzle and the skin, providing an exit path for the gas stream while the gold beads penetrate the skin.

The gun's force is set so that the DNA-coated beads travel less than one hundredth of an inch into the epidermis, the skin's top layer. They go through squamous cells and the stratum corneum to settle in cells of the stratified epithelium.

After the viral DNA in the plasmids has done its job and the epidermal cells in which the plasmids landed have manufactured antigen, these cells are sloughed off. The bare gold beads inside these cells disappear together with them.

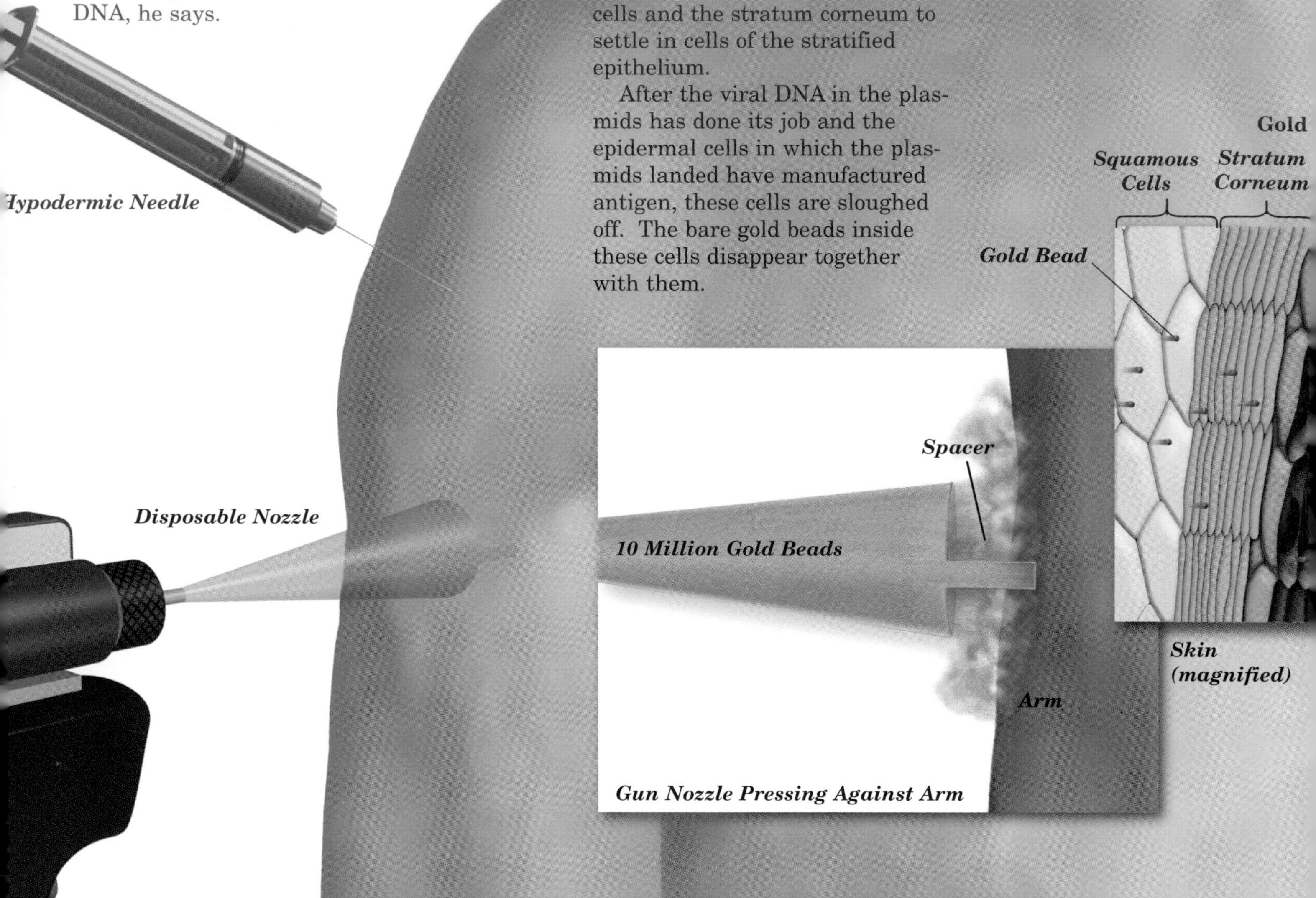

Gold

Squamous Cells *Stratum Corneum*

Gold Bead

Spacer

10 Million Gold Beads

Skin (magnified)

Arm

Gun Nozzle Pressing Against Arm

The first successful test of a DNA vaccine was carried out in 1993, with a hypodermic needle. In a landmark experiment, Margaret Liu, who was then at Merck Research Laboratories in West Point, Pennsylvania, and her colleagues showed that DNA encoding a gene for the internal nucleoprotein of an influenza A virus could protect mice against another strain of flu. This protein is "an attractive target for a vaccine because it is a very conserved (unchanging) portion of the virus," Liu explained. Existing flu vaccines stimulate the body to make antibodies that target only the virus' surface proteins, which mutate rapidly (that's why the vaccines must be updated every year).

Liu's experiment proved that the DNA vaccine worked, at least in mice, and that it aroused both kinds of immune responses—T cells, which can destroy already infected cells, as well as antibodies, which can block a virus from infecting other cells.

Inside the skin cell, the microbial gene is first transcribed into messenger RNA (mRNA), which gets translated into microbial protein. This large antigen is then chopped up into smaller pieces, and each piece picks up an MHC Class I molecule that identifies the cell as part of the person's body. Together the fragment of antigen and the MHC molecule travel toward the cell surface. There, finally, the small antigen binds to a killer T cell, triggering the process of immunity.

Beads Entering Skin

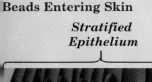

Stratified Epithelium

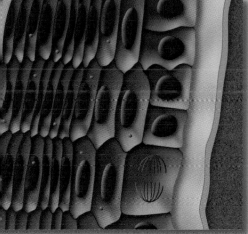

Only a few of the 10 million DNA-covered beads actually land inside the nuclei of skin cells and make antigens. Yet the small amount of protein they produce leads to a strong immune response.

"These results were so surprising that people had a hard time believing them," Liu says. But since then, several groups of researchers have started clinical trials of naked DNA vaccines against a variety of diseases—hepatitis B, AIDS, herpes simplex, influenza, T cell lymphoma, and malaria. DNA vaccines against other diseases are being tried in animals at a rapid pace.

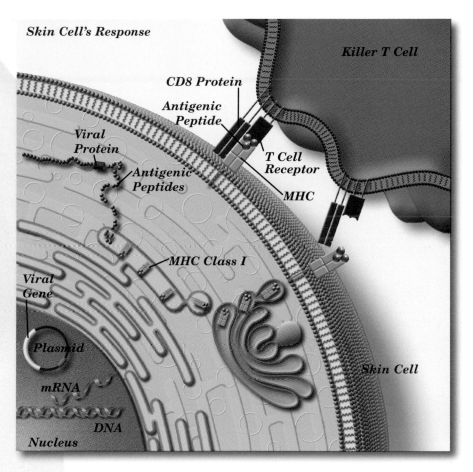

Skin Cell's Response

Killer T Cell

CD8 Protein

Antigenic Peptide

Viral Protein

Antigenic Peptides

T Cell Receptor

MHC

MHC Class I

Viral Gene

Plasmid

Skin Cell

mRNA

DNA

Nucleus

Naked DNA vaccines have yet to be proved safe and effective. If so, they may be applicable to any number of diseases. "The method is the same for any target," Liu points out. "The DNA is almost completely the same, except for the part that codes for the gene. So any other application could come very quickly." Furthermore, vaccines for multiple diseases could be given in a single inoculation.

In the next few years, "we will have sequenced the genomes of most, if not all, of the world's pathogens," declares Stephen Johnston of the University of Texas Southwestern Medical Center. "DNA vaccines probably offer the best way to translate all that sequence information into useful vaccines."

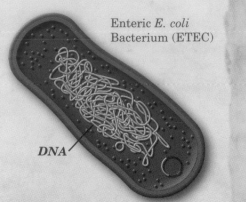

Enteric *E. coli*
Bacterium (ETEC)

DNA

Genetically
engineered soil
bacteria
(white) infect
plant cells,
transferring an
ETEC gene
into the cells.

Edible Vaccines From Bananas and Other Fruit

Imagine a baby eating a spoonful of banana puree that will make her immune to diarrheal diseases or hepatitis B. This may actually happen, perhaps within a decade, says Charles Arntzen, president of the Boyce Thompson Institute for Plant Research in Ithaca, New York, who in 1992 came up with a rather odd but promising idea: producing edible vaccines in transgenic plants.

"Bananas would be both the factories and the packages" for the vaccines, Arntzen says.

Why bananas? Because plant vaccines need to be eaten raw, he explains—heat would inactivate some of the antigens—and few people would like to eat the transgenic tobacco leaves or raw potatoes his group used in its first experiments. By contrast, "bananas are as baby-friendly as they are plentiful," Arntzen says. They would also be very inexpensive. However, they might have to be prepared like baby food to control the dose of the vaccine.

To make vaccine against an infectious *E.coli* bacterium, ETEC (enterotoxigenic *E.coli*), whose toxins cause travelers' diarrhea, Arntzen's group plucked a gene from ETEC and spliced it into a naturally occurring soil bacterium called *Agrobacterium tumefaciens*. The altered *Agrobacterium* infected the cells of potato plants, transferring the foreign gene into them.

The potato cells then started to produce some ETEC protein. When mice ate the potatoes, their immune systems recognized the ETEC protein as alien and began to churn out responses. From then on, the mice were primed to make antibodies specific for the ETEC toxin.

A dozen volunteers at the University of Maryland's Center for Vaccine Development recently ate similar potatoes (raw, but chopped up to make them edible) during a clinical trial of the vaccine, with promising results. Elsewhere,

Norwalk virus vaccine has been grown in potatoes. Rabies and hepatitis B vaccines have been grown in tomatoes. Cholera vaccine has been made in cantaloupes. And European scientists have just proved that a vaccine grown in black-eyed beans can protect mink against a viral disease that attacks cats, dogs, mink, and other animals.

Naked DNA and edible plant vaccines are just the beginning. Many other new vaccines appear to be in store for us, thanks to molecular biology.

FINDING THE

CRITICAL SHAPES

X-ray crystallography and
computer graphics bring
new approaches to the
fight against viral diseases,
cancer, and other ills.

Howard Hughes Medical Institute
6701 Rockledge Drive
Bethesda, Maryland 20817
(301) 571-0200

CRITICAL SHAPES

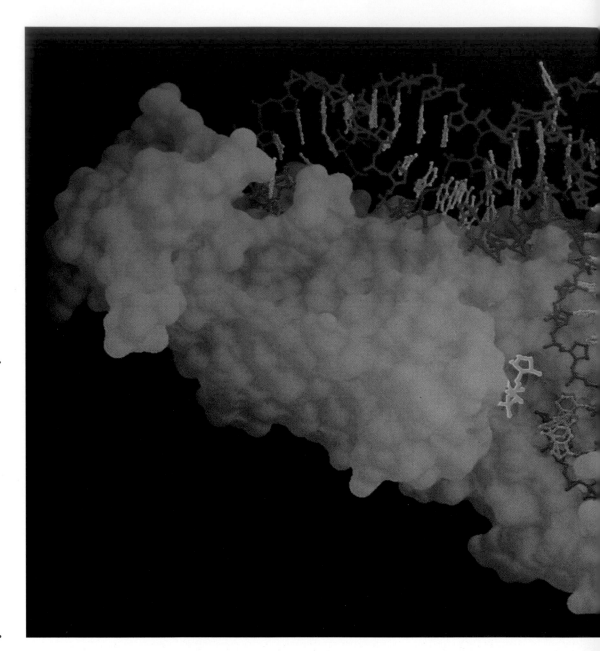

In billions of brief encounters every day, biological molecules such as these run the body's chemical machinery. Here, two molecules are shown recognizing each other by their shapes. Their 3-D structure was deciphered in the laboratory of Thomas Steitz in the HHMI unit at Yale University.

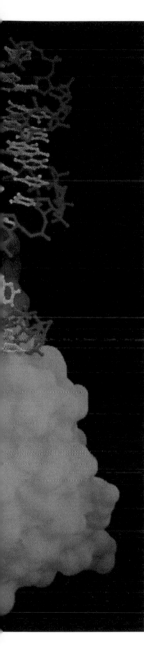

FINDING THE CRITICAL SHAPES

X-ray crystallography and computer graphics
bring new approaches to the fight against viral diseases,
cancer, and other ills.

INTRODUCTION

. . . .

Seeing the 3-D structure of a biological molecule can be a revelation. A look at James Watson and Francis Crick's model of the DNA double helix, for instance, offers immediate insight into how the molecule copies itself. "The minute you see it, you know how heredity works," says Don Wiley, an HHMI investigator at Harvard University.

Since this model's creation in 1953, the study of molecular shapes has grown into a critically important science that is fueling much of biomedical research. Scientists who can view the locations of the individual atoms of a protein and also examine the protein's entire structure, in relation to other proteins or DNA, are no longer flying blind.

That is the lure of structural biology: It reveals the intricate, 3-D structures of molecules in our bodies—and sometimes even their function. Structural biology used to be a slow, painstaking pursuit in which scientists worked for many years, even decades, to decipher a single molecule. But recombinant DNA technology jolted the field by providing abundant quantities of proteins for study. New tools became available, such as the x-ray beam line designed for high-speed experiments on biological molecules at the synchrotron of the Brookhaven National Laboratory. And brightly colored images of newly solved structures began to appear ever more frequently in leading journals.

The 3-D images showed the subtle bumps and hollows by which molecules recognize one another in billions of fleeting encounters every day. They revealed whether two shapes fit snugly enough together, like a key in a lock, to allow nutrients into cells or to receive signals from other cells.

They also clarified certain mechanisms of disease, allowing scientists to create new kinds of therapies—for example, drugs to block viral infections. Without such information, drug designers would be as limited as surgeons who tried to operate on a patient without knowing the location of relevant bones and organs.

The molecules being deciphered were so small, however, that even the most powerful microscopes offered little help in analyzing their structural details. Scientists had to rely on more difficult methods of detection—primarily x-ray crystallography.

Max Perutz, a pioneer of this method, required 22 years of intensive labor to solve the structure of hemoglobin, the oxygen-bearing molecule in red blood cells, which he and his colleagues mapped in 1959. If Perutz had had a synchrotron beam line similar to the one at Brookhaven, he might have mapped the hemoglobin molecule in just a few weeks from the time he obtained a good crystal.

A half-dozen other synchrotrons, even more powerful than Brookhaven's, have opened in various parts of the world, but the HHMI beam line at Brookhaven remains the most productive line for structural studies with the "MAD" method pioneered by Wayne Hendrickson, an HHMI investigator at Columbia University. MAD allows crystallographers to use ever smaller and more delicate crystals, relieving the bottleneck caused in the past by a shortage of large and perfect crystals.

In addition, growing numbers of protein structures are being solved by nuclear magnetic resonance (NMR) spectroscopy, which has become vastly more powerful in recent years. Previously this technique could be used only with very small molecules, but now it can analyze small to medium proteins. This bypasses entirely the long and sometimes erratic process of making crystals of the proteins.

Another change is the extraordinary increase in computer power, which enables scientists to tackle far more difficult problems.

This massive progress is changing the shape of structural biology. In the past, researchers worked on whatever proteins yielded suitable crystals. Now their main goal is to solve interesting biological problems. Increasingly, they define themselves by their subject matter—not as crystallographers or NMR spectroscopists, but as biologists who want to find out how genes are turned on, for instance, or how drugs bind to enzymes.

To experience the 3-D world in which structural biologists operate, pick up the stereo viewer supplied on page 411, unfold it, place it over one of the stereo-pairs on pages 411-415, and stare a few minutes. It often takes this long for the eyes to relax and the brain to put the two images together. But if you persist, the 3-D structures will become clear—and so will the attraction of exploring molecular shapes.

Maya Pines, *Editor*

LIGHT AT THE
BEND
OF THE TUNNEL

*Opening a Powerful X-Ray Beam Line
to Probe the Most Difficult Crystals*

by Natalie Angier

When scientists push a button this spring to let an extremely intense and fine-tuned x-ray beam stream into a new pipeline at the National Synchrotron Light Source in Brookhaven, New York, they will further galvanize a field that is already one of the busiest in biological science: the study of molecular shapes.

The x-ray beam line will help researchers probe the delicate structures of proteins, DNA, and other molecules that keep our bodies alive and thrumming. The beam will sail into an air-locked work station known as a hutch. There the high-energy ray will encounter a tiny speck of a crystallized protein held suspended in a thin glass capillary. For the briefest instant, the beam will pass through the delicate lattice of the protein, the geometric straightness of its course nudged ever so slightly by the atoms in the crystal. The diffracted rays will conclude their micro-odyssey by slamming into an imaging plate perfectly positioned to capture the event and to collect the data crystallographers need for their analyses.

By mapping out the 3-D structures of important biological molecules, scientists can better comprehend how these molecules work. Most of the mapping is done by x-ray crystallographers, who aim x-rays at crystals composed of particular molecules and then analyze the patterns of the diffracted rays for clues to the molecules' structures. Their work is speeded up when they can use the intense x-ray beams spawned by synchrotron accelerators— giant rings where subatomic particles are gunned up to nearly the speed of light to generate bursts of powerful radiation. Synchrotrons have allowed crystallographers to make great strides in recent years; but at times these scientists have labored rather as poor relations, forced to wait in line and make do with equipment and experimental conditions designed for physicists.

. . . .

*Wayne Hendrickson (below) and codesigner
Jean-Louis Staudenmann (above) align a crystal
in the new diffractometer, where the crystal will
be spun to expose different angles to a high-
powered x-ray beam.*

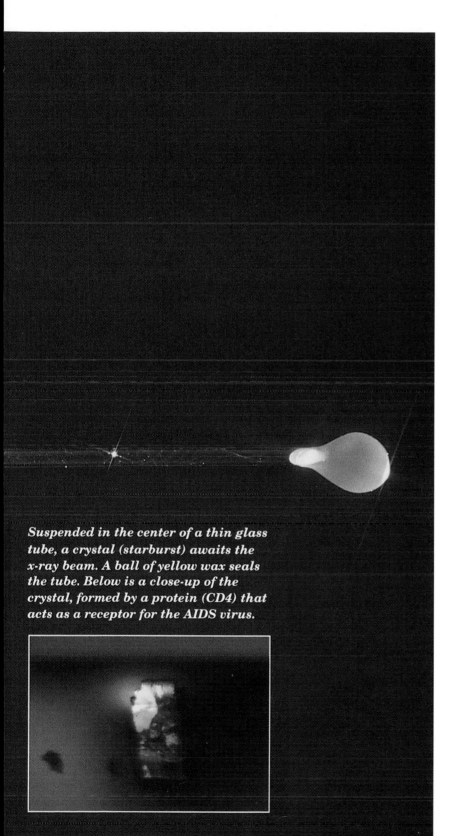

Suspended in the center of a thin glass tube, a crystal (starburst) awaits the x-ray beam. A ball of yellow wax seals the tube. Below is a close-up of the crystal, formed by a protein (CD4) that acts as a receptor for the AIDS virus.

For biologists, HHMI's $3.2 million facility at Brookhaven National Laboratory (of which the Light Source is a part) should be a boon. The newly completed assembly of lenses, grids, detecting devices, and computer graphics displays will form one of the few synchrotron beam lines dedicated to the singular needs of x-ray crystallographers. In addition, it will offer some unique features designed by Wayne Hendrickson, a professor of biophysics in the HHMI unit at Columbia University, to provide information so rapidly and accurately that crystallographers will be able to tackle a large number of previously insoluble problems.

The opening of this beam line—and work on two additional beam lines to open there soon—reflects the current growth of research in structural biology. There is hardly a university, pharmaceutical company, or biotechnology firm around that isn't racing to establish an x-ray crystallography department, or to expand existing facilities. "Whenever I visit a college that doesn't have a crystallography group, they'll pull me aside and ask me how to get one started," says Hendrickson. "I tell them, well, you've got a lot of competition." *Nature, Science, Cell,* and other leading journals feature reports on the proteins that have been analyzed through the techniques of x-ray crystallography. Already the atomic structures of over 500 proteins have been entered into the Protein Data Bank at Brookhaven, and the list lengthens almost monthly.

As perhaps the greatest index of the vitality of the field, graduate students and postdoctoral fellows are flocking to the laboratories of prominent x-ray crystallographers. And they are clamoring for positions despite the admitted difficulty of the science. To excel as a crystallographer, a young researcher must master not only biology, but chemistry, biochemistry, physics, a significant amount of mathematics, and computer skills as well.

The allure of x-ray crystallography is not surprising, given the heady goals of its practitioners. Crystallographers are seeking to determine nothing less than the shape of life itself. They want to know what the vital molecules of the body look like: how each one twists, kinks, curls, and shudders into the particular 3-D configuration that allows it to perform its assigned task in the cell. Crystallographers who study DNA—the double helix that enfolds all the genetic instructions for maintaining life and creating new life—wish to understand how the ultra-compressed bends and coils of the molecule interact with proteins to dictate which genes are switched on and which are shut down at any given time. That switching system determines such basic events as cell division, whether a cell will develop into a muscle or a liver or a brain cell, and when it will die. By exploring the minute details of DNA coiling, investigators hope to unravel the mysteries of gene regulation.

Other crystallographers are tracing the complex curves and pleats of proteins, the end products of the genes that are encoded in DNA. There are about thirty thousand different proteins in the average human cell, and, like a tireless maintenance crew, they keep the house in order. Many are enzymes, catalyzing in moments crucial biochemical reactions—such as the breakdown of foodstuffs into usable fuel—that might take centuries without them. So-called structural proteins serve as beams and girders, lending the cell its shape and resiliency. Proteins known as membrane receptors poke through the surface of the cell, receiving signals from the bloodstream or from neighboring cells. Still other proteins are the hormones or messenger molecules that stimulate those membrane receptors. And some proteins latch on to DNA, enabling the mastermind molecule to translate its genetic instructions into a fresh batch of proteins.

In sum, whatever needs doing is done by proteins; and a protein's form dictates its function. Through analyzing precisely how the individual hydrogen, carbon, nitrogen, and other atoms of the protein are stacked together in 3-D space, crystallographers can better understand why one protein is able, say, to clasp a passing bundle of cholesterol and ferry it into the cell cytoplasm, while an antibody protein, secreted

In response to polarized light, these large crystals of an enzyme from the pancreas glow in bright colors, demonstrating a high degree of internal order.

by cells of the immune system, can zero in on an invading pathogen. The tiniest nuances of shape—a gully here, a side chain there—can prove indispensable to interpreting the protein's character and biological role. In one recent x-ray study of an enzyme that is critical to the inflammatory process, Paul Sigler, a professor of molecular biophysics in the HHMI unit at Yale University, and his colleagues charted out the feature that flicks the enzyme on

Laboratory of Alexander McPherson, University of California, Riverside

and off: a little chamber in the protein's midsection that clutches and releases calcium molecules. "We solved the atomic structure in record time, two to three weeks," he says exultantly, "and now we're looking at information on a computer screen that may hold the key to blocking the inflammatory response."

Crystallographers trust that by deciphering the architecture of enough proteins, they'll spot revealing patterns and consistencies—structural motifs that turn up again and again in proteins that perform similar tasks, or that evolved from a mutual ancestral protein. When Michael Rossmann, a professor of biochemistry at Purdue University in Lafayette, Indiana, recently analyzed the structure of rhinovirus-14—a virus that causes human colds—he was astonished to see that it resembles the shape of certain plant viruses. "That told us how closely related many viruses actually are," he says. So, perhaps, are proteins.

Other crystallographers note that they have already seen similarities among the so-called DNA-binding proteins, which help to turn genes on and off. "We've found that there are certain patterns in the arrangement of the backbones of these DNA-binding proteins," says Carl Pabo, an HHMI investigator at The Johns Hopkins University School of Medicine in Baltimore, Maryland. "For example, we see a type of twisted structure known as 'helix-turn-helix' that is conserved from viruses and bacteria right on up to higher organisms."

The more proteins they can link together into families and super-families, researchers say, the closer they will be to understanding the chemistry of major cellular processes. "The subatomic physicists have their standard model that takes account of all particles and observations out to, what? five significant figures," says Paul Sigler. "Well, we biologists are on the threshold of coming up with a standard theory of our own, one that will account, in chemical terms, for how cells differentiate, develop, and regulate themselves. But that's not going to happen until we figure out what the key players look like."

Crystallographers are seeking to determine nothing less than the shape of life itself.

Knowledge of protein structure promises to have broad medical and practical value. Wayne Hendrickson, for example, is studying the shape of CD4, a cell-surface protein that serves as a receptor for the human immunodeficiency virus, the cause of AIDS. Johann Deisenhofer, a professor of biochemistry in the HHMI unit at the University of Texas Southwestern Medical Center at Dallas, recently won the Nobel Prize in Chemistry together with his former colleagues Robert Huber and Hartmut Michel of West Germany for their structural analysis of a protein complex that orchestrates photosynthesis—a discovery of unsurpassed importance in understanding one of the fundamental processes of nature (see stereo pictures, p. 412). Now, Deisenhofer says, the agricultural industry would like to use his results to create safer and more targeted weed killers. Food chemists are exploiting the tools of crystallography in quest of better food stabilizers. "It may sound presumptuous," says Brian Matthews, an HHMI scientist at the University of Oregon in Eugene, "but we'd like to engineer proteins that are superior to those found in nature. In fact, we have already constructed a protein that is much more stable than its natural counterpart."

GAZING INTO A CRYSTAL WALL

Matching the impetus to decipher protein structure is the significant challenge of the science itself. The average protein consists of several hundred amino acids, and there are 20 different types of amino acids to choose from. Those numbers, however, don't begin to convey the complexity of what biologists call the "protein folding" problem. A protein is produced in the cell as a linear strand of amino acids, but upon construction the chain begins to flop and bunch under a boggling cross fire of chemical and electric forces. Each amino acid can twist into about 10 different shapes. Thus, each type of protein could, in theory, wiggle into 10^{100} possible configurations, a nearly infinite number. Certain factors rule out many of those shapes—for example, two atoms cannot occupy the same space—yet enough potential forms remain to confound easy analysis. Even when biologists know the amino acid sequence of the protein, and even when they have some insight into its biochemical activity, they must look directly at the finished product if they hope to claim intimate acquaintance with their subject. "When you want to know how a car runs," says Paul Sigler, "the best way to go about it is to lift up the hood and take a look."

Of course, with biological molecules, it is not simply a matter of popping open the hood and poking around. DNA and proteins are exceedingly tiny, and their structures are far below the resolving power of any microscope in existence. "You need a ruler commensurate with what's being measured," says Wayne Hendrickson, "and the size we're talking about here is the distance between atoms." That distance is roughly 1.5 angstroms, with one angstrom equalling one 10-billionth of a meter. The only things compact enough to gauge such bantam diameters are x-rays, which run anywhere from 0.5 to 500 angstroms each—about a thousand times shorter than the wavelength of visible light.

The use of x-rays presents problems of its own. Because they are so short, the rays tend to whisk through most compounds entirely unimpeded—that is, they do not scatter well. And it is scattering that allows us to visualize the world. "When you look at an object in a room, the light from the bulb hits the object, scatters off it, and is collected by the lens in your eye," Hendrickson explains. "That light is then focused on the retina and sent on to your brain, where neural processes turn it into a mental image.

"We would like to be able to use x-rays to probe molecular structure analogously, but we can't. To a very good approximation, x-rays travel in straight lines. Essentially there are no lenses that will produce images from x-rays. Moreover, molecules of protein scatter x-rays almost imperceptibly."

Luckily for crystallographers, "almost imperceptibly" doesn't mean the protein has no effect at all; the influence simply is vanishingly small, since very few of the rays in an x-ray beam get scattered by the electrons

"Lift up the hood and take a look."

A CRYSTALLOGRAPHER'S COOKBOOK

Probing the atomic structure of a biological molecule is like preparing a four-star meal: it is part inspiration, part perspiration, part skill, and part sheer good luck. As with gourmet cooking, there's an element of magic. The many pieces of the analysis all come together at the very end, when the researcher suddenly divines what the molecule looks like in three-dimensional space. Herewith, the steps that precede the moment that Wayne Hendrickson refers to as the great "ah-ha!"

■ Perhaps the hardest step is to obtain a good crystal in the first place (see The Art of Growing Crystals, p. 392). A researcher will often create two versions of the crystal, one of the biological molecule itself, another of the same molecule with markers added for the sake of comparison. The markers might be either a heavy metal such as mercury, which will bind tightly to the crystal lattice in a few key spots; or selenium, a sulfur-like element which, for biochemical reasons, is able to replace the sulfur in methionine, one of the amino acids of the protein under study. In either case, the introduced markers are useful because they scatter x-rays effectively. Mercury, for example, is such a dense element that it will diffract an incident beam of energy 200 times more powerfully than will carbon, a standard constituent of biological molecules.

■ Measuring less than 0.5 millimeter on a side, the crystal is then suspended in a thin glass capillary and bombarded with an x-ray beam. Depending on the intensity of the beam and other factors, the crystal might be irradiated for anywhere from a few seconds to 48 hours or longer. The crystal is rotated to get a "snapshot" from every possible angle. Sometimes a researcher must stop in the middle of the experiment and replace the crystal with a fresh sample. Reason: biological molecules quickly decay under the withering force of an x-ray beam, and a crystal that has been irradiated for too long ends up looking something like a rotten tooth.

■ Positioned behind the crystal, a detecting device—either an imaging plate, a photographic film, or an electronic counter—records the pattern of scattered energy waves. The quantity of data collected is staggering and must be reduced to the simplest pattern possible before proceeding. Thus, crystallographers try to determine the smallest slice of information they must decipher in order to calculate the molecule's atomic structure. Scanning the diffraction patterns, they establish the size, shape, and symmetry of what is known as the *unit cell* of the crystal. If a crystal is thought of as a giant wall constructed of millions of bricks, the unit cell is the basic brick pattern—the motif—that is repeated over and over in the wall.

■ Once the unit cell has been determined, crystallographers attempt to recreate, through complex mathematical algorithms, the constituents of the unit cell that produced the observed x-ray pattern. In essence, the researchers use mathematics as their "lens." The retina of our eyes, or the lens of a light microscope, will focus scattered light waves bouncing off an object into a discernible image of that object. No lens exists that can resolve scattered x-rays, so crystallographers rely on mathematics—and computers—to mimic what a lens might do.

Here is where the importance of comparison markers comes in. The strong signal that bounces off, say, a metal atom stands out from the other diffraction signals and thus permits researchers to calculate the position of that atom. From there the scientists can begin "bootstrapping" along to the other, lighter components of the molecule.

■ The results of these elaborate operations is an electron-density map, a graphic representation of the data that resembles a hiker's contour map—but doesn't necessarily give an accurate picture of the terrain. As its name implies, an electron-density map indicates the distribution of electrons in the unit cell, thereby revealing information about the atoms that possess those electrons. But in dealing with crystals of large molecules such as proteins, which are held together by comparatively few, weak bonds and which consist in large part of water, such maps are of relatively low resolution. They seldom show individual atoms. They also tend to contain a number of errors.

■ The success of the analysis depends largely on how the crystallographer interprets the electron-density map to produce a preliminary atomic model of the unit cell—the crucial motif of which the entire crystal is constructed. Even then a lot of hard work remains; the model must be optimized, or refined. Employing yet another series of mathematical algorithms, the researcher painstakingly compares the calculated x-ray diffraction from the rough model with the original observed x-ray diffraction data. Where the model was wrong, it is corrected in successive cycles. "We try to find the best fit between the model and the observed data," explains Axel Brünger, an HHMI investigator at Yale University, who uses an advanced mathematical technique called simulated annealing. "We imagine such things as 'What would happen if you stretch a bond between these two atoms?' 'What would happen if you heated, then cooled, the molecule?' We develop these so-called empirical energy functions to simulate what might be going on in the entire molecule."

■ Most of the final calculations are performed by a computer, and the results are displayed on a high-resolution graphics screen. By manipulating a joystick or mouse, the researcher can turn the image to reveal any side or angle of the model desired. And to the crystallographer, this ultimate image is an indisputable masterpiece. "It's the aesthetics of crystallography that really launched me into this field," says Sherry Mowbray. "To be able to take a molecular 'photograph' of what these things in our body really look like is the most miraculous thing I can imagine."

—*N.A.*

in the atoms of a single molecule. That's why crystallographers build crystals. They create lattices in which the molecules are aligned atop one another like bricks in a wall. And though the crystals are themselves rather small—about 0.3 millimeter on a side—they are virtual Great Walls of China compared with isolated protein molecules. "The number of separate molecules in a crystal is staggering," says Hendrickson. "It's something like 10^{15} molecules, which is one molecule for every penny in the national debt. What happens is that all those molecules scatter in unison, with the result that, while scattering from one protein molecule may be weak, the combined scattering or 'diffraction' from the collection of replicas is quite strong." Sherry Mowbray, an HHMI investigator at the University of Texas Southwestern Medical Center at Dallas, compares a lone protein molecule to a plastic sheet with a very pale tint of pink: "Just one looks clear, but if you had thousands of them stacked together, the bunch would look bright pink."

It is this luminous signal, this splash of crystalline "color" that is the heart of x-ray crystallography. By recording the diffracted waves, either on film or with an electronic counting device, the researchers gain their first handle on the nature of the molecule responsible for the scattering. But capturing the wave pattern is just a start, the first toddling step toward a solution of the molecule's structure—as the earliest x-ray crystallographers realized only too well.

THE PHASE IT'S GOING THROUGH

Chic though x-ray crystallography may be, the science is, in fact, three-quarters of a century old. In 1912, Max von Laue, a German physicist, had the idea that one could determine the atomic structure of molecules in a crystal by passing x-rays through the crystal, since the crystal's repetitively spaced layers would act like a diffraction grating and scatter the x-rays in a pattern. He tried it out with a crystal of zinc sulfide, a simple organic molecule. But von Laue did not fully understand the relationship between the pattern of x-ray spots that he saw on his film and the arrange-

Crystallographers like to work with stable, fast-growing crystals such as these, which are formed by con-canavalin B, a dietary protein found in beans and peas.

ment of atoms in the crystal that produced the pattern. At around the same time, Sir Lawrence Bragg, the English physicist, proposed an equation to explain the key relationship, at least for x-ray patterns generated by neatly ordered crystals such as table salt.

Biologists seized on the technology to attack their molecules. Perhaps the most spectacular use of diffraction patterns occurred in the 1950s, when James Watson and Francis Crick were struggling to decipher the structure of DNA. The two men might well have remained stymied had it not been for Rosalind Franklin's excellent x-ray diffraction data on the molecule. Her photographs helped them figure out that DNA is composed of a double helix of complementary nucleotide subunits which face each other like steps on a spiral staircase and are supported by a backbone of sugar and phosphate.

Solving the basic components of DNA proved far easier than deducing the contours of proteins, which are much less regu-

lar and repetitive than the double helix. As biologists attempted to form images of these molecules, they ran into what is known as "the phase problem." They had a wretched time working backwards from their diffraction data to figure out the type and position of each atom that caused the rays to diffract as they did; they could not simply apply Bragg's equations to their molecules, because proteins are so much more complicated and uneven than salt.

Max Perutz toiled for 22 years to determine the rough structure of hemoglobin, a protein that consists of 10,000 atoms. To underscore the difficulty of the enterprise, hemoglobin isn't even especially large; many proteins have hundreds of thousands of atoms.

As he plied away the decades, Perutz invented a splendid method for cracking the phase problem that is still employed by many crystallographers today: he added a sprinkling of heavy metal atoms to his protein crystals. Heavy metal atoms are densely packed with electrons and thus scatter

This luminous signal, this splash of crystalline "color"...

rays more strongly than do hydrogen, carbon, and other atoms of proteins. The metal inserts bind to certain positions in the crystal and serve as glittering markers, allowing scientists to calculate the position of the other atoms. Joining Perutz at Cambridge University, John Kendrew applied this technique, known as isomorphous replacement, to the analysis of a smaller protein, myoglobin, and the two scientists received the Nobel Prize in Chemistry in 1962.

A "NEAR-RELIGIOUS CONVERSION"

Despite the celebrated triumphs of Perutz and Kendrew, x-ray crystallography was long viewed as an almost impossible field. Who wanted to spend half a career on one enzyme?

"When I entered protein crystallography, in 1958, there were literally only five or six people in the world with expertise in the area," Michael Rossmann recalls. "We were a *very* small family."

By the 1970s, with the advent of recombinant DNA technology, most biology students rushed headlong into the new science of gene cloning and gene sequencing, and crystallography was left behind. "Ten years ago, almost nobody took up crystallography, for the simple reason that there weren't that many jobs available," says Robert Fox, an HHMI investigator at the Yale University School of Medicine. "I think that those who bucked the tide underwent a near-religious conversion. When you see a protein molecule folded for the first time, it's pretty miraculous."

Ironically, genetic engineering helped spark the current renaissance in x-ray crystallography. As the geneticists tweezed apart ever more sequences of DNA, they became increasingly curious about the shape and purpose of the proteins encoded by their cloned genes. For their part, crystallographers saw in recombinant DNA methods a fabulous source of their most coveted raw material: protein.

To harvest a usable crystal, researchers need to start with something like a tenth of a gram of a protein—and in the pioneering days of crystallography they needed far more than that. Most proteins, however, are extremely difficult to extract from tissue. The early crystallographers generally could not get at them or else had to spend months extracting them, a millionth of a gram at a time. With today's gene-splicing protocols, investigators can often insert a cloned gene into bacterial or yeast cells and then spur those prolific organisms to churn out virtually unlimited quantities of the gene's protein product in a matter of days.

Other technical advances have fueled the boom in crystallography. High-speed computers can perform in seconds the thousands or tens of thousands of calculations necessary to settle the position of a single atom. Computer graphics can help translate molecular images into chemically sensible models. New electronic "area detectors" can register x-ray diffraction patterns hundreds of times faster and more precisely than standard x-ray film or the old electronic counters.

And then there is that most exotic new apparatus in the crystallographer's kit pack: the synchrotron accelerator.

A WINDOW OF OPPORTUNITY

Synchrotron radiation began life as an irritating and uninvited guest. In the 1940s, physicists built their first high-energy accelerators, pushing electrons and other subatomic particles to almost the speed of light and then smashing the particles together in their effort to study the fundamental forces of nature. The physicists observed that every time the electrons were accelerated by the giant magnets in the collider rings, the particles emitted bursts of energy that spanned the electromagnetic spectrum. Because the machines were called synchrotrons, researchers dubbed the outbursts "synchrotron radiation," and they viewed the phenomenon as larcenous, for it drained energy from the speeding particles and limited the design of their equipment.

However, other scientists leaped on synchrotron radiation as a potent tool for their work. Solid-state physicists saw it as a means of studying the surfaces of compounds; engineers coveted it for circuit design and imprinting, while weapons researchers were happy to find another fountain of furious energy.

Finding a fountain of furious energy

Window-like ports were constructed at the turning points of collider rings to tap the synchrotron radiation and divert it to various projects. Eventually, the demand for time on the beam lines became so great that accelerator rings were built simply to beget synchrotron radiation—the second-generation machines.

Today there are about a dozen synchrotron light sources around the world: in England, France, Germany, Japan, the Soviet Union, and elsewhere. Here in the United States, facilities at Cornell University in Ithaca, New York, and Stanford University in Palo Alto, California, are of the parasitic first-generation type; the National Synchrotron Light Source at Brookhaven is, as its name implies, a second-generation accelerator.

The benefits of synchrotron radiation for protein crystallography are significant. The x-ray beam from a synchrotron port is at least a thousand times stronger than the beam produced by a conventional laboratory x-ray device.

This means that biologists can gather more data about their crystals while limiting the time the molecules are exposed to x-rays—an important advantage. Solid-state physicists do not need to worry about exposure time, because an inorganic sample such as silicon can bathe in radiation indefinitely and still retain its form. By contrast, a biological crystal deteriorates rapidly when irradiated, and crystallographers often must stop in the middle of an experiment to replace a damaged sample with a fresh one—that is, if another crystal is available.

Synchrotron beams may help conquer the problem of protein decay. "An experiment that might take three days on 'home' x-ray sources takes approximately five minutes at the synchrotron," says Wayne Hendrickson. "The chemical reactions that chew up your crystal don't have a chance to happen."

For this reason, a number of intrepid biologists have taken advantage of synchrotron radiation, and they insist that some of their experiments could only have been done at the accelerators. But other crystallographers have shunned the giant rings. They grouse that, because the facili-

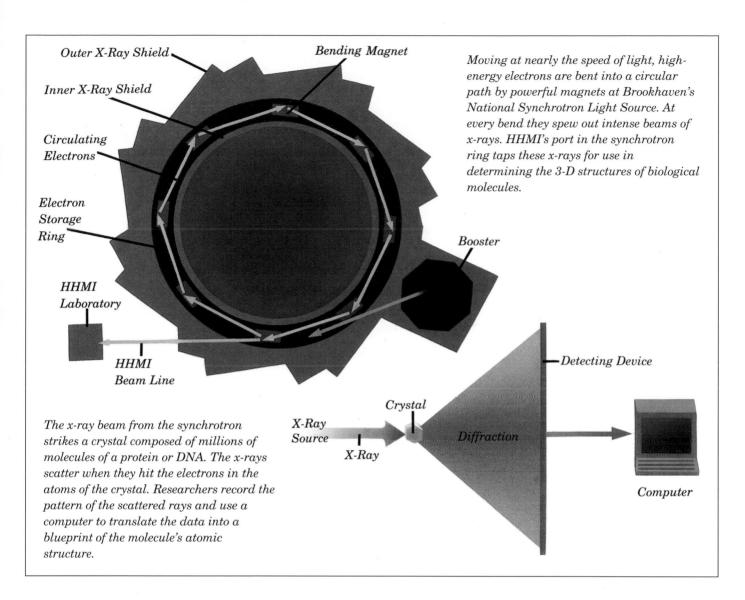

Moving at nearly the speed of light, high-energy electrons are bent into a circular path by powerful magnets at Brookhaven's National Synchrotron Light Source. At every bend they spew out intense beams of x-rays. HHMI's port in the synchrotron ring taps these x-rays for use in determining the 3-D structures of biological molecules.

Outer X-Ray Shield

Inner X-Ray Shield

Bending Magnet

Circulating Electrons

Electron Storage Ring

HHMI Laboratory

HHMI Beam Line

Booster

The x-ray beam from the synchrotron strikes a crystal composed of millions of molecules of a protein or DNA. The x-rays scatter when they hit the electrons in the atoms of the crystal. Researchers record the pattern of the scattered rays and use a computer to translate the data into a blueprint of the molecule's atomic structure.

X-Ray Source

X-Ray

Crystal

Diffraction

Detecting Device

Computer

ties were built for physicists and engineers, there are no amenities for biologists. Doing an experiment at an accelerator means booking time on the beam line far in advance, spending large sums of money to journey to a distant state or nation—and, worst of all, arriving at the ring only to find that it has been shut down for repairs.

THE DREAM BEAM MACHINE

Wayne Hendrickson had such plaints echoing in his ears as he set out to draft the biologist's perfect beam line. He also had plenty of notions of his own. The venture began in early 1985, when his department head at Columbia University told him that HHMI was considering a pro-

gram in the field of structural biology. Could Hendrickson please come up with some proposals—say, in a week or two? Hendrickson dashed off two suggestions: a new beam line at Brookhaven and a supercomputing center at Columbia. "They accepted the beam line idea and rejected the other," says Hendrickson. "I'm glad they made that choice."

Hendrickson marks the true beginning of the project as February 1987. That's when he recruited the man who helped bring it all together: Jean-Louis Staudenmann. An engineer and solid-state physicist and a native of Geneva, Switzerland, Staudenmann was then working at Iowa State University in Ames. He had one particularly outstanding qualification: he had already designed the optics system for another beam line at Brookhaven. Staudenmann was soon joined on the project by technicians Randy Abramowitz and David Cook, computer scientist Gyungock Kim, and machinist Walter Zotterman.

With monomaniacal drive, Hendrickson, Staudenmann, and their co-workers plotted the details of their new machine. Hendrickson had a keen sense of what would work and what would not; after all, he had sampled most of the synchrotron facilities around the globe. "The best way to learn about instrumentation is not to read papers or even just look at it," he says, "but to go ahead and use it."

The goals of the team were threefold: to further compress x-ray exposure time; to offer a candy store of choices; and to incorporate a novel method devised by Hendrickson for cracking the phase problem. The design that emerged is composed of not one beam line but three, each band of x-rays streaming off the accelerator ring from the same port but encased in its own steel conduit. One beam is engineered for what Hendrickson calls the "bread-and butter experiments" of crystallography. With it, biologists will be able to analyze the structure of their molecules using the old heavy metal insertion techniques—but more swiftly than ever before. Another beam, still in the construction stage, is geared for "Laue radiation" assays, which could eventually permit biologists to observe split-second enzymatic reactions as they occur.

The third beam must truly be considered the soul of the new machine. It is intended to execute Hendrickson's structure-solving innovation, the "multiple wavelength anomalous diffraction" technique—MAD for short. A variation on Perutz's metal-doping procedure, MAD is meant for those molecules that, because of peculiarities of their composition, cannot be analyzed through more pedestrian methods. In the elaborate protocol, the crystallographer first studs the molecule with several markers of selenium, a sulfur-like element. Next the researcher must hit the molecule with a beam of energy that can be shifted from one part of the x-ray spectrum to another. Reason: at some point, the changing beam will hit just the right wavelength to cause the embedded selenium atoms to begin pulsing, or resonating. And as they resonate, they will leave characteristic spots on the x-ray imaging film—just the signposts needed to begin analyzing molecular structure.

Conventional x-ray sources, however, cannot be tuned to varying wavelengths or focused in the necessary manner. Such options require a device like the one that Hendrickson has installed at the synchrotron beam: a monochromator. Priced at $200,000, the monochromator looks like a large clothes dryer set down in the middle of the steel conduit that carries the x-rays from the accelerator. Inside the monochromator are two high-precision silicon crystals, which serve to sculpt and define the x-ray beam. When it streams out from the belly of the accelerator, the beam is a hot, wide fan of x-rays, ranging in wavelength from a tenth of an angstrom to many hundreds of angstroms. Only specific wavelengths are wanted at a given time, so the monochromator crystals must dispense with the others and let only the chosen ones pass through.

The tuning and focusing of the beam can be changed to suit the needs of the experimenter and the properties of the protein under study. Whatever parameters are chosen, the monochromator crystal directs the chiseled radiation on a horizontal path through the pipeline and into the hutch, a 300-square-foot radiation enclosure where the experiment proper occurs. After its 24-meter odyssey from giant accelerator to cramped hutch, the beam at last is fit to rendezvous with the crystal under study.

Suspended in a thin glass capillary, the crystal is under the exacting control of a $60,000, half-ton diffractometer, which measures 1.5 meters in length and 1.0 meter in height. "This is one of the largest and most accurate diffractometers that you can find on the market," says Staudenmann, "and Wayne's specifications take it to its maximum capacity." By those

The beam
is a hot,
wide fan of
x-rays.

demanding specifications, the computer-driven diffractometer must orient the crystal so that it deviates from the line of the oncoming beam by no more than five-hundredths of a millimeter. The Brookhaven crew also has configured the diffractometer to spin the crystal along four axes simultaneously, at high speeds. In other words, as the x-ray beam zaps the crystal, the diffractometer rotates the sample, exposing as many angles of the molecule as possible.

Two additional gadgets round out the ensemble. To record the x-ray diffraction patterns, Hendrickson has chosen an imaging system of high-resolution plates, called imaging phosphors, which were invented originally for use in such medical equipment as CT scans. Each $400 plate is covered with a phosphorescent material that will glow when irradiated. Clamped in a cassette about 200 millimeters behind the crystal, the plate can register up to 5 million x-ray data points—"a hell of a lot of information," says Hendrickson. Because of the need to analyze the crystal from different angles, a crystallographer will deploy

Technician Randy Abramowitz inspects the monochromator that tunes and focuses the x-ray beam before it streams into the conduit (right) leading to the crystal.

some 80 to 100 of the re-usable plates to analyze a single molecule (the cassette changes the plates automatically). And each plate will have tens of thousands of spots where x-rays have hit the surface with greater or less intensity—depending on the array of atoms that sent the rays bouncing.

Because there is such a dizzying amount of data, the beam scheme includes a $300,000 "laser reader system," which will scan the completed imaging phosphors at a fleet rate of some 200 seconds per plate and translate the data into a workable, digitized format. (The laser system will also be used for experiments on the other two beam lines.) Hendrickson expects eventually to have on-site computers that will distill the digitized information into a preliminary blueprint of the crystal's atomic structure, ready for a biologist to take home and refine into a publication-worthy report.

In putting together their almost gothically interlocking system, the scientists have logged strings of 20-hour days, fretted over delayed shipments of key parts, ques-

tioned their own judgment in choosing one method over another—and sometimes squabbled for days on end.

Nevertheless, the designers trust that the result will prove well worth the investment of time and angst, and that the beam line will become a sort of crown jewel of crystallography. Hendrickson estimates that, once the network is at full throttle, biologists will be able to wrap up most experiments in half a day. "Our goal is that people will be able to come, throw the crystal on the diffractometer, digitize the plates, and go home with the data processed," he says. He promises an abundance of amenities and frills to ease the biologists' task: a cold room for protein work; a temperature-controlled environment so that samples don't spoil; biological and chemical reagents. Says Hendrickson, "We're also going to do our best to keep the beam line from ever breaking down."

In dispensing time on the three beam lines, researchers associated with HHMI will have priority, but at least 25 percent of the time will be available to outside biologists. Visiting scientists will have to pay only for room and board; the use of all equipment and accoutrements at the synchrotron is free of charge.

Hendrickson hopes that many crystallographers will adopt the MAD method to decipher the 3-D structure of heretofore intractable molecules. He also confesses that part of the incentive for erecting the beam line at Brookhaven was to expand the collection of eight champagne corks sitting on his office bookshelf. "Every time we solve a structure," he says, "we celebrate by opening a bottle of champagne."

"The real reason for building all this equipment is to have a chance to do the experiments we want to do. There are these highs that you get when you finally understand what's going on with a structure. It all comes together in a crescendo, and there's this great 'ah-ha!' of discovery. That's the feeling crystallographers go for again and again." ●

To raise the odds, he is actually trying to levitate the droplet, either in the laboratory or in space.

THE ART OF GROWING CRYSTALS

Some people have a knack for it, a sort of green thumb. But for most x-ray crystallographers, nothing is so humbling—or exasperating—as the process of growing the crystals their work requires.

"Crystals are totally unpredictable," says David Davies, a veteran crystallographer at the National Institutes of Health. Asked how one grows them, he laughs and replies, "If you find out, let me know." A few years ago, when he wanted to study a fragment of an antibody molecule, his laboratory produced a fairly good crystal of it within a few weeks, on the very first try. It seemed like an easy molecule to work with, so the researchers tried to make more and better crystals of the same kind. But after an entire year of effort, "we had nothing at all," Davies recalls. "The first one was the only good one. You never know with crystals. You do exactly the same thing over and over again, and they never come out the same."

"It's always a problem," agrees Michael Rossmann, of Purdue University. "Just recently one of our projects was stuck for two years because a virus crystal wouldn't grow. Getting crystals to grow is the weakest part of the process."

A crystal is a form of perfection, an ideal, beautiful state in which all the atoms and molecules are arranged in precise order and their pattern is repeated regularly in three dimensions. It is particularly difficult to attain with large, complex, and flexible molecules such as proteins. Yet it must develop naturally: crystals may be coaxed to grow, but they cannot be forced.

The trick, then, is to find the specific set of conditions under which a particular protein is most likely to crystallize. One starts out with pure protein (this used to be the stumbling block, but now that recombinant DNA technology allows many proteins to be produced in quantity, it is no longer such a problem). The protein is then dissolved in water, and the water is made to evaporate very slowly, in a friendly environment.

As the water evaporates, the protein solution becomes more concentrated, bringing the protein molecules closer together. This increases the attraction between some of their atoms so that the molecules begin to cluster.

Then, one prays.

In most cases the result is a disappointing mess—either a disordered solid or little grains that look like sand and are called, contemptuously, a "precipitate." Only rarely do the molecules form a crystal nucleus, a perfectly ordered critical mass of identical units around which a crystal can grow. This happy event requires exactly the right combination of temperature, acidity, protein concentration, calm, and other mysterious conditions.

If a crystal nucleus does appear, other molecules will attach themselves to it *if* conditions are suitable. But here again, uncertainty reigns. Nobody knows precisely

. . . .

This seemingly perfect crystal of nerve growth factor, shown inside a hanging drop, proved nearly useless to crystallographers because of a peculiar geometric defect.

why some crystals continue to grow, while others never come close to the size at which they would be useful to crystallographers (about 0.5 millimeter in length). Nor can anyone explain why some crystals form in the shape of needles, while others look more like diamonds (some proteins have been known to crystallize into a dozen different shapes, though their structure remains the same). What is clear is that protein crystals contain at least as much water as protein and should not be exposed to air. They must be kept moist, in their "mother liquor," even while undergoing x-ray analysis.

Assuming that a crystal does reach minimum size, it may not have enough internal order to produce good diffraction patterns that can be interpreted with reasonable accuracy. And in the end, if a crystal has passed all of these hurdles and attained the right state of purity, it may suddenly crack or shatter before any measurements have been taken.

No wonder many researchers still consider crystal-growing a black art. "There are stories out there about phases of the

> "I just like to do it— it's like panning for gold."

Crystals of a bacterial virus protein reflect light from many planes, as in a painting by Braque.

moon and that sort of thing," says HHMI investigator Stephen Sprang, of the University of Texas Southwestern Medical Center at Dallas.

"It's difficult to control the conditions under which a protein crystal will grow. We were very puzzled once because our crystals of tumor necrosis factor would grow in one room, but not in another, as we discovered by accident after we moved. Then we found out that the inner side of the corridor, which adjoined the second room, was warmer than the other side by 2 degrees, and that made the difference! There are good explanations for all these things. But every protein has its own conditions under which it will crystallize, and you have to search them out."

Small molecules, such as salt or sugar, will form crystals easily. "Take one of the simplest examples, a tidal pool of salt water," says Alexander McPherson of the University of California at Riverside, who is widely acknowledged as "the greenest thumb" among crystal growers. "When the sun beats down on the pool, you are left with salt crystals." But proteins, being larger, have fewer atoms on their surface in proportion to their volume and therefore form fewer bonds with their neighbors. (For the same reason, crystals of protein are much more delicate than crystals of smaller molecules.)

"You need to develop an eye for growing crystals of protein," says McPherson. "You need to intuit what the next step should be. But you also need a great deal of experience. It takes five to seven years to learn how to grow them efficiently." Most crystallographers grow one protein crystal "and hope they never have to grow another one," McPherson says. "Then they go off to work on their crystal for the rest of their lives. But I've always had an interest in it. I just like to do it—it's like panning for gold. You take something that has very little value, like a protein solution, and make it valuable for revealing structures."

In the past 20 years his laboratory has crystallized about 70 different proteins, an astounding achievement. Most of these crystals were produced inside "hanging drops"—droplets of protein solution that hang on the underside of a glass cover, held there by surface tension. Eventually the protein in this drop becomes more concentrated as some of the water in the solution evaporates by vapor diffusion, reaching an equilibrium with a small reservoir of more concentrated solution in a well under the drop. This changes the protein concentration very slowly, increasing the chance of producing small nuclei that will turn into crystals. Nevertheless, "we still get a precipitate more often than a crystal—a thousand times more often, in some cases," McPherson says.

To raise these odds, he is actually trying to levitate the droplet, either in the laboratory or in space. The idea sounds far-fetched, but there is some reason to believe that containerless growth, which eliminates contact with solids such as the glass cover, may allow crystals to grow more uniformly.

"We can levitate the drop with electrostatic fields in the lab," McPherson says. "This method does work with very light objects. And of course in space, if you set the droplet free, it will just sit there." Protein solutions suspended in such droplets would not have to contend with the turbulence produced by gravity, which one researcher compared to "putting your thumb into the flow of a river."

The space shuttle has carried crystal-growing experiments on seven flights since 1985, but these flights last only a few days. Other researchers have sent their experiments up on the Russian space station, Mir, which gave them more time in so-called microgravity.

However, "microgravity will not solve the problem," McPherson emphasizes.

"The real problem still remains: Proteins are different from one another—as different as human beings. Each protein must be treated separately, so it's very slow going."

The most important contribution of efforts to grow crystals in space may well be that they forced scientists to develop robots and other kinds of automated devices that can take over various steps in the process.

With the help of automation, researchers are now doing large-scale experiments in which each condition of crystal growth—temperature, acidity, concentration, etc.—is varied systematically, for many types of proteins, and the results are monitored automatically. Some precise rules of crystal growth may emerge from these experiments.

Another kind of systematic variation is being tried by Carl Pabo, an HHMI scientist who studies protein-DNA complexes. Pabo lets his proteins bind with DNA fragments of different lengths—fragments that are easily made in the laboratory. He is building a library of prospective DNA linkers that could be sprinkled in with dissolved proteins and would bind with them. These linkers appear to stack up in the crystal, forming columns of different sizes that may help the crystal to grow.

Although Pabo cannot yet predict which DNA linker will be best for any particular protein, his method allows him to try several dozen protein-DNA combinations in a short period of time, increasing his chances of obtaining a suitable crystal.

X-ray crystallography techniques have improved dramatically in the past five years. At the same time, thousands of proteins have been identified, and more are being discovered every month. But crystallographers can study only a few of the proteins—those for which they have crystals. Until crystal-growing changes from an art to a science, the shortage of crystals will remain a major bottleneck for them.

—M.P.

JAMMING THEIR STRUCTURES:

A NEW WAY TO STOP THE VIRUSES THAT CAUSE INFLUENZA, COLDS, AIDS

by Maya Pines

Viruses travel light, like thieves. Armed with just a few genes of their own, they steal whatever they need from the cells they invade. "Them becomes us," remarked Baruch Blumberg, who won a 1976 Nobel prize for his work on hepatitis, a viral disease. This tendency of viruses to commandeer our cells' chemical machinery for their own use has been a major obstacle to finding treatments against viral infections, since most drugs that could stop viruses would damage our cells as well.

In this artist's conception, an influenza virus lands on a human cell and binds to molecules of sialic acid (red) on its surface. Other viruses approach the cell. If such binding can be prevented, there will be no infection.

"We were boxing blindfolded. It helps to know where to hit."

New clues to treatment are now emerging as the detailed, 3-D structures of viral molecules reveal features that belong to viruses alone. "We were boxing blindfolded," declares a researcher. "It helps to know where to hit."

Scientists who actually see viral structures in three dimensions are in a much better position to understand how viruses function—how these tiny particles that are not quite alive clamp on to living cells, invade them, take over their machinery, reproduce, and then go off by the thousands to invade other cells. At the same time, the structures suggest ways in which specific steps of the viral life cycle might be blocked.

For years we have relied almost entirely on vaccines to guard against viral diseases. Vaccines work by stimulating the immune system to produce antibodies that are highly specific and have long memories. The antibodies act like scouts; they range through the bloodstream, sensing the minute differences that identify foreign substances. When antibodies recognize a virus they have encountered before, they attach themselves to it and tag it as an enemy so that various natural defense systems of the body can attack it. In this way, vaccines have provided magnificent protection against smallpox, polio, measles, yellow fever, and other diseases.

But we still have no reliable defense against viral infections for which vaccines do not exist, such as AIDS. There are extremely few antiviral drugs of any sort, and none that protect against a broad range of viral diseases—nothing comparable to penicillin or the other powerful antibiotics that have been so effective against a variety of bacterial infections for nearly half a century. The reason for this lag is that bacteria offer many more targets for drugs than viruses do. Unlike viruses, bacteria are complete organisms, with reproductive systems of their own and dozens of components that are identifiably different from those of animal cells. Drugs aimed at these components are likely to knock out the bacteria without harming the host.

By focusing on the shapes of key viral molecules, scientists hope to find new therapies that do not depend on the immune system but, rather, fight infection directly. One particularly promising strategy would be to block the initial binding of viruses to cells.

If there is no binding, there can be no infection. And molecules bind together only when they fit together precisely—when the bumps and crevices on certain molecules of one surface (virus, for example) dovetail with those on another surface (cell). The idea, then, is to plug up the cavities in the binding site with decoy molecules and thus prevent the virus from infecting any more cells. But first one needs precise information about the two structures involved.

Deciphering the structure of the viral molecules that bind to cells has proved surprisingly difficult, both because these are relatively large proteins and because they are frequently embedded in the fatty membrane that surrounds many viruses.

A SHIFTY PROTEIN'S SHAPE REVEALS A VULNERABLE SPOT

A long, thin protein that sticks out of the surface of influenza virus was the first human virus molecule to yield its 3-D secrets. The structure of this hemagglutinin (HA) was worked out in 1981 by Don Wiley, a professor of biochemistry and biophysics at Harvard College (now an HHMI investigator there), and Ian Wilson, a postdoctoral fellow who has become an associate member at the Research Institute of Scripps Clinic in La Jolla, California.

Influenza viruses look like little balls of fuzz as they float about in search of cells to invade. The fuzz consists of some 500 spikes made up of HA, as well as shorter spikes made up of another protein called neuraminidase. The HA protein, which binds to human cells, is the key target of antibodies against influenza and also the most important ingredient in influenza vaccines.

In order to analyze HA, Wiley first had to pry it loose from the viral membrane without damaging it and then dissolve it in a watery solution—the only way to get it ready for crystallization so it could be studied by x-ray crystallography. In 1974 he heard that a British researcher, John Skehel—who is now director of the

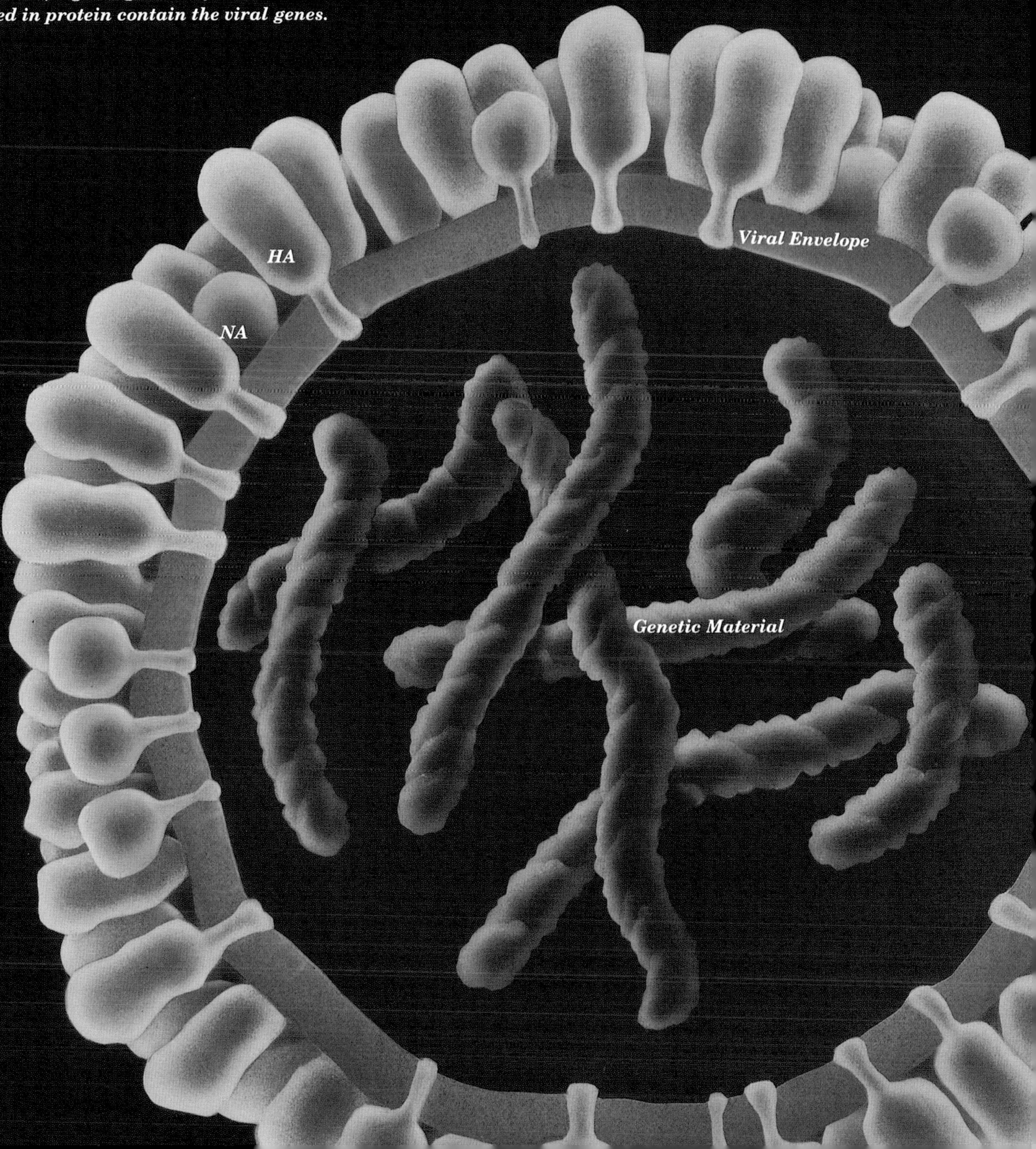

Ready to bind with human cells, molecules of hemagglutinin (HA) cover the surface of an influenza virus, shown here in cross-section. The viral surface is also studded with another protein, neuraminidase (NA). Inside, eight segments of RNA wrapped in protein contain the viral genes.

HA

NA

Viral Envelope

Genetic Material

Deciphering the 3-D structure took six years of work and 1.5 million eggs.

National Institute for Medical Research in London—had lifted the HA protein right off the viral membrane with the aid of bromelain, an enzyme extracted from pineapple. Skehel's method left only a tiny stem of the HA anchored inside the membrane. An additional advantage of the bromelain was that it made the HA soluble. Wiley wrote to Skehel right away, starting a long and fruitful collaboration that continues to this day.

"When Skehel got my letter, he sent me some of the HA he had purified," says Wiley. "I worked on it here for a while. Then I went on sabbatical to his lab in Mill Hill, near London, for a year. There I learned some virology, his specialty, while he learned some crystallography, my specialty. Now, everything we do on the flu, we do together."

The two men soon produced a pure crystal of HA protein. But actually solving the structure of HA at atomic resolution took six years of work, trials with 10 different strains of virus, thousands of crystals, and extraordinary amounts of protein for raw material. The Harvard team obtained the protein from England, where Skehel set up a large-scale supply operation for them.

"Five thousand chicken eggs were infected every week," Wiley recalls. "The virus grew over the weekend. On Monday morning, all five thousand eggs had their tops taken off carefully and their fluid removed. Then the HA protein on the surface of the viruses was extracted and purified. This was done every single week for six years." It would have been almost impossible to get these supplies of protein by the usual form of funding in the United States, Wiley maintains. "Yet to make an overall assault on the structure-function problem, you have to have pure material, and enough of it."

Even before the researchers had completely solved the structure of HA, rumors about their findings—and what these findings meant—began to fly in the Harvard laboratory.

Influenza virus is known for its exceptional ability to mutate rapidly: whenever a large number of people become immune to a particular strain, the virus changes itself into a different strain and regains its power of

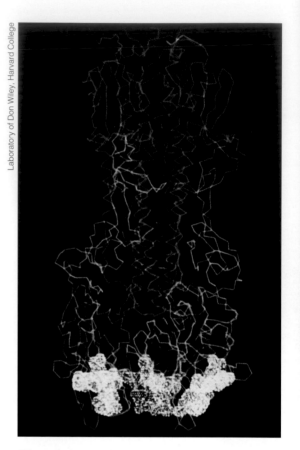

Three pockets (white) at the tip of the HA molecule are its binding sites, shaped to fit with sialic acid on the surface of human cells.

· · · · · · · · · · · · · · · · ·

infection. That is why we never gain any long-lasting immunity to it. Measles or polio generally can be caught only once in a lifetime, and there are effective vaccines against them. But variations of influenza virus keep coming, in unpredictable cycles, to infect us over and over again. The manufacturers of vaccines are hard put to keep up with it.

As Wiley explains, "The antibodies that the immune system sends out to fight disease are made to recognize specific shapes. If the virus changes shape, the antibodies don't recognize it any more."

To avoid the antibodies, however, the virus need not change its entire shape. Altering just a few of the amino acids that make up its key proteins seems enough, judging from the small differences between the HA's amino-acid sequences in all known strains of influenza virus. As the

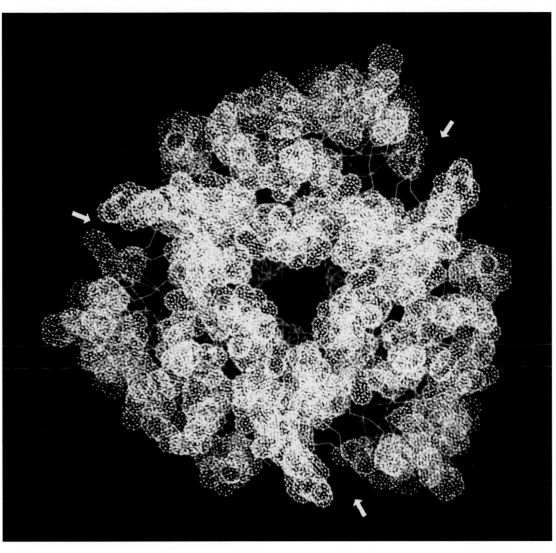

The influenza virus avoids the immune system by changing some amino acids (yellow) on the surface of the HA molecule. But the three binding pockets (blue) on the HA's tip, seen from above, remain unchanged in all known strains of influenza. Three central helices (red) extend downward to stabilize the HA's structure.

Harvard researchers examined the HA structure of different strains, they kept finding the variable amino acids located in the same general area: in highly exposed loops around the tip of the HA molecule, just where the molecule would be expected to make contact with cells. The loops were also at the front lines to meet the antibodies sent by the immune system to fight the virus. This pattern was repeated in each of three identical subunits, or monomers, that make up the HA molecule.

Between the exposed loops the researchers found a small pocket that aroused their curiosity. In each monomer the pocket was lined by amino acids that appeared to remain unchanged in all strains of influenza. Could this protected, unchanging cavity be the binding site through which the virus latched on to receptors on the surface of human cells?

Wilson remembers his excitement "when it became clear that all the amino-acid changes were going on at the surface of the molecule, except in one patch between the variable loops. It was the first indication of how a virus changes its structure to cause disease," he says. After correlating the structural changes with the years of new epidemics, the researchers concluded that epidemics occur whenever at least one amino acid is substituted in each of five specific sites on the surface of each monomer in the HA molecule.

"Every two years or so, the HA monomer gets about 14 amino acid replacements in

> **"D**rugs of this sort—perhaps in the form of a nasal spray —could inactivate the virus."

its structure," Wiley explains. "When you sum them up, you see five different places in each monomer where there is at least one change for each epidemic. So in 1981 the idea developed that maybe human antibodies were binding to those five places"— and that changing these sites was the virus's way of staying one step ahead of the immune system (or, sometimes, ahead of the vaccine makers).

Meanwhile the binding pocket, which does not change, remains a tantalizing feature of the HA structure. Scientists had known for years that sialic, or neuraminic, acid, a sugar molecule found all over cell surfaces, is involved in binding influenza virus. But there was no evidence linking it to the HA pocket. In 1988 William Weis, a graduate student working with Wiley, soaked HA crystals in a solution of small molecules containing sialic acid. As he hoped, the HA and sialic-acid–containing molecules bound together. Next he examined the structure of the "complex" formed by the two molecules and found a perfect fit between them.

"We proved that this section of the HA molecule is indeed the binding site for sialic acid," says Weis. "We also learned which way the sialic acid is oriented in the site." Researchers can now begin to work on drugs that might resemble sialic acid but bind to the HA more tightly. Drugs of this sort— perhaps in the form of a nasal spray—could inactivate the virus, Wiley says.

In addition, Wiley and his associates are exploring ways to stop the virus later in its life cycle, right after it has attached itself to the cell. Specifically, they would like to interfere with membrane fusion, the process through which an influenza virus fuses its membrane with that of the host cell as it prepares to penetrate the cell. A drug might be developed to block this process, and the piece of the HA molecule that is responsible for fusion—a segment far from the binding site—may provide a convenient target.

Besides, membrane fusion is interesting in its own right: it occurs millions of times a day in every human body as various substances that are enveloped in membranes move in and out of human cells.

THE LIFE OF A VIRUS

"A virus is a piece of bad news wrapped up in protein."—Peter Medawar.

Straddling the border between life and non-life, viruses are built so economically that they cannot even reproduce until they enter a living cell (of human, animal, plant, or bacterium). They consist of genetic material—just one or a few short segments of RNA or DNA—covered by a protein shell, which in some cases is further surrounded by a fatty envelope.

The first job of an animal virus is to find a cell and attach to it. Then the virus crosses the cell membrane (or fuses its envelope with the membrane) and disappears into the cell.

Once inside, the virus releases its genes, which proceed to take over the cells' machinery. Like slave drivers, they force the infected cell to work around the clock, neglecting its own needs while producing thousands of new viral subunits.

Eventually the subunits assemble into complete virus particles, or virions. These emerge from the cell either by "budding" out of the cell's membrane (in the case of enveloped viruses) or by bursting out of the cell and destroying it.

There are some exceptions, however. In certain cases (for example, in AIDS infections), the viral genes become integrated into the host cell's chromosomes and quietly replicate as part of the cell.

Some of these long-simmering viruses are retroviruses, so named because they replicate in a contrary manner. Retroviruses defy the usual rule that genetic information moves from DNA to RNA and then to protein. Their genes are made of RNA, and they use an enzyme called reverse transcriptase to copy this RNA into a complementary strand of DNA, thereby gaining the power to make the infected cell produce more virus. Unfortunately, this mechanism may lead to cancer if the viral DNA lands in the wrong place.

In 1956 James Watson and Francis Crick pointed out that viruses contain so few genes—and therefore produce so few kinds of proteins—that their shells must be composed of many copies of identical proteins held together in highly symmetrical configurations. Later on, other scientists showed that "spherical" viruses, such as the tomato bushy stunt virus, are actually made up of multiples of 60 subunits arranged in regular patterns, as in Buckminster Fuller's geodesic domes. The tomato bushy stunt virus, for instance, has 180 subunits and looks like a soccer ball.

Because of their small size, most viruses could not be seen at all until the development of electron microscopes in the 1930s. These instruments, about 100 times as powerful as the best light microscopes,

. . . .

Like a soccer ball, the poliovirus's protein coat is composed of 12 symmetrical sections, each with five subunits. Three kinds of proteins (in blue, red, and yellow) make up each subunit in this computer-generated model.

allowed scientists to determine the general structure of many viruses, including bacteriophages, the tiny viruses that prey on bacteria.

The relative simplicity of viruses made them a favorite subject of biologists, who used them as tools to figure out some of the basic principles of modern biology. Studies of the genetics and chemistry of bacteriophages led directly to the birth of molecular biology.

Most recently, viruses have been used to carry specific genes into living cells. This ability to ferry genes may prove to be of special value in gene therapy.

Despite their awful consequences in such diseases as rabies or AIDS, most viral infections do little or no harm.

Viruses have evolved to get along with their hosts rather than kill them. They need living creatures that sneeze or otherwise spread infection; otherwise they cannot multiply. A disease that proves rapidly lethal may be an accident—a horrible mistake—both for the virus and for the host.

—*M.P.*

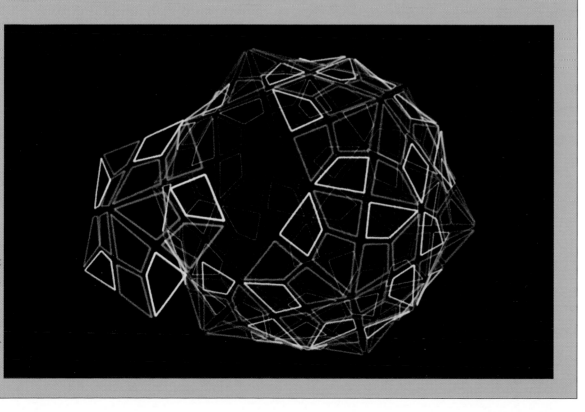

Coming: effective treatments for the common cold.

Seeing Human Viruses in Atomic Detail

Ideally, scientists would like to see the 3-D structure of an entire influenza virus—not just its surface proteins—but so far, this has proved impossible. "Flu virus won't crystallize because it has a lipid [fatty] outer membrane," Wiley explains. The AIDS virus cannot be crystallized for the same reason. Lipid membranes are too squishy to pack into the neat symmetry required for a crystal.

Some other viruses are built without a lipid envelope; they just have a shell of proteins around their nucleic acid, and they can be crystallized. In 1978 Stephen Harrison, now an HHMI investigator, and his colleagues at Harvard University solved the structure of a whole virus (the tomato bushy stunt virus) at atomic resolution for the first time. In 1985 the structures of two human viruses were deciphered: human rhinovirus 14, which causes colds, and poliovirus.

Michael Rossmann and his associates at Purdue University, who solved the structure of rhinovirus 14, believe their work may lead to a long-sought goal: a truly effective treatment for the common cold, or at least for the most common forms of the cold. There are so many varieties of cold viruses (more than a hundred strains of rhinovirus alone) that scientists doubt a useful vaccine can ever be developed against the common cold. But other treatments—drugs that snuff out colds—may well become available within a decade.

When Rossmann looked at the structure his group had produced, he was at once struck by a series of deep clefts, or "canyons," that cut through the surface of the virus in 60 symmetrical places. Antibody molecules had been shown to attach themselves to certain sites on the rims of these canyons, but the antibodies were much too large to fit inside the narrow canyons.

Each canyon was thus ideally shaped to bind with a small receptor protruding from the surface of cells, Rossmann realized. Certain exposed parts of the virus might mutate to evade the immune system's anti-bodies, but other parts (inside the canyon) would "remain constant in various strains and retain their ability to seek out the same receptor on the surface of human cells," he said.

Soon after solving this structure, Rossmann heard that scientists at the Sterling-Winthrop Research Institute in Rensselaer, New York, had synthesized two compounds that, on routine screening, appeared to reduce infections by viruses of the picornavirus family—the family that includes rhinoviruses. The compounds were said to interfere with the "uncoating" process through which viruses release their genetic material into the cell they have invaded. Intrigued, Rossmann asked for samples and began to study how these compounds worked. Soon he was collaborating with the Sterling-Winthrop researchers to develop more effective versions of these drugs.

By 1986 the researchers announced that they had soaked crystals of rhinovirus 14 in solutions containing two of these drugs, WIN 51711 and WIN 52084, and had solved the structures of the two drug-virus complexes. "This was the first time that anyone had been able to see a drug bind to a virus," Rossmann says. The structures showed that the drugs went right into the canyon, then through a hole, like a trap door, on the canyon floor, and into a large pocket.

"The drugs are much smaller than antibodies, so the canyon is no problem for them," Rossmann explains. "It's like putting mules into the Grand Canyon." When the drug slides into the pocket under the canyon, it appears to stiffen the virus structure; this interferes with the virus' ability both to bind with cells and to shed its protein coat, a crucial step in its life cycle. Designing drugs that fit this pocket more tightly than the original WIN compounds—and thereby block the virus more effectively—is complicated by the pocket's oily nature. Marc McKinlay, of Sterling-Winthrop, and Rossmann recently compared this problem to designing a properly shaped "piece of grease" to fit into a "greasy pocket." Nevertheless, they are hopeful that more advanced drugs against colds and other picornaviruses (such as poliovirus, Coxsackie viruses, or hepatitis A) will go into clinical trials soon.

The structures of poliovirus and rhinovirus 14 turned out to be very similar to one another, as well as to the structures of certain viruses of tomatoes, beans, and other plants.

The tomato virus work in 1978 had shown the value of seeing viral structures as a whole, points out James Hogle of the Research Institute of Scripps Clinic, a co-discoverer of the structure of poliovirus. "If Steve Harrison had solved the individual subunits in isolation, they wouldn't have made much sense, because they have a loop structure that interacts with that of neighboring proteins and plays a critical role," he says. Both the rhinovirus and the polio structures have similar loops, which surround and protect their binding sites.

Meanwhile, at Harvard University, Harrison solved the structure of the entire simian virus 40, or SV40, a virus that causes tumors in hamsters. This was the first time the structure of a whole virus that contains DNA had been deciphered (other solved viruses contain the related nucleic acid, RNA).

New Leads in the Fight Against AIDS

Research on viral structures has not ignored today's most frightening epidemic—AIDS. Rapid progress is being made in understanding the human immunodeficiency virus (HIV), which causes AIDS, and in finding new targets for drugs against the disease.

The first effective treatment for AIDS, azidothymidine (AZT), was developed without benefit of information about the 3-D structure of the virus. But the drug has limited value and is quite toxic. AZT was originally synthesized 25 years ago by a chemist who hoped to produce an anti-cancer agent. It failed to stop cancer and also failed to kill bacteria, so it was abandoned. However, in 1984, when the Burroughs Wellcome Co. began screening all of its available compounds for activity against the newly discovered AIDS virus, AZT suddenly showed some promise. The National Cancer Institute confirmed that AZT blocks replication of the virus by interfering with reverse transcriptase. This key enzyme converts the virus's genetic material into DNA so that its genes can direct infected cells to make more virus.

Clinical trials with AZT began in 1985. A larger, double-blind study that started in early 1986 was halted six months later because the usefulness of AZT treatment was so clear that it no longer seemed fair to give half the patients a placebo. From then on, all patients in the trial were offered the drug. Nevertheless, AZT is far from a cure-all.

New drugs against AIDS are now being created in a more logical manner, often on the basis of structural information. In a particularly promising development, scientists deciphered the structure of a key enzyme in the AIDS virus: a protease (an enzyme that cuts through proteins) that snips apart a large polyprotein and liberates some essential viral proteins, including reverse transcriptase. Until the protease has freed them, these viral proteins cannot function and the virus cannot reproduce.

Manuel Navia and his associates at Merck & Co., Inc., reported the structure of the AIDS protease in February 1989 and suggested that it might be used to design drugs to jam the virus's reproductive machinery—for example, a drug might be fashioned to slide into the spot where protease releases the proteins, preventing this release. Another report on the enzyme's structure appeared shortly afterwards, when Alexander Wlodawer and his colleagues at the Frederick Cancer Research Facility, a branch of the National Cancer Institute, corrected some details of the Merck model. Their version brought the AIDS protease structure more in line with those of pepsin, renin, and other better-known enzymes of the same family, the aspartyl proteases.

Because of this similarity, researchers could tap into a wealth of information accumulated by chemists who had experimented with thousands of inhibitors of renin, for instance, in their search for better medications against hypertension. This was a major step towards developing better inhibitors of the protease. ●

"It's like putting mules into the Grand Canyon."

WHY
THEY ALL WANT COMPUTER GRAPHICS

by Frederic Golden

Invited to speak before the august Royal Society in London a few years ago, Robert Langridge arrived with twin projectors, handed out stereo glasses to the audience, and unreeled a 3-D film of animated, almost Disneyesque biological molecules dancing across the screen.

"It was a very unusual approach to take at a scientific meeting," admits Langridge, head of the Computer Graphics Laboratory at the University of California, San Francisco (UCSF). He still shudders slightly at the reaction of the meeting's chairman, an eminent biochemist. Introducing Langridge, the chairman said archly, "And finally, we have our last talk. It's Bob Langridge and his Hollywood show!"

Langridge's embarrassment was short-lived, however. Today almost everyone who works with complex biological molecules uses the sort of computer graphics he showed off at the meeting—even the scornful biochemist who introduced him. Langridge's UCSF colleague, crystallographer

Robert Stroud, considers computer graphics as essential to analyzing proteins as an automobile is to getting around. "You could walk, but it would take forever," he says. Robert Fletterick, another UCSF protein scientist, sees it soon becoming "the biochemist's scratch pad, as common as paper and pencil."

Computer graphics reveals information and stimulates insights that might otherwise remain buried forever in long columns of dreary numerical data. Says Langridge, "If you have seven pages of coordinates, you can't make much of them unless you can display them as pictures."

Even more important, computer graphics is interactive. Almost as easily as youngsters operate video games, scientists can move a molecule in any direction across the screen, examine it from any angle, and strip away parts of it to expose its inner recesses. They can slice through any section of the molecule, check its fit against another molecule, and by reshuffling components or perhaps adding entirely new ones, transmute the molecule into completely novel forms.

This electronic wizardry has become indispensable to scientists, not only in deciphering the structure of proteins, whose intricate loops and twists determine their biological roles, but also in understanding proteins more deeply and even in redesigning them. With its help, they are tooling up new drugs and vaccines for hitherto intractable diseases.

Developing new drugs used to be largely hit or miss. Molecular graphics cannot guarantee success, but it offers an almost instantaneous electronic testing ground. Scientists can try out new ideas on a video screen without ever picking up a pipette or reagent. Also, because the computer screen

. . . .

Computer graphics is "indispensable," says Florante Quiocho as he fits his model of a protein's atomic structure into an electron-density map (shown in red on the screen) that was produced by x-ray crystallography. The fitting is done electronically, with the aid of a program called FRODO.

is such a powerful communications tool, researchers can easily trade thoughts at the terminal. Whatever flaws are discovered at such times can be corrected long before a projected design is ever exposed to the less forgiving environs of a living system.

For scientists who study proteins the old way—and they are getting scarce—nothing is quite so exhilarating as their first encounter with molecular graphics. Just a tap on a few buttons, a push on a joystick, a tweak of a mouse, or a twist of some knobs—and, presto, the action begins. A protein appears on the screen. The molecule expands, then shrinks. Next, it flies across the screen and docks smoothly in a hollow of a neighboring molecule. The operator is in total command of this molecular world.

"You feel like you're flying a plane," says Langridge, himself a licensed pilot and a designer of MIDAS (Molecular Interactive Display and Simulation), one of the computer graphics programs that have become so important to investigators seeking to unearth the secrets of proteins. Most such programs display images not only in color, but also in three dimensions, conveying a lifelike sense of reality. Donning stereo glasses, the viewer watches as one part of a molecule seems to float over or behind another, drifting in and out of sight. With a nudge of the joystick, the world begins to turn dizzily, almost as if the observer were looking out from the cockpit of an aircraft that has begun to bank.

Before computer graphics systems started appearing in laboratories in the late 1970s, scientists had to construct large models of molecules with rods, balls, and wire scaffolding—a difficult task, especially when it came to large molecules. Florante Quiocho, of the HHMI unit at Baylor College of Medicine, remembers how he struggled to build a large protein that was held together by screw-type connections and also required taut wire anchors at top and bottom to keep it in place. (Now he plans to replace the screws with permanent connections and place the structure in his garden as an outdoor sculpture. "What else is it good for these days?" he asks.) Tolerances had to be close; parts could be off by no more than a small fraction of an inch or degree in size or angle of placement. As components were added, the model quickly swelled in size, sometimes occupying the better part of a room. It also became more fragile and often started to sag. Small vibrations or careless handling could bring the whole edifice tumbling down, wrecking weeks of labor.

In one electronic swoop, molecular graphics has eliminated these perils. As Langridge quips, "It subtracts gravity from the equation." Still, Langridge, who did plenty of old-fashioned model building as a fledgling scientist under the distinguished x-ray crystallographer Maurice Wilkins at the University of London, looks back on the old technique somewhat nostalgically. "I always think that before they go into computer graphics, graduate students ought to be required as an exercise to build a protein," he says. "I'd give them all the coordinates (for the positions of the atoms) and the sticks and wire for the job. 'Here are the parts for your model,' I'd say. 'Go build your molecule.'"

THE REWARDS OF MODEL BUILDING

When everything slips into place for the model builder, the results can be aesthetically as well as scientifically rewarding. The classic example is that of James Watson and Francis Crick, who toyed for months with wire and sticks before uncovering the elegant double helical structure of DNA. Their model at once explained how a cell's genetic messages are encoded (along the sequences of nucleotides that form the steps of the staircase). It also suggested how DNA replicates (by separating into two complementary single-stranded chains, each of which pairs with matching nucleotides to form a new double helix).

DNA, however, is a model of simplicity compared to proteins, even though it directs their production. Among nature's largest molecules, proteins consist of long chains of amino acids crumpled into bewildering shapes—"like folded-up sausages," says Stroud. The birth of a protein begins when little factories inside the cell—the ribosomes—respond to instructions from the nucleus (brought to them by messenger RNA) and link up amino acids one after another, like beads on a string, into segments known as polypeptides. This linear chain of amino acids is the protein's primary structure. Almost immediately, the newly formed polypeptides twist into a jumble of structures.

Their most common component is the alpha helix, discovered more than 40 years ago by Linus Pauling in a brilliant bit of pioneer model making. Pauling used as his clues the known angles and lengths of the hydrogen bonds between amino acids; in this case, they dictated a tight coil. He also found another common configuration, the beta-pleated sheet, in which a strand of polypeptide folds into side-by-side ribbons that look like a wrinkled sheet. Another shape is the hairpin turn. Diverse shapes, or motifs, such as these make up the protein's secondary structure.

As these motifs interact, they are drawn together or apart by different forces. Some components, for example, are hydrophilic (water-attracting), others are oily and hydrophobic (water-repelling). These forces lead to strange contortions and coiling, sometimes even linkups of amino acids at opposite ends of the chain. The result is the protein's complex 3-D, or tertiary, structure—but not necessarily its final one. Often, several polypeptide chains combine into a single large protein molecule, as in hemoglobin, producing a quaternary structure.

Why proteins fold in particular ways remains profoundly puzzling. Clearly, the

What to do with the old stick and wire models of molecules? Some are exhibited in museums. Quiocho plans to turn his into an outdoor sculpture.

sequence of amino acids determines what the protein eventually will look like; this has been shown experimentally by unfolding a protein chemically and then watching it refold into its old configuration. But even when scientists know the line-up of amino acids in a certain protein—easily determined these days with automatic sequencing machines—they cannot predict its final 3-D form.

Special graphics programs that interpret the data obtained by x-ray crystallography can help researchers decipher these structures. For example, FRODO, developed by T. Alwyn Jones, a Welsh-born scientist at the University of Uppsala, Sweden (who whimsically named his program after a character in a Tolkien story), turns the x-ray data into brilliantly colored computer images that resemble a contour map, with an important difference: the information is shown three-dimensionally. Parts of the map seem to lie in the foreground. Others appear to be at a greater distance, like the vanishing parallel lines in a Renaissance painting. The illusion, visible without stereo glasses, becomes even more realistic as the model rotates.

To look at the protein from a different perspective and expose previously hidden aspects, one can rotate the plane of observation in any direction along any of the three axes of 3-D space. Then one can move along another axis and begin the rotation anew. Called rotations and translations, these maneuvers provide the same six

The illusion becomes even more realistic as the model rotates.

degrees of freedom of motion that a jet pilot has. Against this mobile 3D canvas, the model builder begins to search for the molecule's basic structure.

"You use your knowledge of the protein," says Arthur Olson, director of the molecular graphics laboratory at the Research Institute of Scripps Clinic in La Jolla, California. "You know how it is connected, by what types of bonds. You know what amino acids are involved. You know their sequence. Then you use all that information to bring all the parts of the protein together." Where once an initial fitting might have taken many months, with such sophisticated electronic help it can now be done in weeks.

After the basic structure is completed and refined—also with the help of computer graphics—scientists want to learn how the protein works. As Olson explains, "You need to understand the protein as a three-dimensional machine: how it catalyzes, how it recognizes other molecules, how it self-assembles. You have to understand how this complex array of amino acids functions."

This means looking more closely at the protein's properties, such as its surface, the electrostatic forces at work within the molecule, the location of its binding sites. For this purpose, Olson and his Scripps colleague Michael Conolly, who is now at New York University, designed a graphics program called GRANNY (because it interacts with a graphics software language named GRAMPS). GRANNY "knows about chemistry and biochemistry," says Olson. It can display the electrostatic forces around a particular atom or amino acid and even show whether a portion of the molecule is hydrophilic or hydrophobic, an important clue to how the protein might work.

Experiments using GRANNY can reveal how a particular protein goes about its business and help researchers to check some of their hunches. Recently biochemist Elizabeth Getzoff and her colleague, John Tainer, at Scripps used this program to explore how a particular enzyme defends the body against toxic ions. They found a cavity in the enzyme's structure that acted

as a deadly trap, not only capturing the toxic ions but also changing them into harmless molecules.

Pleased with their results, Getzoff and Tainer joined with Olson to make a film of how the enzyme did its job. The film—"Terms of Entrapment"—has been used to show off the power of molecular graphics to many scientific groups. Olson now plans to establish a graphics center at Scripps that will help researchers turn their molecular structures into animated form. "These are such complex three-dimensional objects that even showing one or two detailed images doesn't do them justice," he says. "By animating them, you can add significantly to the information presented."

Just down the hall from Getzoff and Tainer, another Scripps scientist is using computer graphics to tackle a bigger problem—the rules of protein folding. Instead of studying simulations of real molecules on the screen, mathematical chemist Jeffrey Skolnick creates very simplified models of his own. He begins with only two types of amino acids—one assumed to be hydrophobic, the other hydrophilic—which look no more complicated than beads. Both types are given the propensity to form helices or other motifs. Then Skolnick lets the computer take over.

Like atoms and molecules in real life, Skolnick's amino acids are kept in a state of constant agitation. Fluttering about on the screen, they seem at first to be moving chaotically. But after a few seconds, a hint of order appears. Some of the amino acids approach each other, others scoot apart as their hydrophilic and hydrophobic preferences come into play. A string of amino acids begins to form, its ends waving restlessly. Finally, the thrashing amino acids reach a minimum free energy state and settle into place. All motion on the screen stops. To Skolnick's delight, his amino acid beads have formed a hairpin turn, one of the basic shapes of proteins. Next they go on to form more complicated structures characteristic of real proteins.

Do such simulations even vaguely resemble real life? Is this how certain

(Continued on page 416)

ENTERING
THE WORLD
OF 3-D IMAGES

I t takes
a few minutes to get the hang of viewing 3-D images. Pick up the
stereo viewer that is supplied next to this page, unfold it, and
place it directly between these two pictures of New York. Then put your eyes over the
ses and relax. Since the pictures were shot from two different angles, as if from the left and right
eyes, your brain will fuse them into a single 3-D image if you give it enough time.
Slowly the depth of New York City's canyons will emerge.
You are now ready to see the biological structures shown on the next four pages in 3-D.
All were photographed off the computer graphics screens of researchers in structural biology.

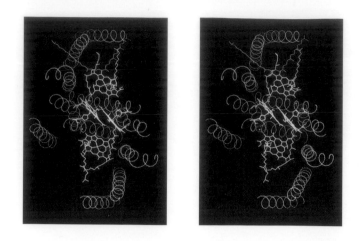

Nobel Prize–winning structure.
Eleven helices in blue, purple, or brown go right through the surface membrane of a purple bacterium in this model of a large protein—the photosynthetic reaction center—viewed from above the membrane. The first membrane-spanning protein whose 3-D structure was solved, it earned a Nobel Prize for Johann Diesenhofer (an HHMI investigator), Robert Huber, and Hartmut Michel. [From the laboratory of Johann Deisenhofer, University of Texas Southwestern Medical Center at Dallas.]

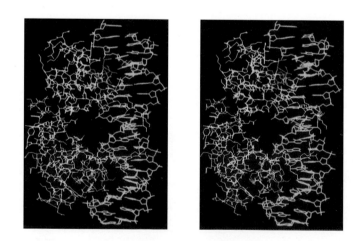

Turning genes on and off.
A repressor protein (C-shaped structure, left) binds to a control site on viral DNA (blue double helix, right). Fitting precisely into two of the DNA's major grooves, it turns off most of the viral genes. However, it also turns on the gene that codes for repressor protein, ensuring that more of it will be made. [From the laboratory of HHMI investigator Carl Pabo, The Johns Hopkins University School of Medicine.]

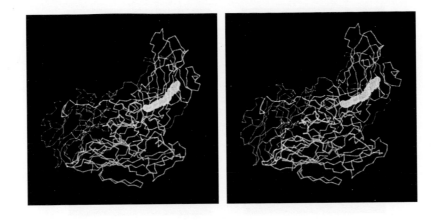

Drug jams virus.

An antiviral drug (yellow) binds to a rhinovirus, a common cold virus. The drug fits inside a pocket that never changes, though other parts of the virus frequently mutate. By stiffening the virus's structure, the drug prevents the virus from releasing its genetic material inside an infected cell. [From the laboratory of Michael Rossmann, Purdue University.]

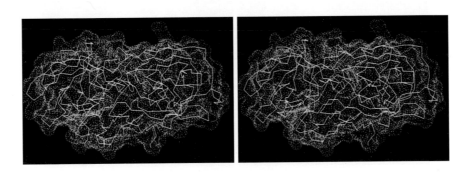

Treatment for AIDS?

A man-made inhibitor (red) is shown binding to, and inactivating, a protease (an enzyme that cuts through protein—in this case, a fungal enzyme called rhizopuspepsin). In AIDS, a similar protease (which could be inactivated in the same way) snips apart a large molecule of polyprotein and liberates some proteins that are essential to the viral life cycle. [From the laboratory of David Davies, National Institute of Diabetes and Digestive and Kidney Diseases.]

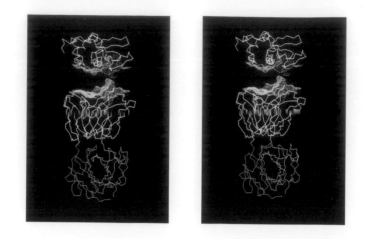

Antibody meets foreign substance.
Ready for a meeting, the bumps on the surface of an antibody (green) complement the
depressions on the surface of an antigen (pink), an antibody-stimulating substance—in this
case, the enzyme lysozyme. [From the laboratory of David Davies, National Institute of
Diabetes and Digestive and Kidney Diseases.]

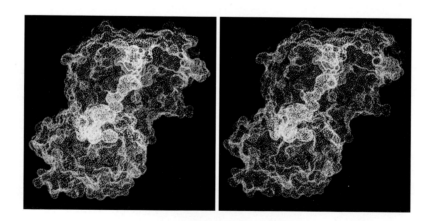

Discovering a tunnel.
A tunnel in which two chemical reactions can take place efficiently is revealed in this 3-D view
of tryptophan synthase, an enzyme of Salmonella bacteria. In this tunnel, the product of one
reaction is channeled directly to the site of the next reaction as in a coordinated machine,
without any loss to the outer environment. [From the laboratory of David Davies, National
Institute of Diabetes and Digestive and Kidney Diseases.]

More powerful computers and more flexible graphics systems are spurring the pioneers of computer graphics to become more ambitious. They would like to develop ways of showing complex biological phenomena—for instance, a protein "breathing," or opening up its structure as it binds with another molecule. Such goals would require the computer to integrate and interpret data from various sources and then present the results in easily readable form, using color and shadowing to deal with various kinds of information simultaneously. Here are two examples of new techniques from the laboratory of Arthur Olson, Research Institute of Scripps Clinic.

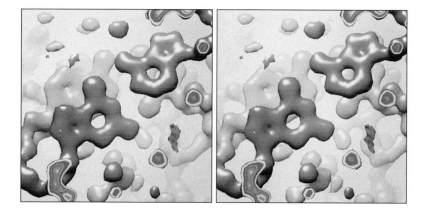

New way to show volume.

Here Olson bypasses the use of contour-like maps to show the electron densities calculated from x-ray diffraction data. Instead, he produces a solid-looking picture by a new technique called volume rendering. Two DNA nucleotides (green) that form one step in a DNA staircase are highlighted on the screen. Different intensities of color represent varying levels of electron density; other nucleotide pairs along the staircase can be seen in the background, in paler color.

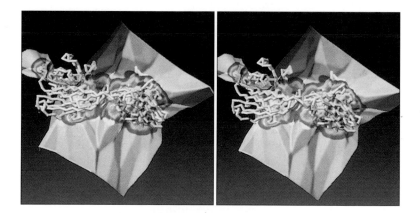

The interface between molecules.

The crumpled sheet represents the surface formed by the mid-points between two chains (greenish-blue tubes) of an antibody's Fab (antigen-binding) region. Only one of the chains can be seen in its entirety. The parts of the sheet that are closest to the two chains are colored red; those farther from the chains are shown in yellow, blue, and finally white. The image reveals a twisted interface between the two chains rather than a flat boundary.

(Continued from page 410)

sequences of amino acids come to prefer one shape over another? Skolnick admits that his computerized protein folding, involving a small number of amino acids, represents only "a fuzzy cartoon of the real thing." But he is confident that he will be able to build more complicated models with perhaps as many as several hundred amino acids and find out some of the answers.

Skolnick's work illustrates the increasing interdependence of the computer's image-making and number-crunching capabilities. This marriage of raw computational force and delicate graphics is even more evident when scientists use nuclear magnetic resonance spectroscopy (NMR)—and especially a form called two-dimensional NMR—to analyze protein structure. Unlike x-ray crystallography, NMR does not require crystallization of target molecules—an important advantage, since so few proteins can be crystallized.

NMR holds atoms in strong magnetic fields while subjecting them to pulses of high-frequency radio waves. In response, the atomic nuclei—protons—broadcast little signals of their own. These resonances, like fingerprints, are uniquely characteristic of their sources. They can be used not only to identify individual atoms in a molecule, but also to approximate their relative positions. However, it takes a tremendous amount of data processing to summon a molecule's structure on the screen, fit together its pieces, and fine-tune it.

Recently, Peter Wright, a specialist in NMR analysis, and his colleagues at Scripps used a combination of NMR and molecular graphics to solve the 3-D structure of the binding portion of a so-called zinc finger, one of a class of regulatory proteins containing zinc. Zinc fingers lock into nucleic acids, turning genes on and off, and may play important roles in disease.

The down side to NMR is that it now works only on smaller proteins—up to about a hundred amino acids. But Wright and other NMR spectroscopists are hopeful that it will extend its range. Says he,

"We're now at the stage where crystallographers were 20 years ago."

Another rapidly emerging field in which imaging systems play a role is protein dynamics. X-ray crystallography and even NMR give the impression that protein structures are essentially static—although in fact the molecules are constantly vibrating. When one looks at the computer screen, one does not see a snapshot of any individual molecule, but rather an "average" picture of many molecules spread over a particular time and space. By mathematical simulations, however, Martin Karplus of Harvard University and other protein dynamicists have been able to stop the fidgety proteins in their tracks.

Through their calculations of the internal movements of the atoms in a molecule, these researchers are able to create "stop-action" exposures of that molecule. The exposures are of astonishingly short duration—only one femtosecond, or a quadrillionth of a second (a quadrillion is 1 followed by 15 zeros). Even though these are only mathematical exercises, they are based on real protein data. By selectively displaying the computer's simulated stop-action pictures on the screen, the investigators can inspect a molecule at virtually any instant in its history.

They can also study a molecule's changes over time. Using a sequence of these stop-action images—perhaps one out of every hundred created by the computer—and photographing them individually off the screen, they are able to assemble what is in effect a time-lapse movie of the molecule in motion. "You won't see electrons or protons moving from one place to another, but you might be able to see the movement of an amino acid as the molecule rearranges itself," says Arthur Olson.

MORE THAN LUCK: "RATIONAL" DRUG DESIGN

The newest and perhaps most exciting application of molecular graphics is in the growing arena of computer-aided drug design. "Most of the drugs you're aware of in the marketplace were found serendipitously," says Irwin Kuntz, a physical chemist who heads a drug-design team at UCSF. "Penicillin mold flew in through the window; aspirin came from a willow tree."

To make the search for new drugs less haphazard, pharmacologists in the 1960s and 1970s turned to a more directed, "rational" approach to drug design: they applied their broad knowledge of the body's chemistry to create drugs that would block or enhance a specific molecular activity. A notable product of this method is the anti-ulcer drug cimetidine, which reduces gastric acidity by disrupting the histamine molecules that stimulate the production of digestive juices. Still, these pharmacologists usually did not know how their creations worked in every detail—where, for instance, a drug locked on to a particular molecule.

The study of molecular structures is now rapidly filling in these details, allowing Kuntz and his colleagues, as well as other scientists, to carry rational drug design one step further. Using the precise information unearthed by such tools as x-ray crystallography and NMR, they look for a binding site—either on a protein or on a stretch of DNA whose action they want to block—and try to plug it up with some complementary molecular fragment.

"The old way was just to sit and look at possible structures," Kuntz says. "You would use your chemical intuition and your biochemical knowledge to identify sensitive points and then go after them in ways that you thought might somehow attack the target structure." But this is extremely difficult, especially with complex objects like proteins, because it is hard for the mind to visualize whether one complicated 3-D object will mate neatly with another. "That's not just a shortcoming of chemists, it's a human failing," Kuntz says.

On the other hand, the computer is superbly equipped to make such geometric matchups because it can handle many variables at once. To take advantage of it, Kuntz and his colleagues have developed a program called DOCK, which pharmaceutical houses are already using as a drug discovery service. Drug designers who work with computer graphics generally display a protein on the screen and then make repeated attempts to "fly" a molecular fragment—the potential blocking agent—into the grooves or cavities that might be possible landing sites, carefully dodging any obstacles along the way. DOCK automates this time-consuming procedure and identifies the most likely sites. Then it tries out many different "keys" to fit these biochemical "locks."

The keys come out of an international chemical inventory known as the Cambridge Crystallographic Database, which contains structural information for some fifty thousand small compounds—peptides, steroids, benzene rings, and so on—crystallized over the past two decades. A scientist might spend years trying out all the database's keys manually on the screen, but DOCK can whip through the inventory in a few hours.

Even with such high-powered assistance, site-directed drug design, as it is called, requires the collaborative efforts of a number of different specialists. At UCSF, it involves chemists, biologists, and computer scientists who meet regularly to discuss their progress, often convening around the computer screen. "Like the old cracker barrel," says Kuntz, "it's our common meeting ground. It provides geometric images that everyone can understand." If the chemists run into difficulties when attempting the synthesis of a particular design, the other team members weigh in with their own ideas—switching to another target, say, where they may be able to "land" with an easier-to-make chemical structure.

After a possible compound has been tested successfully in vitro, it needs further analysis. "You would probably recrystallize the entire complex (the mated lock and key) to see if the drug and the original protein fit in the way you intended them to," says Kuntz. Several years may well elapse before the team actually produces a molecule for live testing.

These methods are so promising that many drug companies are betting on computer-aided design as their best weapon in the fight against viral infections, cancer, and other ills. ●

"Penicillin mold flew in through the window; aspirin came from a willow tree."

The HHMI beam line at Brookhaven has proved extremely productive. During just one year, 1998, it enabled researchers to solve dozens of recalcitrant structures, including 18 analyzed with Hendrickson's "MAD" method.

More Beam Lines for "MAD"

This success led to the opening of other synchrotron beam lines designed specifically to take advantage of the MAD method, both in the United States and abroad. Some of them, including several at the Advanced Photon Source at Argonne National Laboratory and the ESRF in Grenoble, France, have intrinsically greater power than Brookhaven's.

A second x-ray beam line now being started at Brookhaven under HHMI auspices will focus its beam even more effectively than the first and will allow researchers to work with even smaller crystals.

Crystals: Improving One's Luck

Meanwhile, the art of crystal growing has changed radically. One technique described by Stephen Burley, an HHMI investigator at The Rockefeller University, greatly improves the chances of obtaining crystals from either individual proteins or "complexes" of proteins bound to DNA. By using this "dynamic light scattering" technique, "we can test in five minutes whether or not a particular form of a protein is likely to crystallize. This saves a lot of time," Burley says.

Another advance came from the discovery that freezing will protect crystals against damage from the x-rays used in studying them. "The marriage of freezing crystals with the MAD techniques is a real dream," Hendrickson observes. "Even a poor crystal may work fine."

A Catalogue of Basic 3-D Structures

The crystallographers' hope of spotting some basic structural motifs in proteins has also come true. "Nature uses a limited set of protein motifs," says HHMI investigator John Kuriyan of The Rockefeller University. "When we determine a new protein structure now, about three-fourths of the time we recognize its units. So there is a feeling that we are converging on a more or less complete set of 3-D structures. When we reach a complete cataloguing of these structures, the connection between gene sequence and protein structure will become clearer."

New Tools for Larger Structures

Some of the structures being studied are so large that it has been almost impossible to solve them with x-ray crystallography. But there are new ways to examine them.

To decipher the shape of chromatin (the DNA-protein complex that condenses into chromosomes), for example, several scientists now use the scanning force microscope, which has an extremely fine and sensitive needle that can trace the outline of such complexes in samples of water under varying conditions.

Other scientists have chosen a different method to solve the structure of the ribosome, a large complex of some 50 proteins and 3 RNA chains that is the site of protein synthesis. They use cryoelectron microscopy, which involves an electron microscope, extremely low doses of radiation (so as not to damage the structure, which is frozen in a layer of ice), and new methods of 3-D visualization and structural analysis developed by Joachim Frank of the State University of New York at Albany.

With all these techniques, as well as improvements in nuclear magnetic resonance spectroscopy, which can show moment-to-moment changes in a protein's shape, "structures are now coming out fast and furious," says Hendrickson. The Protein Data Bank at Brookhaven, which contained about 500 protein structures at the time of our report, had accumulated more than 8,000 entries by early 1999 and had been transformed into a highly sophisticated system to keep up with the flood of new structures being solved not just each month, but each day.

Forty Champagne Corks

Every time Hendrickson's lab solves a new structure, the team celebrates by opening a bottle of champagne. Back in 1990, writer Natalie Angier reported that there were eight champagne corks on Hendrickson's office bookshelf. Having made a new count, Hendrickson is happy to provide an update. As of early 1999, he says, "we now have 40 corks."

Fusing With the Host

Don Wiley's work on the shape of HA, a protein that sticks out from the surface of the flu virus, has had a life of its own. His original structure showed how HA binds to human cells, but researchers knew that the HA molecule had an added role: After the virus docks with the cell, HA helps the virus to fuse with the host cell membrane. The scientists reasoned that in order to do this, HA would have to alter its shape.

In 1993, Peter Kim, an HHMI investigator at the Whitehead Institute for Biomedical Research in Cambridge, Massachusetts, predicted that HA's shape change would be dramatic—as if a "spring-loaded" coil had sprung out. A year later, Wiley and Skehel announced they had solved HA's second structure by x-ray crystallography. They showed that, like a jack-in-the-box, HA pops out and penetrates a cellular membrane, allowing the virus to dump its genes inside the cell.

This led directly to an understanding of how HIV, the virus that causes AIDS, penetrates its target cell. In 1997, Kim and Wiley independently solved the structure of gp41, the needle-like HIV protein that jabs into the cell membrane with the same type of spring-loaded mechanism as HA, giving the virus access to the cell's reproductive machinery.

The virus' first job, however, is to make contact with a cell. In 1998, Hendrickson and his colleagues succeeded in solving the structure of gp120, a protein that pokes out from the surface of HIV and grasps the target cell. They analyzed a complex of gp120 bound to CD4, a protein on the target cell. Hendrickson believes this structure may help researchers design new drugs or a vaccine against HIV.

Meanwhile, the protease inhibitors, which were just a dream at the time the structure of protease was solved, have become the most powerful drugs available against HIV. Used in combination with other medications, they have greatly extended the lives of many people with HIV.

A Cure For the Common Cold?

Though there is still no cure for the common cold, Michael Rossmann's discovery of the structure of rhinovirus 14 has led to an experimental drug. A derivative of the two compounds mentioned in our report, the drug is manufactured by ViroPharm of Philadelphia and has gone through three phases of FDA testing. It is already in use in selected hospitals against other viruses, such as coxsackie, that are in the same family but have more serious consequences. "As long as no side effects show up, it will be used against the common cold," says Rossmann. "This is a good example of how structure can be a help in the design of drugs."

New Uses For Computer Graphics

Faster and more powerful computers, better software, and more sophisticated methods of predicting how specific shapes will fit together have made computer graphics even more popular than before.

It also serves new purposes. With scores of small molecules emerging as candidates for drug development, researchers want to go beyond seeing how well these molecules fit into a particular protein. "There is a big effort now to select compounds not just for complementarity of shapes, but also for how good they would be as drugs," says Irwin Kuntz. "We can evaluate their drug-like character. For example, computer programs now exist that can predict a compound's oral availability"—that is, how well the compound would work if swallowed.

Arthur Olson has developed a computer program that uses what he calls "a co-evolution technique" to figure out how the HIV protease becomes resistant to drugs. "We model both the HIV protease and its inhibitors," he explains. "We study how different inhibitors fit into the protease molecule's active site and how the active site changes its shape to avoid them. The virus evolves—but so can the inhibitors. We plan to design inhibitors that will attack the viral mutants. It's a kind of cat-and-mouse game."

The new programs are written in a flexible manner and can be run on a variety of different computers. There is new richness, too, in available databases. The Cambridge Crystallographic Database now contains structural information for more than 190,000 small compounds, rather than 50,000 as before. The Available Chemical Directory of San Leandro, California, which gets even more traffic, not only shows the structure of 154,000 small molecules but also makes all of these compounds available for sale to researchers.

CONTRIBUTING WRITERS

Natalie Angier is a Pulitzer prize-winning science reporter for the *New York Times*. Her third book, *Woman: An Intimate Geography,* was published in 1999. She lives with her husband and daughter just outside of Washington, D.C.

Sandra Blakeslee is a science correspondent for the *New York Times* who writes about the life sciences; her specialty is neuroscience. She lives in Santa Fe, New Mexico.

Steven Dickman is a writer and consultant in Cambridge, Massachusetts. He is the founder of Creative Biotech Consulting, a firm that specializes in communications and strategy for new biotech companies.

Deborah Franklin, who writes about science and medicine, is a senior editor for *Health Magazine* in San Francisco.

Jeff Goldberg is the author of *Anatomy of a Scientific Discovery*. He lives in New Jersey, where he is a senior writer on HIV products at Integrated Communications.

Frederic Golden is a freelance science writer. He contributes regularly to *Time,* where he was formerly a science editor, as well as to other publications. He writes on a wide variety of topics in science, medicine, and technology.

Stephen S. Hall is the author of *A Commotion in the Blood: Life, Death and the Immune System*. He is a contributing writer for the *New York Times Magazine* and writes a column for *Technology Review* magazine.

Robert Kanigel, who teaches science writing in the Program in Writing and Humanistic Studies at the Massachusetts Institute of Technology, is the author of *Apprentice to Genius, The Man Who Knew Infinity,* and *The One Best Way.*

Beverly Merz is the editor of *Harvard Women's Health Watch,* a monthly consumer newsletter published by Harvard Medical School.

Geoffrey Montgomery is writing a book about the discovery of a half-billion-year-old cluster of genes—the *Hox* genes—that play a central role in patterning animal and human forms.

David Noonan is a feature writer for the Sunday *Daily News* in New York City. He writes often about science and medicine. He is the author of *Neuro-,* a book about neurological medicine, as well as the novel *Memoirs of a Caddy.*

Maya Pines is the editor and principal writer for this volume. She was the senior science editor at the Howard Hughes Medical Institute for more than 10 years, beginning in 1987. Before joining HHMI, she wrote four books in the fields of science and education and contributed more than 50 articles to national publications.

Peter Radetsky is a contributing editor of *Discover* and the content developer for BBH Exhibits, a producer of traveling museum exhibits about science, history, culture, and art. He teaches science communications at the University of California, Santa Cruz.

Harold M. Schmeck, Jr. was a science writer for the *New York Times* for many years. Since he retired in 1989, he has written two science books: *Turning Medical Progress into Print* and *The Hostage Brain* (with Bruce McEwen), as well as a novel. He regularly contributes articles to national magazines.

Larry Thompson, who wrote *Correcting the Code,* is the former science editor of the *Washington Post* health section. Since earning a master's in fine arts in film at American University, he has joined the Food and Drug Administration as a television producer.

Shawna Vogel is a freelance writer and author of *The Skinny on Fat: Our Obsession with Weight Control;* she also was the author of *Naked Earth: The New Geophysics*. She writes for various popular science and health magazines. She lives in Boston.

· ·

Editor . . . . .Maya Pines   *Production Manager*
Editorial Consultant . . . . .Gertrude Kelly   *and Permissions Editor* . . . . . .Kimberly A. Blanchard

Art Direction and Design: RCW Communication Design Inc.
Rodney C. Williams, Leon Lawrence III
William Jensen, Jessica Dawn Cohen

Editorial Director: Robert A. Potter

· ·

*We wish to thank the many scientists who so generously contributed their time and knowledge to help us
produce this book and the reports on which it is based.*

This publication is a project of the Institute's Office of Communications, Robert A. Potter, director, (301) 215-8855.
Published September 1999. Printed in the USA.

ILLUSTRATION AND PHOTOGRAPHY CREDITS

Page	Source
1	**Photograph**—Kay Chernush Human chromosomes-Photo Researchers, Inc. ©Biophoto Associates/Science Source **DNA**—Robert J. Driscoll, Michael G. Youngquist, and John D. Baldeschwieler, California Institute of Technology, *Nature* 346:294-296, 19 July 1990, ©1990 Macmillan Magazines Limited **Injected mouse eggs**—Ralph Brinster, University of Pennsylvania School of Veterinary Medicine, *Harvey Lectures, Series 80, 1986*, pp. 1-38 ©1986 Wiley-Liss, Inc., reprinted by permission of Wiley-Liss, a division of John Wiley & Sons, Inc. **Mapping of human muscle glycogen gene**—Peter Lichter et al., cover, *Science* 247:64-69, 5 January 1990, ©1990 American Association for the Advancement of Science **Double helix**—Rob Wood-Stansbury Ronsaville Wood, Inc.
2-3	Peter Lichter et al., cover, *Science* 247:64-69, 5 January 1990, ©1990 American Association for the Advancement of Science
6-7, 8-9, 10	Kay Chernush
13, 14	Karen Barnes-Stansbury Ronsaville Wood, Inc.
15	Burt Glinn/Magnum
16	Karen Barnes-Stansbury Ronsaville Wood, Inc.
18-19	Kay Chernush
22-23	Mary E. Challinor
25-32	**Foldout (Chromosomes)** **Double helix**—Karen Barnes-Stansbury Ronsaville Wood, Inc. **Human chromosome photos**—Photo Researchers, Inc. ©Biophoto Associates/Science Source **Genetic Diseases**—Design: Stansbury Ronsaville Wood, Inc.—Illustrations: Collagen protein, Richard J. Feldmann, National Institutes of Health; Antibody, Richard J. Feldmann, National Institutes of Health; Hemoglobin ©Leonard Lassin, FBPA, in collaboration with Hans Dijkman, Ph.D., and Waldo Feng, Mt. Sinai Medical Center, Department of Physiology and Biophysics; Molecular Modeling Facility, Peter Arnold, Inc. **Abnormal Chromosomes and Cells**—Peter Lichter et al., *Molecular Genetics of Chromosome 21 and Down Syndrome*, 1990 by David Patterson and Charles J. Epstein, pp. 68-79, ©1990 Wiley-Liss, Inc., reprinted by permission of Wiley-Liss, a division of John Wiley & Sons, Inc.
35	©1983 Steve Uzzell
36	©1986 Steve Uzzell
40-41	Antonio Baldini and David Ward, *Genomics* 9:770-774, April 1991 ©1991 Academic Press, Inc.
42-43	Ralph Brinster, University of Pennsylvania School of Veterinary Medicine, *Harvey Lectures, Series 80, 1986*, pp. 1-38 ©1986 Wiley-Liss, Inc., reprinted by permission of Wiley-Liss, a division of John Wiley and Sons, Inc.
45	Ralph Brinster, University of Pennsylvania School of Veterinary Medicine, cover *Nature*, 300:611-615, 16 December 1982,
46	©1982 Macmillan Magazines Limited.
50	Burt Glinn/Magnum
54-55	**Cell**—Lennart Nilsson, *A Child Is Born*, pp. 58-59, ©1990 Dell Publishing Company; **Photo**—Kay Chernush **Cross-section of fly's eye**—Nicholas E. Baker; **Axon pathways**—Nipam H. Patel and Corey S. Goodman **Worm**—Erik Jorgensen; **Fly embryo**—James Langeland, Stephen Paddock, and Sean Carroll
56-57	T. M. Jessell and M. A. Hynes
60-61	Lennart Nilsson, *A Child Is Born*, pp. 52-53, ©1990 Dell Publishing Company
62	E. B. Lewis
63	A. C. Spradling
64	Lennart Nilsson, *A Child Is Born*, p. 60, ©1990 Dell Publishing Company
67	Kay Chernush
70-71	After Romanes, 1901
72-73	Kay Chernush
74	E. B. Lewis
76	Stansbury Ronsaville Wood, Inc
79	Carl O. Pabo, *Cell*, 63:581, 2 November 1990 ©1990 Cell Press
81	**Egg chambers**—D. Montell and A. C. Spradling **Diagram**—*From Egg to Embryo*, second edition, figure 7.2, page 215 ©1991 Cambridge University Press (adapted by Mary E. Challinor)
82	W. Driever and C. Nüsslein-Volhard
83	James Langeland, Stephen Paddock, and Sean Carroll
84-85	Erik Jorgensen
86	Erik Jorgensen
88-89	**Lineage map**—John Sulston (adapted by Mary E. Challinor) **Worm**—H. Robert Horvitz (adapted by Stansbury Ronsaville Wood, Inc.)
93	Gerald M. Rubin, *Cell* 67:709, figure D, 15 November 1991 ©1991 Cell Press
94	Nicholas E. Baker
95	Gerald M. Rubin, *Cell* 67:708, figure B, 15 November 1991, ©1991 Cell Press
96	©Leonard Lessin/Peter Arnold, Inc.
97	Photo Researchers, Inc., ©Biophoto Associates/Science Source
100-101	A. Ramoa, G. Campbell, C. Shatz
103	Carol Donner
105	Nipam H. Patel and Corey S. Goodman **Illustration**—Sally J. Bensusen
106	*Molecular Biology of the Cell* by Alberts, Bray, Lewis, Raff, Roberts, and Watson, page 1129 ©1989 Garland Publishing Co (adapted by Mary E. Challinor)
108-109	**Yeast**—©Manfred Kage/Peter Arnold, Inc. **Sea urchin**—©Fred Bavendam/Peter Arnold, Inc. **Worm**—Erik Jorgensen **Zebrafish**—Carolina Biological Supply Company **Frog**—E. R. Degginger/Folio **Fruit fly**—E. B. Lewis **Mouse**—The Jackson Laboratory **Humans**—©Lori Adamski Peek/Tony Stone Worldwide
113	**Head**—A. Gevins and M. E. Smith, EEG Systems Laboratory, San Francisco **Retinal cells**—Richard H. Masland, Massachusetts General Hospital **Nerve cell**—Guang Yang and Richard H. Masland, Massachusetts General Hospital **Hair cell**—John Assad, Gordon Shepherd, and David Corey, Massachusetts General Hospital **Retina**—Photographic Laboratory of Bascom Palmer Eye Institute, University of Miami **Profile**—Michael Freeman **Sensory receptor cells**—Eade Creative Services, Inc./George Eade illustrator (**cone cell** adapted from *Scientific American* 256:42, 1987)
114-115	Guang Yang and Richard H. Masland, Massachusetts General Hospital
116-117	**Photo of head**—Michael Freeman **MRI scan of head**—John Belliveau, NMR Center, Massachusetts General Hospital
118-119	Mark R. Holmes © National Geographic Society
120	**Brain**—John Belliveau, NMR Center, Massachusetts General Hospital **Homunculus**—Reprinted with permission from Simon & Schuster, Inc., from *The Cerebral Cortex of Man* by Wilder Penfield and Theodore Rasmussen, ©1950 Macmillan Publishing Co.; renewed ©1978 Theodore Rasmussen
122	Eade Creative Services, Inc./George Eade illustrator (**rod and cone cells** adapted from *Scientific American* 256:42, 1987)
123	Eade Creative Services, Inc./George Eade illustrator (**rod cell** adapted from *Scientific American* 256:42 1987 **Transduction** adapted from figure 28-3, page 404, *Principles of Neural Science* by Eric R. Kandel, James H. Schwartz, and Thomas M. Jessell, ©1991 Elsevier Science Publishing Company, Inc.).
124-125	Kay Chernush
126	Instituto Cajal. CSIC. Madrid
127	Eade Creative Services, Inc./George Eade illustrator (**visual system** adapted from a drawing by Laszlo Kubinyi and Precision Graphics on page 9 of *Images of Mind* by Michael I. Posner and Marcus E. Raichle, Scientific American Library ©1994 **Retinal cells** adapted from a drawing by Carol Donner/Tom Cardamone Associates on page 37 of *Eye, Brain and Vision* by David H. Hubel, Scientific American Library, ©1988)
128	Richard H. Masland, Massachusetts General Hospital
129	Eade Creative Services, Inc./George Eade illustrator (adapted from figure 28-4, page 405, *Principles of Neural Science*, by Eric R. Kandel, James H. Schwartz, and Thomas M. Jessell ©1991 Elsevier Science Publishing Company, Inc.
130	Photographic Laboratory of Bascom Palmer Eye Institute, University of Miami
131	Ann H. Milam, University of Washington, Seattle
133	Fritz Goro, Life Magazine ©Time Inc.
135	Joe McNally/Sygma
138	Eade Creative Services, Inc./George Eade illustrator (adapted from a drawing by Carol Donner on page 60 of *Eye, Brain and Vision*, by David H. Hubel, Scientific American Library ©1988)
139	Simon LeVay, *Journal of Comparative Neurology*, vol. 159, 1975. Reprinted by permission of Wiley-Liss, Inc., a subsidiary of John Wiley & Sons, Inc.
141	Marcus Raichle, *Human Brain Mapping*, 1:1, 1993. Reprinted by permission of John Wiley & Sons, Inc.
142	Vincent Clark and James Haxby, Section on Functional Brain Imaging, Laboratory of Psychology and Psychopathology, NIMH, NIH
142-143	Kay Chernush
144-145	Kay Chernush
146-147	John Assad, Gordon Shepherd, and David Corey, Massachusetts General Hospital
149	Michael J. Mulroy and M. Charles Liberman, Massachusetts Eye and Ear Infirmary
150	Jennifer Jordan/RCW Communication Design Inc. (adapted from a sketch by James Hudspeth, HHMI, University of Texas Southwestern Medical Center)
152-153	Masakazu Konishi, California Institute of Technology
154-155	Kay Chernush
155	Eric Knudsen, Stanford University
156	Randall R. Benson and Thomas Talavage, NMR Center, Massachusetts General Hospital

158-159	Kay Chernush
161	Scott T. Barrows ©National Geographic Society
164	Kerry J. Ressler, Susan L. Sullivan, and Linda B. Buck, *Cell* 73:601, figure 4(5),7 May 1993 ©Cell Press
166	*The Lovers* by Pablo Picasso, National Gallery of Art/Chester Dale Collection
168	**Photo**—*The Journal of NIH Research*, vol. 6, January 1994. Reprinted with permission. **Illustration**—Eade Creative Services, Inc./George Eade illustrator (adapted from *The Journal of NIH Research* vol. 6, January 1994)
169	Urs Ribary, Rodolfo Llinás et al., from *Proceedings of the National Academy of Sciences* 90:3594, April 1993 ©National Academy of Sciences
170,170-171	A. Gevins and M. E. Smith, EEG Systems Laboratory, San Francisco
175	**Background photograph**—Lennart Nilsson, *The Body Victorious* ©Boehringer Ingelheim International GmbH **Photograph**—Kay Chernush **Stem cell symbol**—RCW Communication Design Inc. **Red blood cells**—Lennart Nilsson, *The Incredible Machine* ©Boehringer Ingelheim International GmbH **Fat-clogged artery**—Lennart Nilsson **Red cells flowing through capillaries**—Lennart Nilsson, *The Incredible Machine* ©Boehringer Ingelheim International GmbH **Macrophage eating a red blood cell**—Marcel Bessis, *Corpuscles*, chapter on Red Cell Death, ©1974 by Springer-Verlag **Line illustration**—Tim Phelps, CMI, FAMI **Fat cells**—Lennart Nilsson, *The Incredible Machine* ©Boehringer Ingelheim International GmbH
176-177	Stuart H. Orkin, *Nature Genetics*, 1:93, figure 1, 1 May 1992
180-181	Kay Chernush
182	Courtesy of David A. Williams
183	Lennart Nilsson, *The Body Victorious* ©Boehringer Ingelheim International GmbH
184-185	The Studio of Wood Ronsaville Harlin, Inc.
186	Tim Phelps, CMI, FAMI
187	David Hockley, National Institute for Biological Standards and Control, S. Mimms, UK
188	T. D. Allen, Christie Hospital, Manchester, UK, *Atlas of Blood Cells-Function and Pathology*, by D. Zucker-Franklin, M. F. Greaves, C. E. Grossi, A. M. Marmont, 2nd edition, Edi. Ermes, Milano, 1988
191	Rafael Espinosa, laboratory of Janet D. Rowley
192-193	Kay Chernush
194	G. P. Rodgers, C. T. Noguchi, A. N. Schechter, National Institutes of Health
195	M. E. Challinor
196	Kay Chernush
199	M. E. Challinor
201-208	**Foldout (Genetic Diseases)** **A Molecule to Breathe With**—The Studio of Wood Ronsaville Harlin, Inc. **Red Blood Cell photos**—Lennart Nilsson, *The Incredible Machine* ©Boehringer Ingelheim International GmbH **Spread**—**Design and illustrations** of stem cells and erythropoietin, of developing red cells, of human circulation-Tim Phelps, CMI, FAMI; **Photos** of developing red cells (4 photos)-David A. Williams; Red cell island and bone marrow photos-Y. Yawata, Kawasaki Medical School, Japan, from Sysmex 1993 calendar; Heart pumping red cells-Lennart Nilsson, *The Incredible Machine*, ©Boehringer Ingelheim International GmbH; macrophage engulfing red cell-Marcel Bessis, *Corpuscles*, chapter on Red Cell Death ©1974 Springer-Verlag **Photo of Bob Massie**—Kay Chernush
211	The Studio of Wood Ronsaville Harlin, Inc.
212	Lennart Nilsson, *The Body Victorious* ©Boehringer Ingelheim International GmbH
215	M. E. Challinor
216	Gernsheim Collection, Harry Ransom Humanities Research Center, University of Texas at Austin
217	**Illustration**—M. E. Challinor **Photo**—FPG International
218, 219	Kay Chernush
220-221	Lennart Nilsson, *The Body Victorious* ©Boehringer Ingelheim International GmbH
222-223	**(1)** Cheng-Teh Wang and Yuh-Ling Lin, Institute of Life Science, National Tsing-Hua University, Hsinchu, Taiwan, *Blood* 80:12, 15 December 1992 ©W. B. Saunders Co. **(2, 3, 4)** Lennart Nilsson, *Living with Haemophilia* ©Boehringer Ingelheim International GmbH
224-225	(Drawings) (top and middle) National Library of Medicine (bottom left and right) from *Bloodletting Instruments in the National Museum of History and Technology*, A. Davis and T. Appel, Smithsonian Institution, Washington, D.C., 1979 **Photograph**—Courtesy of Alexander Tulinsky
226	Lennart Nilsson, *The Incredible Machine* ©Boehringer Ingelheim International GmbH
227	Lennart Nilsson
228-231	**Helix illustration**—Stansbury Ronsaville Wood, Inc.
230	Kay Chernush
235	**Background**—Peter Miller, The Image Bank **Trypanosomes**—courtesy of Clara Alarcon, laboratory of John Donelson **Flies**—F. L. Lambrecht, Visuals Unlimited **Researcher photo**—Kay Chernush **Mycobacterium**—cover, *Tuberculosis: Pathogenesis, Protection, and Control*, edited by Barry Bloom ©American Society for Microbiology 1994 **Blood cells with malaria parasites**—©Boehringer Ingelheim International GmbH Photo-Lennart Nilsson/Bonnier Alba AB **HIV**—Eade Creative Services, Inc./George Eade illustrator **Schistosoma**—CNRI, Science Photo Library, Photo Researchers Inc. **VSG Structure**—courtesy of Michael Blum, Harvard University
236-237	Daniel Goldberg and David Russell, cover, *Parasitology Today*, 11:8 [122], August 1995
240	Courtesy of Valerie Hale, National Jewish Center for Immunology & Respiratory Medicine
241	Susie Fitzhugh
242-243	Kay Chernush
244	Courtesy of Valerie Hale, National Jewish Center for Immunology & Respiratory Medicine
244-245	Nadia Borowski, Zuma Images
246	**Upper**—Mary Challinor (adapted from *New England Journal of Medicine,* 330:1707) **Lower**—Brian Edlin et al., *New England Journal of Medicine*, 326:1519, 1992. Reprinted by permission.
247	Bettmann
248-249	Kay Chernush
251	Cover, *Tuberculosis: Pathogenesis, Protection and Control*, edited by Barry Bloom, ©American Society for Microbiology 1994
252	Courtesy of Ilan Rosenshine, University of British Columbia
255	Courtesy of James Sacchettini, *Science* 267:1639, 17 March 1995 ©1995 American Association for the Advancement of Science
256-257	Mary Challinor
258	Bob Thomas, Tony Stone
259-266	**Foldout (Microbes)** **Microbes That Threaten Us**—Design: Eade Creative Services, Inc./George Eade illustrator **Microbe photos**—**(1)** David M. Phillips, Visuals Unlimited **(2)** Arthur M. Siegelman, Visuals Unlimited **(3)** David Chase, CNRI, Phototake **(4)** ©Boehringer Ingelheim International GmbH Photo-Lennart Nilsson/Bonnier Alba AB **(5)** Gopal Nurti, CNRI, Phototake **(6)** Institut Pasteur, CNRI, Phototake **(7)** David M. Phillips, Visuals Unlimited **(8)** Custom Medical Stock Photo **(9)** Institut Pasteur, Phototake **(10)** ©Boehringer Ingelheim International GmbH Photo Lennart Nilsson/Bonnier Alba AB **(11)** Institut Pasteur, CNRI, Phototake **(12)** A. B. Dowsett, Science Photo Library, Photo Researchers Inc. **(13, 14)** David M. Phillips, Visuals Unlimited **(15)** ©Boehringer Ingelheim International GmbH Photo-Lennart Nilsson/Bonnier Alba AB **(16)** Stanley Flegler, Visuals Unlimited **(17, 18)** A. B. Dowsett, Science Photo Library, Photo Researchers Inc. **(19)** Alfred Pasieka, Science Photo Library, Photo Researchers Inc. **(20)** ©Boehringer Ingelheim International GmbH Photo-Lennart Nilsson/Bonnier Alba AB **(21)** David M. Phillips, Visuals Unlimited **(22)** Barry Dowsett, Science Photo Library, Photo Researchers, Inc. **(23, 24)** CNRI, Science Photo Library, Photo Researchers Inc. **(25)** ©Boehringer Ingelheim International GmbH Photo-Lennart Nilsson/Bonnier Alba AB **(26)** Photo Researchers Inc. **(27, 28)** Institut Pasteur, CNRI, Phototake **(29)** CNRI, Science Photo Library, Photo Researchers Inc. **Spread**—**Gymnast photos:** Kay Chernush; **Mosquito:** Courtesy of Robert Gwadz, NIAID, NIH; **Syringe:** Custom Medical Stock Photo, Inc.; **Bacterium on cell surface:** Courtesy of Brett Finlay, University of British Columbia; **Cluster of staphylococcus:** David M. Phillips, Visuals Unlimited; **AIDS virus:** Eade Creative Services, Inc./George Eade illustrator; **Macrophage:** ©Boehringer Ingelheim International GmbH Photo-Lennart Nilsson/Bonnier Alba AB; **Antibody molecule:** courtesy of Arthur Olson ©The Scripps Research Institute, 1985; **T cell:** ©Boehringer Ingelheim International GmbH Photo-Lennart Nilsson/Bonnier Alba AB; **Staphylococcus before and after antibiotic:** Courtesy of F. O'Grady and D. Greenwood, *Science*, 163:1076-1078, 1969 ©1969 by the American Association for Advancement of Science; **Polio virus:** ©Boehringer Ingelheim International GmbH Photo-Lennart Nilsson/Bonnier Alba AB; **Trypanosomes:** Klaus Esser, Smith, Kline & French, *Science* 229:190-193, 1985 ©1985 American Association for the Advancement of Science; **Bacteria sharing resistance:** Dennis Kunkel, Phototake.

268-276	**Flies**—F. L. Lambrecht, Visuals Unlimited
269	Kay Chernush
270	RCW Communication Design Inc. (adapted from *Scientific American*, February 1985, p. 48)
272-273	Thomas C. Boyden, Visuals Unlimited
274	James Dennis, CNRI, Phototake
275	Courtesy of Michael Blum, Harvard University
277	Courtesy of Robert Gwadz, NIAID, NIH
278-279	©Boehringer Ingelheim International GmbH Photo-Lennart Nilsson/ Bonnier Alba AB
280-287	**Virus**—©Boehringer Ingelheim International GmbH Photo-Lennart Nilsson/Bonnier Alba AB **Border**—Eade Creative Services, Inc./George Eade illustrator
282-283	**Immune system cell**—Eade Creative Services, Inc./George Eade illustrator
286	Courtesy of Joachim Jäger, Yale University
288	**Background**—Hans Gelderbloom, Visuals Unlimited **Smallpox victim**—courtesy of J. Wickett, WHO
289	**Bacterium diagram**—Robert Fleischman et al., *Science* 269:507, 28 July 1995 ©1995 by the American Association for the Advancement of Science
293	**Background**—Kay Chernush **Illustration**—Eade Creative Services, Inc./George Eade illustrator **Tick**—©1989 Herb Charles Ohlmeyer/Fran Heyl Associates **3-D structure**—J. Choi and J. Clardy, *Science* 273:241, fig. 2c, 12 July 1996 ©1996 American Association for the Advancement of Science **Spirochetes**—Aruvindah de Silva et al., *Journal of Experimental Medicine*, 183:273, 1 January 1966 ©The Rockefeller University Press **MHC molecule**—P. J. Bjorkman, M. A. Saper, B. Samraoui, W. S. Bennett, J. L. Strominger, and D. C. Wiley, *Nature* 329:509, 8 October 1987 ©Macmillan Magazines Limited **Macrophage**—©Boehringer Ingelheim International GmbH Photo-Lennart Nilsson, Bonnier Alba AB
294-295	J. Choi and J. Clardy, *Science* 273:241, fig. 2c, 12 July 1996 ©1996 American Association for the Advancement of Science
298	©Boehringer Ingelheim International GmbH Photo-Lennart Nilsson/Bonnier Alba AB
300	Keith Brofsky
302	Eade Creative Services, Inc./George Eade illustrator
303	William K. Geiger
304, 307	Eade Creative Services, Inc./George Eade illustrator
308-309	J Young and Z. Cohn, *Scientific American*, 258:38-39, January 1988
311	©Boehringer Ingelheim International GmbH Photo-Lennart Nilsson/Bonnier Alba AB
312, 314, 315	Eade Creative Services, Inc./George Eade illustrator
317	Kay Chernush
318	Eade Creative Services, Inc./George Eade illustrator
321	David Powers
322-323, 325	Kay Chernush
326	P. J. Bjorkman, M. A. Saper, B. Samraoui, W. S. Bennett, J. L. Strominger, and D. C. Wiley, *Nature* 329:509, fig. 6, 8 October 1987 ©Macmillan Magazines Limited
330	Eade Creative Services, Inc./George Eade illustrator
332	©Edmonson/Gamma Liaison
338-339	Kay Chernush
341	L. Chen, M. Glover, P. Hogan, A Rao, and S. Harrison, *Nature* 392:44, fig. 3, 5 March 1998 ©Macmillan Magazines Limited
342	J. Choi and J. Clardy, *Science* 273:183, 12 July 1996 ©1996 American Association for the Advancement of Science
345	Eade Creative Services, Inc./George Eade illustrator (source: Gerald Crabtree)
347	RCW Communication Design Inc. (source: Jamie Simon, The Salk Institute)
348-349	©1989 Herb Charles Ohlmeyer/Fran Heyl Associates
350	Stuart Franklin/Magnum
351	Eade Creative Services, Inc./George Eade illustrator (source: *Lyme Disease* by Alan G. Barbour, M.D. ©1966 The Johns Hopkins University Press) **Spirochete**—Russell Johnson, University of Minnesota **Human**—©1997 Smith/Picture Perfect
352	©1989 Bernard Furnival/Fran Heyl Associates
352-353	Russell Johnson, University of Minnesota
353	Timothy Lepore
355	Kay Chernush
356	Aruvindah de Silva et al., *Journal of Experimental Medicine*, 183:273, 1 January 1996 ©The Rockefeller University Press
358, 362-363	Kay Chernush
364-365	©Science VU-NLM Visuals Unlimited
366-369	**Foldout (Vaccines)** Eade Creative Services. Inc./George Eade illustrator (p. 76, source, *New England Journal of Medicine*) 369 **Illustration**—Eade Creative Services. Inc./George Eade illustrator **Upper photo**—Charles Arntzen **Lower photo**—Kay Chernush
372	**Photograph**—Kay Chernush **Structure**—Johann Diesenhofer and Hartmut Michel ©1989 The Nobel Foundation **Crystal**—Alexander McPherson **Painting**—Rob Wood, Stansbury Ronsaville Wood, Inc.
374-375	Mark Rould, John Perona, Paul Vogt, and Thomas Steitz, *Science* 246:1135-1142, 1 December 1989 ©1989 American Association for the Advancement of Science
378-389	Kay Chernush
380-381	Kay Chernush
382	Alexander McPherson
386-387	Alexander McPherson
389	Yvonne Gensurowski, Stansbury Ronsaville Wood, Inc.
391	Kay Chernush
393	Andrzej Joachimiak
394	Alexander McPherson
396-397, 399	Rob Wood, Stansbury Ronsaville Wood, Inc.
400, 401	William Weis, Jerry Brown, Stephen Cusack, James Paulson, John Skehel, and Don Wiley
403	Dan Bloch and Arthur Olson, ©1989 Research Institute of Scripps Clinic
406-407, 409	Kay Chernush
411	David Burder/3-D Images
412	**Top**—Johann Deisenhofer and Hartmut Michel, ©1989 The Nobel Foundation **Bottom**—Steven Jordan and Carl Pabo, *Science* 242:893-899, 11 November 1988 ©1988 American Association for the Advancement of Science
413	**Top**—Thomas Smith, Marcia Kremer, Ming Luo, Gerrit Vriend, Edward Arnold, Greg Kamer, and Michael Rossmann **Bottom**—K. Suguna, Eduardo Padlan, Clark Smith, William Carlson, and David Davies
414	**Top**—David Davies, Steven Sheriff, Eduardo Padlan, Enid Silverton, Gerson Cohen, and Sandra Smith-Gill **Bottom**—Craig Hyde, Ashrafudin Ahmed, Eduardo Padlan, Edith Miles, and David Davies
415	David Goodsell and Arthur Olson, ©1989 Research Institute of Scripps Clinic

Abramowitz, Randy, Brookhaven National Laboratory, Upton, NY

Adelson, Edward, Massachusetts Institute of Technology, Cambridge, MA

Albright, Thomas D., HHMI, Salk Institute, San Diego, CA

Allman, John, California Institute of Technology, Pasadena, CA

Anderson, W. French, University of Southern California School of Medicine, Los Angeles, CA

Arntzen, Charles, Boyce Thompson Institute for Plant Research, Ithaca, NY

Atweh, George, Mount Sinai School of Medicine, New York, NY

Axel, Richard, HHMI, Columbia University College of Physicians and Surgeons, New York, NY

Barbour, Alan, University of Texas Health Sciences Center, San Antonio, TX

Barrell, Bart, Sanger Center, Cambridge, England, UK

Barthold, Stephen W., Yale University, New Haven, CT

Baum, Charles, Searle, Skokie, IL

Baylor, Denis, Stanford University, Palo Alto, CA

Beaudet, Arthur, Baylor College of Medicine, Houston, TX

Behringer, Richard, University of Texas M.D. Anderson Cancer Center, Houston, TX

Belliveau, John, Massachusetts General Hospital, Boston, MA

Belshe, Robert, St. Louis University School of Medicine, Saint Louis, MO

Berg, Paul, Stanford University, Palo Alto, CA

Biddison, William E., NINDS, National Institutes of Health, Bethesda, MD

Billings, Paul, VHA Heart of Texas Veterans Health Care System, Grand Prairie, TX

Bjorkman, Pamela J., HHMI, California Institute of Technology, Pasadena, CA

Blaese, Michael, NIHGR, National Institutes of Health, and Kimeragen, Inc., Newton, PA

Blanchard, John, Albert Einstein College of Medicine of Yeshiva University, Bronx, NY

Bloom, Barry, Harvard School of Public Health, Boston, MA

Bosma, Gayle, Fox Chase Cancer Center, Philadelphia, PA

Bosma, Melvin, Fox Chase Cancer Center, Philadelphia, PA

Botstein, David, Stanford University, Palo Alto, CA

Boyse, Edward, University of Arizona, Tucson, AZ

Breakefield, Xandra, Massachusetts General Hospital, Boston, MA

Brenner, Sydney, Molecular Sciences Institute, Berkeley, CA

Brinster, Ralph, University of Pennsylvania, Philadelphia, PA

Brown, Donald, Carnegie Institution of Washington, Baltimore, MD

Brown, Michael, University of Texas Southwestern Medical Center at Dallas, Dallas, TX

Brown, Patrick O., HHMI, Stanford University School of Medicine, Palo Alto, CA

Broxmeyer, Hal, Indiana University School of Medicine, Indianapolis, IN

Brünger, Axel T., HHMI, Yale University, New Haven, CT

Buchwald, Manuel, University of Toronto, Toronto, Ontario

Buck, Linda B., HHMI, Harvard Medical School, Boston, MA

Burley, Stephen K., HHMI, The Rockefeller University, New York, NY

Capecchi, Mario R., HHMI, University of Utah School of Medicine, Salt Lake City, UT

Caskey, Thomas, Merck Research Labs, West Point, PA

Chien, Yueh-hsiu, Stanford University, Palo Alto, CA

Chisaka, Osamu, Kyoto University, Kyoto, Japan

Cline, Thomas, University of California–Berkeley, Berkeley, CA

Clipstone, Neil, Northwestern University, Evanston, IL

Cole, Stewart, Pasteur Institute, Paris, France

Collins, Francis, NCHGR, National Institutes of Health, Bethesda, MD

Conneally, Michael, Indiana University, Bloomington, IN

Conolly, Michael, Computer Consultant, Menlo Park, CA

Cook, David, Brookhaven National Laboratory, Upton, NY

Cooper, Max D., HHMI, University of Alabama at Birmingham, Birmingham, AL

Corey, David R., HHMI, University of Texas Southwestern Medical Center at Dallas, Dallas, TX

Cowman, Alan, Walter & Eliza Hall Institute of Medical Research, Melbourne, Australia

Crabtree, Gerald R., HHMI, Stanford University School of Medicine, Palo Alto, CA

Cresswell, Peter, HHMI, Yale University School of Medicine, New Haven, CT

Crick, Sir Francis, Salk Institute, San Diego, CA

Cross, George, The Rockefeller University, New York, NY

Cullen, Bryan R., HHMI, Duke University Medical Center, Durham, NC

Davidson, Eric, California Institute of Technology, Pasadena, CA

Davie, Earl, University of Washington, Seattle, WA

Davies, David, NIDDK, National Institutes of Health, Bethesda, MD

Davis, Mark M., HHMI, Stanford University School of Medicine, Palo Alto, CA

Davis, Ronald, Stanford University, Palo Alto, CA

Davison, Owen, Retired

Deisenhofer, Johann, HHMI, University of Texas Southwestern Medical Center at Dallas, Dallas, TX

DeRobertis, Eddy, HHMI, University of California, Los Angeles, CA

Desplan, Claude, HHMI, The Rockefeller University, New York, NY

Desrosiers, Ronald, New England Regional Primate Research Center, Southborough, MA

Dexter, Michael, Paterson Institute, Manchester, UK

Dieffenbach, Carl, NIAID, National Institutes of Health, Bethesda, MD

Doherty, Peter, St. Jude Children's Research Hospital, Memphis, TN

Donelson, John, University of Iowa, Iowa City, IA

Dozy, Andrée, Retired

Dryja, Thaddeus, Massachusetts Eye & Ear Infirmary, Harvard Medical School, Boston, MA

Dulac, Catherine, HHMI, Harvard University, Cambridge, MA

Dulbecco, Renato, Salk Institute, San Diego, CA

Duyk, Geoffrey, Exelixis Pharmaceuticals, Inc., South San Francisco, CA

Ebert, James, Marine Biology Laboratory, Woods Hole, MA

Edelman, Gerald, Scripps Research Institute, La Jolla, CA

Egeland, Janice, University of Miami, Coral Gables, FL

Esmon, Charles T., HHMI, Oklahoma Medical Research Foundation, Oklahoma City, OK

Evans, Martin, Cardiff University, Cardiff, Wales, UK

Faller, Douglas, Boston University School of Medicine, Boston, MA

Fan, Qing R., Harvard University, Cambridge, MA

Fauci, Anthony, NIAID, National Institutes of Health, Bethesda, MD

Feinstein, Paul, The Rockefeller University, New York, NY

Feldman, Dan, NINDS, National Institutes of Health, Bethesda, MD

Fikrig, Erol, Yale University, New Haven, CT

Finlay, Brett, University of British Columbia, Vancouver, Canada

Flavell, Richard A., HHMI, Yale University School of Medicine, New Haven, CT

Fletterick, Robert, University of California–San Francisco, San Francisco, CA

Fox, Robert, Yale University, New Haven, CT

Frank, Joachim, HHMI, Health Research, Inc., at Wadsworth Center, Albany, NY

Garbers, David, HHMI, University of Texas Southwestern Medical Center at Dallas, Dallas, TX

Garboczi, David, NIAID, National Institutes of Health, Rockville, MD

Gearhart, John, The Johns Hopkins University, Baltimore, MD

Gerhard, Daniela, Washington University School of Medicine, Saint Louis, MO

Getchell, Marilyn, Sanders Brown Center on Aging, University of Kentucky, Lexington, KY

Getchell, Thom, Sanders Brown Center on Aging, University of Kentucky, Lexington, KY

Getzoff, Elizabeth, Scripps Research Institute, La Jolla, CA

Gevins, Alan, EEG Systems Lab, San Francisco, CA

Ghosh, Partho, University of California–San Diego, La Jolla, CA

Gilbert, Walter, Harvard University, Cambridge, MA

Ginsburg, David, HHMI, University of Michigan Medical School, Ann Arbor, MI

Gitschier, Jane M., HHMI, University of California, San Francisco, San Francisco, CA

Goldberg, Daniel E., HHMI, Washington University School of Medicine, Saint Louis, MO

Goldstein, Joseph, University of Texas Southwestern Medical Center at Dallas, Dallas, TX

Goodman, Corey, HHMI, University of California–Berkeley, Berkeley, CA

Greenberg, Philip, University of Washington–Seattle, Seattle, WA

Grey, Howard, La Jolla Institute for Allergy & Immunology, San Diego, CA

Gusella, James, Massachusetts General Hospital, Boston, MA

Hajduk, Stephen, University of Alabama, Tuscaloosa, AL

Hammer, Robert, University of Texas Southwestern Medical Center at Dallas, Dallas, TX

Handschumacher, Robert, Yale University, New Haven, CT

Harrison, Stephen C., HHMI, Harvard University, Cambridge, MA

Haxby, James, NIMH, National Institutes of Health, Bethesda, MD

Heimfeld, Shelly, Fred Hutchinson Cancer Research Center, Seattle, WA

Hendrickson, Wayne A., HHMI, Columbia University College of Physicians and Surgeons, New York, NY

Herskowitz, Ira, University of California, San Francisco

Ho, David, Aaron Diamond AIDS Research Center, The Rockefeller University, New York, NY

Hogle, James, Harvard University, Cambridge, MA

Hood, Leroy, University of Washington, Seattle, WA

Horvitz, Robert, HHMI, Massachusetts Institute of Technology, Cambridge, MA

Housman, David, Massachusetts Institute of Technology, Cambridge, MA

Howard, Russell, Maxygen, Redwood City, CA

Hozumi, Nobumichi, Tokyo Science University, Tokyo, Japan

Hubel, David, Harvard University, Cambridge, MA

Hudspeth, A. James, HHMI, The Rockefeller University, New York, NY

Isberg, Ralph, Tufts University School of Medicine, Boston, MA

Iseman, Michael, National Jewish Medical and Research Center, Denver, CO

Jacobs, William R. Jr., HHMI, Albert Einstein College of Medicine of Yeshiva University, Bronx, NY

Jafek, Bruce, University of Colorado, Denver, CO

Jan, Yuh Nung, HHMI, University of California–San Francisco, San Francisco, CA

Janeway, Charles A. Jr., HHMI, Yale University School of Medicine, New Haven, CT

Jardetzky, Theodore, Northwestern University, Evanston, IL

Jessell, Thomas, HHMI, Columbia University, New York, NY

Johnston, Stephen, University of Texas Southwestern Medical Center at Dallas, Dallas, TX

Jones, T. Alwyn, Univ of Uppsala, Uppsala, Sweden

Kaas, Jon, Vanderbilt University Medical Center, Nashville, TN

Kan, Yuet Wai, HHMI, University of California–San Francisco, San Francisco, CA

Kantor, Fred, Yale University, New Haven, CT

Kappler, John, National Jewish Medical and Research Center, Denver, CO

Karplus, Martin, Harvard University, Cambridge, MA

Kauer, John, Tufts University School of Medicine, Boston, MA

Kaufman, Thomas, HHMI, Indiana University, Bloomington, IN

Kim, Peter S., HHMI, Massachusetts Institute of Technology, Cambridge, MA

Kirchhoff, Louis, University of Iowa, Iowa City, IA

Knudsen, Eric, Stanford University, Palo Alto, CA

Konishi, Masakazu ("Mark"), California Institute of Technology, Pasadena, CA

Koshland, Daniel E., University of California–Berkeley, Berkeley, CA

Kunkel, Louis M., HHMI, The Children's Hospital, Boston, MA

Kuntz, Irwin, University of California–San Francisco, San Francisco, CA

Kuriyan, John, HHMI, The Rockefeller University, New York, NY

Lal, Altaf, Centers for Disease Control and Prevention, Atlanta, GA

Lalouel, Jean-Marc, HHMI, University of Utah School of Medicine, Salt Lake City, UT

Lamb, Trevor, University of Cambridge, Cambridge, UK

Lander, Eric, Whitehead Institute for Biomedical Research, Massachusetts Institute of Technology, Cambridge, MA

Langridge, Robert, Retired

Laughon, Allen, University of Wisconsin–Madison, Madison, WI

Lawler, Ann, Johns Hopkins University, Baltimore, MD

Leder, Philip, HHMI, Harvard Medical School, Boston, MA

Lehmann, Ruth, HHMI, New York University, New York, NY

Leiter, Edward, Jackson Laboratory, Bar Harbor, ME

Leonard, Warren, NHLBI, National Institutes of Health, Bethesda MD

Lepore, Timothy, Nantucket Hospital, Nantucket, MA

Leppert, Mark, University of Utah School of Medicine, Salt Lake City, UT

Lewis, Edward, California Institute of Technology, Pasadena, CA

Liu, Jun, Massachusetts Institute of Technology, Cambridge, MA

Liu, Margaret, Chiron Vaccines Research, Chiron Corp., Emeryville, CA

Llinás, Rodolfo, New York University Medical Center, New York, NY

MacNichol, Edward, Boston University, Boston, MA

Macrides, Foteos, retired

Majerus, Philip, Washington University, Saint Louis, MO

Mak, Tak, Ontario Cancer Institute, Toronto, Ontario

Malawista, Stephen, Yale University, New Haven, CT

Markowitz, Martin, Aaron Diamond AIDS Research Center, New York, NY

Marrack, Philippa, HHMI, National Jewish Medical and Research Center, Denver, CO

Martin, Gail, University of California–San Francisco, San Francisco, CA

Martin, Malcolm, NIAID, National Institutes of Health, Bethesda, MD

Massie, Robert, Ceres, Boston, MA

Matthews, Brian W., HHMI, University of Oregon, Eugene, OR

Maxam, Allan, Biolinguistics Institute, Cambridge, MA

McClintock, Martha, University of Chicago, Chicago, IL

McCune, Michael, University of California–San Francisco, San Francisco, CA

McGinnis, William, University of California–San Diego, La Jolla, CA

McKinlay, Marc, ViroPharma, Inc., Exton, PA

McKusick, Victor, The Johns Hopkins University, Baltimore, MD

McPherson, Alexander, University of California–Irvine, Irvine, CA

Meredith, Michael, Florida State University, Tallahassee, FL

Miller, Louis, NIAID, National Institutes of Health, Bethesda, MD

Mitchison, Tim, Harvard Medical School, Boston, MA

Mombaerts, Peter, The Rockefeller University, New York, NY

Moran, David, Human Pheromone Sciences, Inc., Fremont, CA

Movshon, J. Anthony, HHMI, New York University Medical Center, New York, NY

Mowbray, Sherry, University of Uppsala, Uppsala, Sweden

Muller, James, Massachusetts General Hospital, Boston, MA

Nathans, Jeremy, HHMI, The Johns Hopkins University School of Medicine, Baltimore, MD

Navia, Manuel, The Althexis Co., Inc., Boston, MA

Nesto, Richard, Deaconess Hospital, Boston, MA

Newsome, William T., HHMI, Stanford University School of Medicine, Palo Alto, CA

Nüsslein-Volhard, Christiane, Max Planck Institute, Tubingen, Germany

O'Farrell, Patrick, University of California, San Francisco, CA

Olson, Arthur, Scripps Research Institute, La Jolla, CA

Olson, Maynard, University of Washington, Seattle, WA

Orkin, Stuart H., HHMI, Children's Hospital, Boston, MA

Owen, Whyte, Mayo Clinic, Rochester, NY

Pabo, Carl O., HHMI, Massachusetts Institute of Technology, Cambridge, MA

Page, David, HHMI, Massachusetts Institute of Technology, Cambridge, MA

Palmiter, Richard D., HHMI, University of Washington School of Medicine, Seattle, WA

Pantaleo, Guiseppe, Centre Hospitalier Vadois, Lausanne, Switzerland

Parham, Peter, Stanford University, Palo Alto, CA

Pastan, Ira, NCI, National Institutes of Health, Bethesda, MD

Perrine, Susan, Boston University Medical Cancer Center, Boston, MA

Perutz, Max, Retired

Pevsner, Jonathan, Johns Hopkins University, Baltimore, MD

Pinard, Jeff, Grand Rapids, MI

Platt, Orah, Children's Hospital, Boston, MA

Prasad, Vinayaka, Albert Einstein College of Medicine, Bronx, NY

Priess, James, HHMI, Fred Hutchinson Cancer Research Center, Seattle, WA

Quiocho, Florante A., HHMI, Baylor College of Medicine, Houston, TX

Ramachandran, Vilayanur, University of California–San Diego, La Jolla, CA

Ratnoff, Oscar, Case Western Reserve University, Cleveland, OH

Ready, Donald, Purdue University, West Lafayette, IN

Reed, Randall R., HHMI, The Johns Hopkins University School of Medicine, Baltimore, MD

Relman, David, Stanford University, Palo Alto, CA

Rickles, Fred, George Washington University Medical Center, Washington, DC

Riddell, Stanley, Fred Hutchinson Cancer Research Center, Seattle, WA

Riordan, John, Mayo Clinic, Scottsdale, AZ

Rosenberg, Henry, Thomas Jefferson University Hospital, Philadelphia, PA

Rossmann, Michael, Purdue University, West Lafayette, IN

Rowley, Janet, University of Chicago, Chicago, IL

Rubin, Gerald, HHMI, University of California–Berkeley, Berkeley, CA
Ruggeri, Zaverio, Scripps Research Institute, La Jolla, CA
Ryba, Nicholas, NIDR, National Institutes of Health, Bethesda, MD
Sacchettini, James, Texas A&M University, College Station, TX
Sadler, J. Evan, HHMI, Washington University School of Medicine, Saint Louis, MO
Sanger, Frederick, Retired
Schechter, Alan, NIDDK, National Institutes of Health, Bethesda, MD
Schoolnik, Gary, Stanford University, Palo Alto, CA
Schreiber, Stuart L., HHMI, Harvard University, Cambridge, MA
Schwan, Tom, Rocky Mountain Laboratory, Hamilton, MT
Scott, Matthew, HHMI, Stanford University, Palo Alto, CA
Sejnowski, Terrence J., HHMI, Salk Institute, San Diego, CA
Shalala, Donna, U.S. Department of Health, Education, and Welfare, Washington, DC
Sharp, Philip, Massachusetts Institute of Technology, Cambridge, MA
Shatz, Carla, HHMI, University of California–Berkeley, Berkeley, CA
Shaw, George M., HHMI, University of Alabama at Birmingham School of Medicine, Birmingham, AL
Shepherd, Gordon, Yale University, New Haven, CT
Shimonkevitz, R., Rocky Mountain Multiple Sclerosis Center, Englewood, CO
Sigler, Paul B., HHMI, Yale University School of Medicine, New Haven, CT
Simpson, Larry, HHMI, University of California–Los Angeles School of Medicine, Los Angeles, CA
Sinsheimer, Robert, University of California–Santa Barbara, Santa Barbara, CA
Skehel, John, National Institute for Medical Research, London, UK
Skolnick, Jeffrey, Scripps Research Institute, La Jolla, CA
Skolnick, Mark, Myriad Genetics, Inc., Salt Lake City, UT
Small, Rochelle, NIDCD, National Institutes of Health, Bethesda, MD
Smith, Brian, Screaming With Pleasure Productions, East Hampton, MA
Smith, Hamilton, Johns Hopkins University, Baltimore, MD
Smithies, Oliver, University of North Carolina, Chapel Hill, NC
Snyder, Solomon, The Johns Hopkins University, Baltimore, MD
Spencer, David, Baylor College of Medicine, Houston, TX
Spielman, Andrew, Harvard School of Public Health, Boston, MA
Spradling, Allan, HHMI, Carnegie Institution of Washington, Baltimore, MD
Sprang, Stephen R., HHMI, University of Texas Southwestern Medical Center at Dallas, Dallas, TX
Stary, Herbert, Louisiana State University, Baton Rouge, LA
Staudenmann, Jean-Louis, U.S. Department of Commerce, Washington, DC
Staudt, Lou, NCI, National Institutes of Health, Bethesda, MD
Steere, Allen, New England Medical Center, Boston MA
Steitz, Thomas A., HHMI, Yale University, New Haven, CT
Stern, Kathleen, University of Chicago, Chicago, IL
Sternberg, Paul, HHMI, California Institute of Technology, Pasadena, CA
Stevens, Charles, HHMI, Salk Institute, San Diego, CA
Strom, Charles, Illinois Masonic Hospital, Chicago, IL
Strominger, Jack, Harvard University, Cambridge, MA
Strong, Theresa, University of Alabama at Birmingham, Birmingham, AL
Stroud, Robert, University of California at San Francisco, San Francisco, CA
Struhl, Gary, HHMI, Columbia University, New York, NY
Stryer, Lubert, Stanford University, Palo Alto, CA
Suga, Nobuo, Washington University School of Medicine, Saint Louis, MO
Sukata, Satoshi, Osaka University, Osaka, Japan
Sulston, John, Sanger Center, Cambridge, UK
Sung, Ching-Hwa, Weill Medical College of Cornell University, New York, NY
Swain, William, PowderJect Vaccines, Inc., Madison, WI
Tainer, John, Scripps Research Institute, La Jolla, CA
Takahashi, Terry, University of Oregon, Eugene, OR
Telford, Sam III, Harvard School of Public Health, Boston, MA
Thomas, E. Donnall, Fred Hutchinson Cancer Research Center, Seattle, WA
Thomas, Kirk, University of Utah, Salt Lake City, UT
Tomlinson, Andrew, Columbia University, New York, NY
Tonegawa, Susumu, HHMI, Massachusetts Institute of Technology, Cambridge, MA
Townes, Tim, University of Alabama, Birmingham, AL
Tsui, Lap-Chee, Hospital for Sick Children, Toronto, Canada

Unanue, Emil, Washington University, St. Louis, MO
Utz, Ursula, U.S. Food and Drug Administration, Rockville, MD
Valle, David L., HHMI, The Johns Hopkins University School of Medicine, Baltimore, MD
Venter, Craig, President, Cetera Genomics Corporation, Rockville, MD
Verlinsky, Yury, Illinois Masonic Hospital, Chicago, IL
Walker, Bruce, Massachusetts General Hospital, Boston, MA
Wallace, Douglas, Emory University, Atlanta, GA
Watanabe, Kathe, University of Washington, Seattle, WA
Watson, James, Cold Spring Harbor Laboratory, Cold Spring Harbor, NY
Weis, William, Stanford University, Palo Alto, CA
Weissman, Irving, Stanford University, Palo Alto, CA
Wellems, Thomas, NIAID, National Institutes of Health, Bethesda, MD
Welsh, Michael J., HHMI, University of Iowa School of Medicine, Iowa City, IA
Wexler, Milton, Hereditary Disease Foundation, Los Angeles, CA
Wexler, Nancy, Columbia University, New York, NY
White, Raymond, University of Utah School of Medicine, Salt Lake City, UT
Wiesel, Torsten, The Rockefeller University, New York, NY
Wiley, Don C., HHMI, Harvard University, Cambridge, MA
Williams, David A., HHMI, Indiana University School of Medicine, Indianapolis, IN
Williamson, Robert, The Murdoch Institute, Royal Children's Hospital, Melbourne, Australia
Wilson, Ian, Scripps Research Institute, La Jolla, CA
Wilson, James, Institute for Human Gene Therapy, University of Pennsylvania, Philadelphia, PA
Witte, Owen N., HHMI, University of California–Los Angeles, Los Angeles, CA
Wlodawer, Alexander, Frederick Cancer Research Facility, Frederick, MD
Wolpert, Lewis, University College, London, UK
Wright, Peter, Scripps Research Institute, La Jolla, CA
Wysocki, Charles, Monell Chemical Senses Center, University of Pennsylvania, Philadelphia, PA
Yau, King-Wai, HHMI, The Johns Hopkins University School of Medicine, Baltimore, MD
Zeki, Semir, University College, London, UK
Zinkernagel, Rolf, University of Zurich, Zurich, Switzerland
Zipursky, Larry, HHMI, University of California–Los Angeles, Los Angeles, CA
Zotterman, Walter, Brookhaven National Laboratory, Upton, NY
Zuker, Charles S., HHMI, University of California–San Diego School of Medicine, San Diego, CA